Understanding and combatting African Swine Fever

Understanding and combatting
African Swine Fever

A European perspective

edited by:

Laura Iacolina, Mary-Louise Penrith,

Silvia Bellini, Erika Chenais,

Ferran Jori, Maria Montoya,

Karl Ståhl and Dolores Gavier-Widén

Wageningen Academic
Publishers

ASF - STOP

This publication is based upon work from COST Action CA15116, ASF-STOP, supported by COST (European Cooperation in Science and Technology).

COST (European Cooperation in Science and Technology) is a funding agency for research and innovation networks. Our Actions help connect research initiatives across Europe and enable scientists to grow their ideas by sharing them with their peers. This boosts their research, career and innovation.

www.cost.eu

EUROPEAN COOPERATION IN SCIENCE & TECHNOLOGY

Funded by the Horizon 2020 Framework Programme of the European Union

EAN: 9789086863570
e-EAN: 9789086869107
ISBN: 978-90-8686-357-0
e-ISBN: 978-90-8686-910-7
DOI: 10.3920/978-90-8686-910-7

First published, 2021

For any comments or questions, please contact the publisher, Wageningen Academic Publishers, P.O. Box 220, NL-6700 AE Wageningen, The Netherlands. www.WageningenAcademic.com copyright@WageningenAcademic.com

Acknowledgements

This book is based on scientific knowledge and results derived from the COST Action CA15116, ASF-STOP, 'Understanding and combating African swine fever in Europe'. The Action worked towards the main challenge of stopping African swine fever (ASF) from spreading further in Europe and protecting the European pig industry through a comprehensive, multi- and interdisciplinary approach.

I am enormously grateful to COST for supporting ASF-STOP and in this way providing a unique opportunity to bring together individual scientific efforts and expertise into a European-wide strong and integrated network of scientists with a common vision: combatting African swine fever. I thank especially the COST Scientific Officers Dominique Vandekerchove and Ioanna Stavridou for their assistance and advice and the Administrative Officers Christophe Peeters and Cassia Azevedo for the management of the finances. Their professional, timely and continuous support are greatly appreciated. ASF-STOP made special efforts to achieve high excellence and inclusiveness targets. Good gender balance and active participation and integration of young scientists and scientists from less research-intensive countries were accomplished. I am proud of these achievements and thankful to COST for promoting inclusiveness and providing the financial means to make them possible.

The network of ASF-STOP was established during the four years of the Action through the implementation of COST tools, such as scientific conferences and other dissemination means, working group meetings, workshops, training schools, short term scientific missions and inclusiveness target country conference grants. Collaborations were initiated or reinforced and knowledge on many aspects of ASF was exchanged and strengthened. This knowledge is assembled and presented in this book.

The incessant work of the working group leaders of ASF-STOP, Maria Montoya, Marie- Frédérique Le Potier, Ferran Jori, Chris Walzer, Paulius Busauskas, Tomaz Podgorski, Silvia Bellini, Frank Koenen, Marco Tamba, Karl Ståhl, Erika Chenais, Fernando Boinas, and the scientific communication managers Lisa Yon and Laura Iacolina is most appreciated, their work was key to reach the final achievement that this book represents. I also express my sincere gratitude to all participants of ASF-STOP for their contributions and in particular to our colleagues from the Global African Swine Fever Research Alliance (GARA) for their close collaboration, which contributed to a higher and fruitful interaction between ASF-STOP and countries beyond Europe and enhanced links within Europe.

I would also like to acknowledge the National Veterinary Research Institute of Poland, Pulawy for hosting the ASF-STOP International Launch Conference in 2016 and the Istituto Zooprofilattico Sperimentale della Lombardia e dell'Emilia Romagna (IZSLER), Brescia, Italy for hosting the Final International Conference of ASF-STOP in 2020. I also acknowledge the training schools' coordinators, Arvo Viltrop and Sofie Rossi, and the organisers and hosts of the training schools, which were conducted at the Estonian University of Life Sciences, the Friedrich-Loeffler Institute in Germany, and Universidad de Castilla-La Mancha, Ciudad Real, Spain. I am grateful to all the local organisers who hosted working group meetings and short-term scientific missions in the many different countries, the missions were kindly coordinated by Carlos das Neves and Willie

Loeffen. I am also indebted to the coordinator of the inclusiveness target country conference grants, Karoly Erdelyi.

I wish to express my sincere thanks to Susanna Westerberg, National Veterinary Institute of Sweden and Grant Holder of ASF-STOP for supporting the running of the Action with tireless dedication, patience, and professionalism. I am greatly indebted to Francisco Ruiz-Fons, vice-chair of ASF-STOP, for his continuous hard work during the four years of the Action. The coordination of the Action was largely facilitated by his competent, friendly and unlimited support.

I would like to thank all the authors and editors of this book for their essential contributions as experts in the many topics addressed in the chapters. This collected expertise reflects the high excellence and multi-disciplinarity of ASF-STOP and makes this book comprehensive and unique. I would like to express my heartfelt gratitude to Mary-Louise Penrith, globally recognised and long-experienced expert on African swine fever and co-editor of this book, for reviewing and editing the whole book. Her very helpful and wise comments created constructive and enjoyable discussions, provided a more international perspective and significantly improved the scientific quality of the book. Finally, a very special thanks goes to Laura Iacolina, main editor of this book. Without her incessant dedication, encouraging remarks and endless endurance this book would not have been written.

Prof. D. Gavier-Widén
Chair of ASF-STOP

Table of contents

1. African swine fever (ASF), the pig health challenge of the century

C. Martins[1*], F.S. Boinas[1], L. Iacolina[2,3], F. Ruiz-Fons[4] and D. Gavier-Widén[5,6]
[1]CIISA – Centro de Investigação Interdisciplinar em Sanidade Animal, Faculdade de Medicina Veterinária, Universidade de Lisboa, Avenida da Universidade Técnica, 1300-477 Lisboa, Portugal; [2]Department of Chemistry and Bioscience, Aalborg University, Frederik Bajers Vej 7H, 9220 Aalborg, Denmark; [3]Faculty of Mathematics, Natural Sciences and Information Technologies, University of Primorska, Glagoljaška 8, SI-6000 Koper, Slovenia; [4]Health & Biotechnology (SaBio) Group, Spanish Game and Wildlife Research Institute (IREC; CSIC-UCLM-JCCM), Ronda de Toledo 12, 13071 Ciudad Real, Spain; [5]Department of Pathology and Wildlife Diseases, National Veterinary Institute (SVA), 751 89, Uppsala, Sweden; [6]Department of Biomedical Sciences and Veterinary Public Health, Swedish University of Agricultural Sciences (SLU), Box 7028, 750 07 Uppsala, Sweden; cmartins@fmv.ulisboa.pt

Abstract

More than one hundred years ago African swine fever (ASF) was first diagnosed in Kenya. Since then, diverse approaches have been applied to the study of the causative virus, the sole member of the family *Asfarviridae*, aimed at characterising its properties, genome organisation and replication, its antigenic and biological properties as well as to develop treatment and a vaccine. The disease evolved and has persisted in Africa in a sylvatic cycle involving wild suids and soft ticks for a long time, but was introduced, usually through contaminated waste food, into other regions on multiple occasions since 1957. The most recent introduction, into Georgia in 2007, resulted in the spread of the disease to the European Union in 2014 and to the establishment of an international and multidisciplinary network of scientists funded by the European Cooperation in Science and Technology (COST) two years later. The network included a broad variety of scientific fields, animal health and food safety authorities, hunting associations, wildlife managers and food and livestock industries with the goal of increasing preparedness and attempting to stop ASF spread. This book represents the summary of the collective and integrated work of almost 300 dedicated participants in tackling the complex challenge posed by ASF. Here we summarise the state-of-the-art knowledge on this lethal disease, with a focus on the European situation, and identify areas that still need to be explored.

Keywords: African swine fever, history, multidisciplinary, Europe

This publication is based upon work from COST Action CA15116, ASF-STOP, supported by COST (European Cooperation in Science and Technology).
www.cost.eu

EUROPEAN COOPERATION IN SCIENCE & TECHNOLOGY

Funded by the Horizon 2020 Framework Programme of the European Union

DOI 10.3920/978-90-8686-910-7_1,

1.1 Where ASF-STOP comes from

More than one hundred years ago, R. Eustace Montgomery diagnosed for the first time, in Kenya, the occurrence of a swine disease characterised by clinical and pathological features similar but not related to classical swine fever (CSF) or hog cholera. His observations and studies, developed from 1910 up to 1917, showed relevant aspects of the 'East African swine fever' now known as African swine fever (ASF) (Montgomery, 1921). The author's findings are still considered to be pillars for the knowledge attained since then on this disease, that currently inflicts alarming outbreaks with devastating economic and social consequences in numerous countries in Africa, Europe, Asia and Oceania. Hundreds of researchers and specialists in different fields of knowledge, of private and public National and International institutions, have produced, along the years, countless documents and scientific papers dedicated to explore and deepen key aspects on intricate details of the virus biology, disease maintenance and dissemination, complex viral-host interactions, strategies to prevent and control disease outbreaks and dissemination, and efforts to develop safe and successful vaccines. Recent reviews, including the chapters of this book, fully cover different aspects of ASF and it would be superfluous to describe facts already available for the likely already informed readers. Thus, the main ground-breaking contributions that have provided new insights into the current knowledge on ASF are presented here, to reinforce strategies towards prevention and control of ASF.

Since Montgomery's findings, African swine fever virus (ASFV) has been studied through diverse approaches aiming at characterising its structural and physicochemical properties, genome organisation and replication, as well as its antigenic and biological properties. This complex virus is a large double stranded DNA virus (170-190 kb) that encodes more than 150 open reading frames (ORFs), depending on the virus isolate. It is the only known DNA arbovirus and the sole member of the family *Asfarviridae*, classified into the Nucleocytoplasmic Large DNA Viruses superfamily (Takamatsu *et al.*, 2011). The virus infects all members of the *Suidae* family and it is maintained in different epidemiological cycles related to distinct ecosystems. It was diagnosed in Africa for the first time in 1910, when occurrence of ASF outbreaks was related to the presence of warthogs (*Phacochoerus* sp.) and bush pigs (*Potamochoerus* sp.), both shown to survive experimental inoculations with the causative virus without manifesting clinical signs. However, their blood collected up to five to seven days post inoculation was infectious and caused ASF experimentally in domestic pigs. The infected wild suids did not, however, appear capable of disseminating the virus by contact, urine, and faeces although he considered possible the involvement of an arthoropod vector, and was able to exclude fleas and lice (Montgomery, 1921).

Many years later, in the 1960s, when ASF was spreading in Europe, Sanchez Botija found that ASFV infected soft ticks (*Ornithodoros erraticus*), collected in pig sties, were acting as the biological vectors of ASFV in Spain (Sanchez-Botija, 1963). This finding opened new insights regarding the study of ASF epidemiology in Africa. Plowright and colleagues (1969) described thereafter that soft ticks of the *Ornithodoros moubata* complex were vectors and reservoirs of ASFV in Africa. This new knowledge was further pursued by the identification of the main role of *O. moubata* ticks in the ASF sylvatic epidemiological cycle. This cycle was shown to be connected to common warthogs (*Phacochoerus africanus*) in Eastern and Southern Africa (Penrith *et al.*, 2004), the niche of the disease whose aetiological agent has been described as having evolved over

300 years with a time to the most recent common ancestor in the early 18th century (Michaud *et al.*, 2013).

ASF has persisted in Africa in a sylvatic cycle involving wild suids and soft ticks, spreading through infected *O. moubata* ticks to domestic pig populations in which, afterwards, it readily spreads through direct and indirect transmission from infected to healthy pigs (Penrith *et al.*, 2004). The disease was introduced, probably from Angola, to Portugal in 1957 (Manso Ribeiro *et al.*, 1958) in ASFV contaminated waste food from planes that was fed as swill to pigs.

The disease was considered eradicated the following year but was again introduced in 1960 (Manso Ribeiro and Rosa Azevedo, 1961) and it remained endemic in Portugal and Spain up to the early 1990s. Besides being maintained in the domestic pig populations through direct and indirect transmission, it persisted in populations of free ranging Iberian pigs, kept in premises, built mostly with stones and adobe, during the night periods, allowing the establishment of the epidemiological cycle including the soft ticks *Ornithodoros erraticus* as biological reservoirs (Caiado *et al.*, 1988; Sanchez-Botija, 1963). In Europe, the distribution of this soft tick is limited to the Iberian pig extensive production areas, in Central Spain and Southwestern quarter of Iberia (Montado/Portugal and Dehesa/Spain), the last locations from which the disease was eradicated. The occurrence of a sporadic outbreak, in 1999 in the South of Portugal, has been epidemiologically connected with the presence of *O. erraticus* in the pig premises, as these arthropods are very resistant to fasting and may harbour active ASFV for up to 5 years (Boinas *et al.*, 2011). As a consequence, the repopulation of the infested premises in the area became limited since it was only allowed based on the result of a compulsory risk evaluation. Most of them were depopulated, which led to a significant decrease in the prevalence of *O. erraticus* in more recent years (Wilson *et al.*, 2013).

Although the epidemiological role of the Eurasian wild boar (*Sus scrofa*) (WB) during the endemic period in Iberia was considered negligible, epidemiological surveys done at the time showed that WB caused between 5-6% of the ASF outbreaks (Ordas *et al.*, 1983; Perestrelo-Vieira, 1993). Later on, towards the end of the previously mentioned period, serological surveys in Portugal showed only 2.3% seroprevalence in the hunted WB population in 1990 with no positives in 1992 (Commission of the European Communities-Directorate General for Agriculture, 1994). The significant increase of WB in recent years in the whole of Iberian Peninsula, and the increase of free-ranging pig farms, mainly in the Southern areas, certainly would play a relevant role in case of ASF reintroduction.

The WB is known to be as susceptible as the domestic pig to ASF (Blome *et al.*, 2013; Polo Jover and Sanchez-Botija, 1961), and is currently considered to be the main driver of a new epidemiological cycle in disease maintenance and spread since it emerged in the Caucasus, Russia, Eastern Europe and Baltic States (Guberti *et al.*, 2020). This is further described in detail later in this book (Chapters 8 and 9). In the Italian island of Sardinia, ASF has been endemic since 1978, although eradication is likely to be achieved soon. Its persistence for more than 30 years has frequently been associated with the large populations of free-range 'brado' pigs that live in the mountainous area in close contact with WB, although WB is considered of minor importance in this epidemiological scenario (Laddomada *et al.*, 2019).

Studies developed on the role of several species of blood-sucking invertebrates found in pig or warthog habitats have shown that they do not transmit the virus after having an infected blood meal. Although not acting as biological vectors in nature like the argasid ticks, it was experimentally shown that stable flies (*Stomoxys calcitrans*) may act as mechanical vectors of ASFV (Mellor *et al.*, 1987; Olesen *et al.*, 2018).

In the vertebrate host, the aetiological agent replicates preferentially in monocyte and macrophage cells of the mononuclear phagocytic system (van Furth and Cohn, 1968) and causes a range of syndromes and lesions, from peracute to chronic and unapparent forms of disease (Gallardo *et al.*, 2015; Gómez-Villamandos *et al.*, 2013). The clinical evolution of ASF is often related to the diverse virulence of the aetiological agent, which still remains to date one of the most intriguing details of ASFV biology. When ASF was confined to Africa and Iberian countries up to the early '60s, viral isolates in natural occurring outbreaks were shown to be highly virulent, inducing haemorrhagic disease with mortality approaching 100% of infected animals (Manso Ribeiro *et al.*, 1958; Montgomery, 1921). However, even though several authors claimed that ASFV virulence would attenuate in time after the initial outbreaks, possibly through viral adaptation to the host (Plowright *et al.*, 1994), this was not the case. For example, in Portugal highly virulent haemadsorbing isolates (see Chapter 2) were identified up to the end of the endemic period, or even when an isolated and controlled outbreak of disease occurred in 1999 (Wilson *et al.*, 2013).

However, while outbreaks caused by the highly virulent haemadsorbing isolate ASFV Lisbon 60 (L60) prevailed, infections with low virulent non-haemadsorbing isolates were also identified in apparently healthy pigs showing chronic pulmonary lesions (Vigário *et al.*, 1974). This parallel occurrence of both low and highly virulent ASFV isolates in nature is not fully understood. However, it is unlikely that genetic changes have naturally occurred in the highly virulent L60 to give origin to the non-haemadsorbing, low virulent virus. The latter was recently shown to have significant genetic differences when compared to the virulent isolate (Portugal *et al.*, 2015). The haemadsorption effect caused by some virus isolates, and demonstrated by Malmquist and Hay (1960), was thought for a while to be a virulence marker. However, findings in Africa during natural occurring outbreaks have shown that non-haemadsorbing ASFV isolates can be both virulent and non-virulent (Pini and Wagenaar, 1974; Thomson *et al.*, 1979). The origin of non-haemadsorbing isolates, showing different virulence, is not clearly understood. It is accepted that they may occur in natural conditions, originating from parental virus composed of a heterogeneous mix of different viruses, as demonstrated *in vitro* by Pan and Hess (1984, 1985).

The development of molecular methods has been a major advance to study different aspects of ASFV biology. A landmark for ASFV genome characterisation was achieved through the analysis of the complete viral genome sequence of BA71V, an avirulent ASFV isolate, adapted to grow in Vero cells (Yáñez *et al.*, 1995). This work has prompted further research towards the functional characterisation of viral genes that has been pursued by different authors using other viral isolates. As recently reviewed, ASFV encodes for at least 150 proteins and, so far, 38 of them are associated with known or predicted functions in nucleotide metabolism, transcription, replication and repair; more than 24 proteins are involved in virion structure and morphogenesis, and at least 8 are likely involved in host cell interactions, although the functions of a large number of ASFV encoded proteins still remain unknown (Gaudreault *et al.*, 2020). Genome sequencing has allowed molecular characterisation of small conserved DNA regions enabling the detection of origin and traceability of ASFV during outbreaks. The current approach is based on the analysis of the C-terminal end of gene B646L, encoding the major protein p72 (Bastos *et al.*, 2003). More recently, this procedure has been enhanced through the sequencing of the Central Variable Region (CVR) within the B602L gene, or other regions (e.g. E183L encoding p54 protein, CP204L encoding p30 protein), to distinguish between geographically and temporally constrained p72 genotype viruses (Gallardo *et al.*, 2015). This approach has allowed 24 different p72 genotypes to be identified among the currently known virus isolates. Out of these, genotype I is predominant in West Africa and was the only one found outside Africa (Europe, America, and the Caribbean) until the introduction of genotype II from East Africa into Georgia in 2007. This latter genotype is the one currently present in parts of the Caucasus, Europe, the Russian Federation, Southeast Asia and Oceania (Gallardo *et al.*, 2015; Gaudreault *et al.*, 2020). Although important for determining epidemiological details of ASFV, the above-mentioned genotyping does not correlate to the virulence of ASFV isolates as often wrongly stated. The quest for the identification of ASFV virulence factors or markers has been pursued for a long time.

As mentioned above, ASFV virulence was initially correlated to the haemadsorption phenomenon (Malmquist and Hay, 1960). Progressively, studies have been devoted to compare genome sequencing of different virus, for example through the use of isolates of different virulence from the same origin (Portugal *et al.*, 2015). Such approaches are likely relevant to identify genes that may be of importance to be used in vaccine designing. Development of efficient and safe vaccine(s) against ASF is the ultimate goal pursued for more than one hundred years (see Chapter 6) with the first attempts done by Montgomery, who unsuccessfully tried to vaccinate domestic pigs with inactivated ASFV and who also found that anti-ASF serum did not neutralise the virus (Montgomery, 1921). Research on vaccine development against ASF has been extensively reported by many authors covering, among others, details on the complexity of the ASFV particle, the large number of proteins encoded by its genome, the ASFV-host interactions, the mechanisms of protective immunity, and approaches for vaccine development (Arias *et al.*, 2017). Herein, a few aspects will be emphasised regarding viral-host interactions that may be considered as a cautionary route when evaluating vaccine design against ASF. First of all, the preferential replication of ASFV in pig macrophages is important. Numerous studies have been conducted using pig macrophages to characterise aspects of viral modulation of innate immune responses potentially relevant *in vivo*, thus ideally to be used as markers to test efficacy of different vaccine models. For instance, *in vitro* studies have shown that the viral infection did not alter the expression of Complement Fixation (Fc) receptors or the ability to mediate antibody cellular cytotoxicity; in contrast, phagocytosis, antibody mediated phagocytosis and chemotaxis were abrogated independently of the virulence of ASFV isolate (Martins *et al.*, 1987). However, a low virulent ASFV (ASFV/NH/P68 – NHV) was shown to induce an enhanced expression and production of relevant regulatory cytokines interferon (IFN) (IFNalpha), tumour necrosis factor (TNF) (TNF-α) and interleukin (IL) (IL-12p40) on porcine macrophages *in vitro*, when compared to the effect of the virulent ASFV Lisbon 60 (L60) (Gil *et al.*, 2008). More recently, other studies have suggested that both L60 and NHV were able to inhibit interferon I (IFN I) production in macrophages, through a mechanism not dependent on IRF3 modulation (Portugal *et al.*, 2018), while macrophage responses to NHV have been associated with enhanced sensitivity to type I IFN and cytokine responses from classically activated macrophages (Franzoni *et al.*, 2020). Modulation of cellular IFN responses to infection with attenuated virus and induction of different protective immunity has been demonstrated through deletion of genes from multigene families (MGF) on a virulent ASF isolate (Reis *et al.*, 2016).

ASF evolves frequently in domestic pigs as a contagious and usually acute haemorrhagic and lethal disease, although some pigs may survive, which is common when they are infected with low virulent ASFV isolates although, in such cases, chronic type infections may develop (Gallardo *et al.*, 2015; Leitão *et al.*, 2001). In general, pigs surviving ASF are resistant to challenge with homologous or, in some cases, with closely related virus isolates, suggesting the presence of a not yet fully understood efficient protective immunity against infection. On one hand, anti-ASF antibodies may interfere with disease development, as seen by reduced mortality, reduced virulence and a delayed onset of infection in pigs treated with those antibodies (Onisk *et al.*, 1994; Wardley *et al.*, 1985). However, conversely to what occurs in many different animal viral infections, antibodies do not neutralise the pathogenic capacity of ASFV so that vaccines that stimulate the development of the host humoral immune responses are not efficient. On the other hand, cellular immune mechanisms as ASF-specific cytotoxic T (cluster of differentiation) $CD8^+$ lymphocyte and natural killer (NK) cell activities were shown to occur in pigs surviving experimental infection with low virulent NHV isolate (Leitão *et al.*, 2001; Martins *et al.*, 1988, 1993). The cytotoxic T $CD8^+$ lymphocyte activity was seen to play an important role in ASF protective immunity in animals exposed to the low virulent OUR/T88/3 strain that were no longer protected from challenge with the virulent OUR/T88/1 isolate when depleted of $CD8^+$ lymphocytes (Oura *et al.*, 2005). The above mentioned immune responses suggest that ASFV specific antibody alone is not sufficient for protection against ASFV infection, and that there is an important role for the $CD8^+$ lymphocyte subset in ASFV protective immunity (Takamatsu *et al.*, 2013).

1.2 Where ASF-STOP stands

Soon after ASF emerged in Georgia in 2007 and carefully following the development of the situation in Russia, the international scientific community became aware of not only the approaching threat to the European pig industry but also the increased importance of wild boar in this epidemic. With the aim to understand the role of wild boar in ASF epidemiology, improve diagnosis and surveillance in this species and develop management practices that would decrease the role of wild boar in ASF epidemiology, a network of wildlife scientists, under the umbrella of the European Wildlife Disease Association (EWDA), came together to tackle the challenge of ASF expansion into new territories. A core group met at a workshop in Uppsala, Sweden, on 6-7 March 2014 to plan a proposal to the European Cooperation in Science and Technology (COST), a funding organisation under the Horizon 2020 programme of the European Union. The COST Action (Action is the term used by COST and it means 'networking project') ASF-STOP was launched on 1 May 2016. The main objective of the Action was 'to stop African swine fever from spreading further in Europe and protecting the European pig industry by combating ASF through a comprehensive, multi- and interdisciplinary approach'. The network included a broad variety of fields comprising, among others, virology, wild boar management, pathology, biosecurity, vaccinology and communication (Figure 1.1). The inclusiveness of the network led to the involvement of research institutions like universities and veterinary institutes side by side with animal health and food safety authorities, hunters' associations, wildlife managers and food

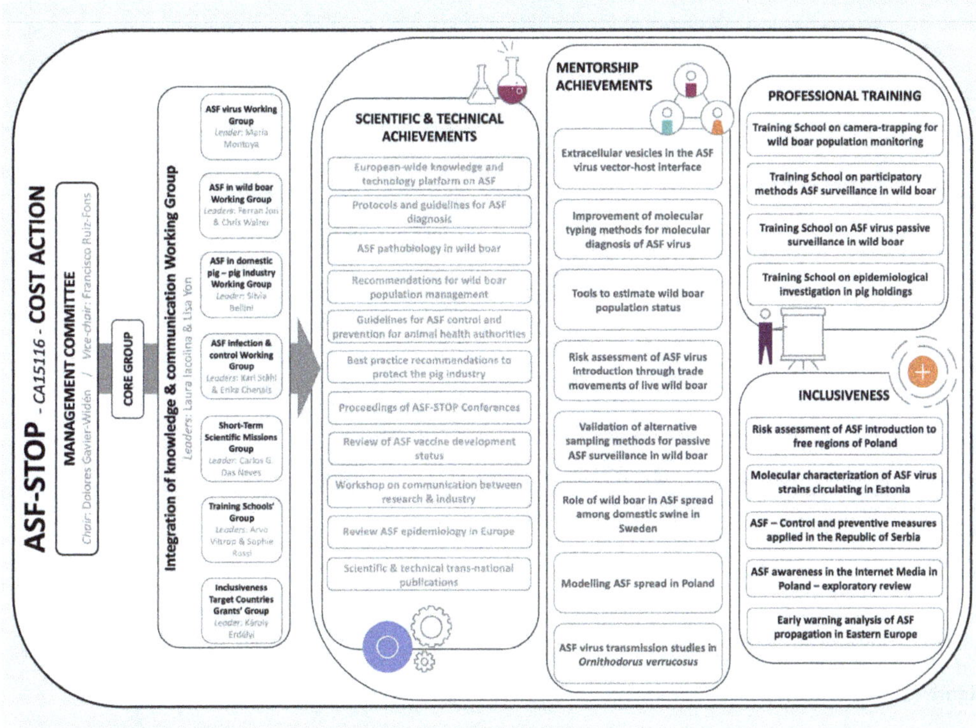

Figure 1.1. Structure of ASF-STOP and overview of the main outputs.

and livestock industries. Non-European partners also joined the network, shared their experience (from work in African and previously ASF infected countries) and contributed to the exchange of knowledge. Most European institutions conducting research on ASF participated in ASF-STOP, as well as many ASF national and international reference laboratories. Overall, the network included 37 countries; 17 of them were inclusiveness target countries in Europe, and three international Agencies, with a total of almost 300 participants (Figure 1.2). All worked together in a coordinated and integrated effort to collectively address the multiple challenges posed by ASF, moving towards the common goal of increasing preparedness and attempting to stop its spread. ASF-STOP worked in close collaboration with the Global African Swine Fever Research Alliance (GARA) and participated in the creation of a global research agenda (GARA https://www.ars.usda.gov/GARA/). In collaboration with GARA, to facilitate the flow of information and, at the same time, avoid unnecessary overlap or duplication of efforts, ASF-STOP created an inclusive environment where experts from different fields came together to complement each other's knowledge and build solutions.

The network worked together on coordinated objectives, which included all aspects connected to ASF (summarised in this book). High scientific excellence, active collaboration and integration of disciplines led to the achievement of the objectives. Special focus areas were the development of standardised methods, the identification of best practices applicable to different field conditions and the development of preventive measures, including advancing in vaccine development. Multidisciplinary teams worked in close collaboration to understand the role of wild boar in

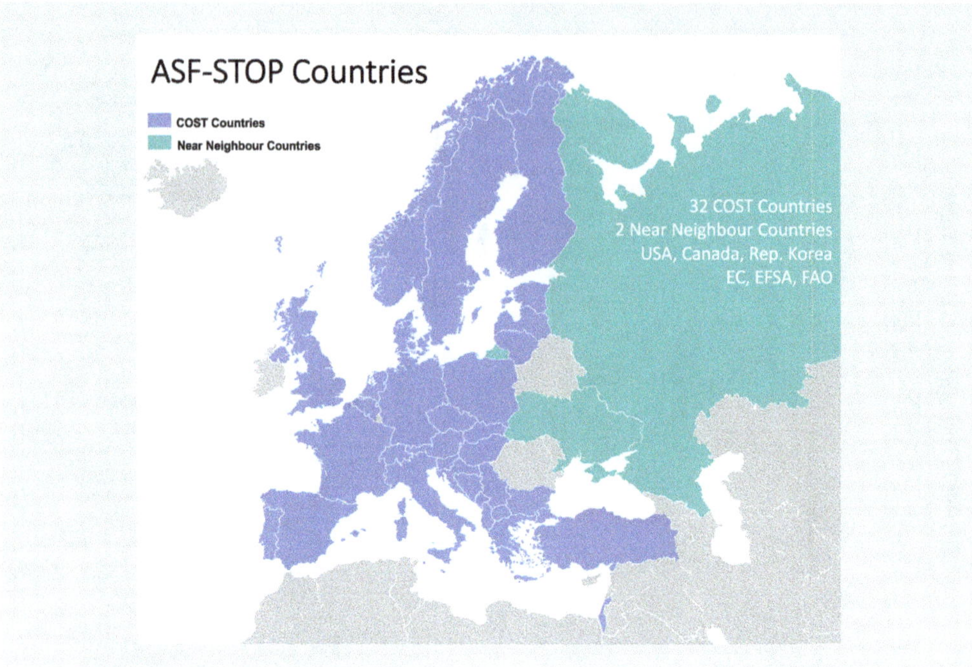

Figure 1.2. Map of the ASF-STOP participating countries.

virus transmission in different conditions and management practices. The teams designed control measures adapted to the different European habitats and wild boar densities. Leaders in virology, vaccinology, pathology, diagnostics and immunology came together to address aspects of the virus biology, its effects on the host, potential therapies and the best ways to detect infection in animals with varying levels of immunological status as well as in samples with different preservation conditions. This latter aspect is particularly relevant in the current epidemiological context where wild boar, as proven by the work of epidemiologists and wildlife biologists in the network, constitute a new epidemiological cycle for the virus in Europe and passive monitoring of wild boar carcasses has proven crucial for early detection of outbreaks. Close monitoring of the health status of wild boar is important for the prevention of introduction of ASFV into the pig production system. The pig industry and food production system are extremely variegated in terms of size, practices, needs and cultural importance. To protect this economically important sector and the welfare of the animals, experts in biosecurity, pig production systems and cleaning and disinfection collaborated with the pig and food industry and animal health and food safety authorities to identify potential risks for ASFV introduction and define guidelines specific for different branches of pig production, from industrial to backyard.

ASF-STOP also worked towards the following capacity building objectives: (1) creation of a Europe-wide scientific knowledge and technology exchange platform to prevent, control and mitigate ASF; (2) enabling informed decision-making based on current scientific data; (3) building capacity in Europe and strengthening trans-national collaboration; (4) achieving inclusive collaboration; (5) achieving wide dissemination of the outcomes and results of ASF-STOP; and (6) establishing a Pan-European research agenda.

Without the collective and integrated work of ASF-STOP the impact of all these efforts in research and development would have been mostly confined to specific fields. The continuous commitment to share this information within and beyond the network, communicating and engaging policy makers, breeders and farmers, hunters, wildlife managers and society as a whole largely contributed to the implementation of the knowledge and the impact of ASF-STOP. As a whole, ASF-STOP made important contributions towards the control and eradication of ASF in Europe, developed and communicated new knowledge, built an extended international network of scientists working on ASF and contributed to capacity building in Europe.

1.3 What ASF-STOP paved the way for

Within the four years of its life span, ASF-STOP has faced many challenges and resulted in even more success stories. A lot has been learnt but several gaps of knowledge have been identified in the process. This book aims to present the overview of the currently available knowledge on ASF, with special focus on the European situation, including but not limited to the results of the Action. This book is also a call for action for the scientific community to investigate those areas that still need to be explored, for policy makers and funding bodies to support those efforts and for society to actively engage in the global challenge of stopping ASF.

Recent research on ASFV (Chapter 2) has shown how the virus inhibits different immune responses and how multiple factors can lead to different lethality rates and that, although knowledge on the architecture of ASFV has progressed over time, more light needs to be shed on this aspect. At the same time, Chapter 2 shows how genome and computational methods could prove fundamental for the development of antiviral drugs and a vaccine.

The immune system is an organism barrier against harmful pathogens and while the host immune system reacts to ASFV infection, the virus interactions with macrophages or dendritic cells remain largely unknown, as is the immune response to ASFV in natural infections compared to experimental settings (Chapter 3).

Knowledge on pathology provides the foundation for understanding pathogenesis and host response to the pathogen. Disease caused by ASFV may range from peracute to subclinical or inapparent depending on different factors, among which the virulence of the isolates is particularly important. The presence and severity of lesions are highly variable and may affect multiple organs (Chapter 4).

Building on this body of knowledge on the lesions caused by ASF, the immunological processes and the characteristics of the virus several diagnostic tools have been developed with good sensitivity and specificity (Chapter 5). However, it remains of paramount importance to carefully follow guidelines for the choice of the most appropriate tool and sampling procedure based on the situation at hand, the field conditions and the requirements.

An aspect that has gained renewed attention over the past few decades is the development of a vaccine. Over time multiple approaches have been tested but most of them were either not successful or led to undesired side effects; however, a number of promising candidates with good efficacy are currently being tested (Chapter 6), with the whole pig sector awaiting the results of large-scale trials. This industry is not only economically important but is also highly relevant for food security, which may be threatened by the ability of ASF to significantly reduce both the availability and affordability of adequate supplies of pork. To prevent the ravages of ASF, the industry has already undergone major rearrangements and changes, particularly in small size pig holdings that, together with backyard farms, are those most at risk (Chapter 7). For this reason, specific guidelines have been developed for the different farming premises with biosecurity information (Chapter 10) and best practices focusing on cleaning and disinfection procedures (Chapter 11) to prevent the introduction of ASFV into the pig holding. Both biosecurity and cleaning and disinfection procedures, if carefully and routinely implemented, have proven to be effective preventive measures even when the virus is circulating in the wild boar population around the farm.

The density and widespread distribution of wild boar in Europe has led to the identification of a new ASF epidemiological cycle between the wild boar and its habitat (Chapter 9). Several studies have reported the same epidemiological principles in both wild boar and domestic pigs and highlighted the important role played by humans in transmitting ASF to both wild and domestic pigs, thus confirming the paramount importance of correct implementation of biosecurity practices. At the same time, the identification of the wild boar-habitat cycle brought renewed attention to the need to manage the increasing European wild boar population (Chapter 8). Expertise developed in wildlife biology and management has been complemented with all aspects of ASFV transmission, survival and pathology to develop integrated approaches that combine management methods and pathogen transmission prevention practices. Positive field implementations in countries, such as Czech Republic and Belgium showed early detection and swift collaborative action involving all stakeholders can lead to control and, ultimately, eradication of ASF from wild boar populations.

Overall, the present summary of different relevant aspects of ASFV biology and viral-host interactions, acquired by different authors up to date, reinforces that to overcome the main challenges highlighted, certain major efforts should be pursued to prevent and control ASF. In Europe the following aspects are particularly important: improvement of sanitary and biosecurity measures in all operations regarding production of domestic pigs; control of WB populations responsible for maintaining and spreading the disease and above all, the development of an efficient vaccine.

Acknowledgements

This publication is based on work from 'Understanding and combating African swine fever in Europe (ASF-STOP COST action 15116)' supported by COST (European Cooperation in Science and Technology). LI received funding from the Horizon 2020 research and innovation programme under the Marie Sklodowska-Curie Action (Grant Agreement no. 656697). Additionally, we are grateful to all ASF-STOP participants for the productive, engaged and friendly exchange of ideas and information (Figure 1.3).

Figure 1.3. ASF-STOP final conference group picture (by G. Spalenza).

References

Arias, M., De la Torre, A., Dixon, L., Gallardo, C., Jori, F., Laddomada, A., Martins, C., Parkhouse, R.M., Revilla, Y., Rodriguez, F. and J.-M., Sanchez-Vizcaino and Sanchez-Vizcaino, J.M., 2017. Approaches and perspectives for development of African swine fever virus vaccines. Vaccines 5: 35. https://doi.org/10.3390/vaccines5040035

Bastos, A.D.S., Penrith, M.L., Crucière, C., Edrich, J.L., Hutchings, G., Roger, F., Couacy-Hymann, E. and Thomson, G.R., 2003. Genotyping field strains of African swine fever virus by partial p72 gene characterisation. Archives of Virology 148: 693-706. https://doi.org/10.1007/s00705-002-0946-8

Blome, S., Gabriel, C. and Beer, M., 2013. Pathogenesis of African swine fever in domestic pigs and European wild boar. Virus Research 173: 122-130. https://doi.org/10.1016/j.virusres.2012.10.026

Boinas, F.S., Wilson, A.J., Hutchings, G.H., Martins, C. and Dixon, L.J., 2011. The persistence of African swine fever virus in field-infected *Ornithodoros erraticus* during the ASF endemic period in Portugal. PLoS ONE 6: e20383. https://doi.org/10.1371/journal.pone.0020383

Caiado, J.M., Boinas, F.S. and Louza, A.C., 1988. Epidemiological research of African swine fever (ASF) in Portugal: the role of vectors and virus reservoirs. Acta Veterinaria Scandinavica. Supplementum 84: 136-138.

Commission of the European Communities-Directorate General for Agriculture, (CEC-DGCI), 1994. Programme to eradicate African swine fever from Iberian Peninsula. Commission of the European Communities, Brussels, Belgium.

Franzoni, G., Razzuoli, E., Dei Giudici, S., Carta, T., Galleri, G., Zinellu, S., Ledda, M., Angioi, P., Modesto, P., Graham, S.P. and Oggiano, A., 2020. Comparison of macrophage responses to African swine fever viruses reveals that the NH/P68 strain is associated with enhanced sensitivity to type I IFN and cytokine responses from classically activated macrophages. Pathogens 9: 209. https://doi.org/10.3390/pathogens9030209

Gallardo, M.C., Reoyo, A. de la T., Fernández-Pinero, J., Iglesias, I., Muñoz, M.J. and Arias, M.L., 2015. African swine fever: a global view of the current challenge. Porcine Health Management 1: 21. https://doi.org/10.1186/s40813-015-0013-y

Gaudreault, N.N., Madden, D.W., Wilson, W.C., Trujillo, J.D. and Richt, J.A., 2020. African swine fever virus: an emerging DNA arbovirus. Frontiers in Veterinary Science 7: 215. https://doi.org/10.3389/fvets.2020.00215

Gil, S., Sepúlveda, N., Albina, E., Leitão, A. and Martins, C., 2008. The low-virulent African swine fever virus (ASFV/NH/P68) induces enhanced expression and production of relevant regulatory cytokines (IFNα, TNFα and IL12p40) on porcine macrophages in comparison to the highly virulent ASFV/L60. Archives of Virology 153: 1845-1854. https://doi.org/10.1007/s00705-008-0196-5

Gómez-Villamandos, J.C., Bautista, M.J., Sánchez-Cordón, P.J. and Carrasco, L., 2013. Pathology of African swine fever: the role of monocyte-macrophage. Virus Research 173: 140-149. https://doi.org/10.1016/j.virusres.2013.01.017

Guberti, V., Khomenko, S., Masiulis, M. and Kerba, S., 2020. African swine fever in wild boar ecology and biosecurity. OIE and EC, Rome, Italy. https://doi.org/10.4060/ca5987en

Laddomada, A., Rolesu, S., Loi, F., Cappai, S., Oggiano, A., Madrau, M.P., Sanna, M.L., Pilo, G., Bandino, E., Brundu, D., Cherchi, S., Masala, S., Marongiu, D., Bitti, G., Desini, P., Floris, V., Mundula, L., Carboni, G., Pittau, M., Feliziani, F., Sanchez-Vizcaino, J.M., Jurado, C., Guberti, V., Chessa, M., Muzzeddu, M., Sardo, D., Borrello, S., Mulas, D., Salis, G., Zinzula, P., Piredda, S., De Martini, A. and Sgarangella, F., 2019. Surveillance and control of African swine fever in free-ranging pigs in Sardinia. Transboundary and Emerging Diseases 66: 1114-1119. https://doi.org/10.1111/tbed.13138

Leitão, A., Cartaxeiro, C., Coelho, R., Cruz, B., Parkhouse, R.M.E., Portugal, F.C., Vigário, J.D. and Martins, C.L.V., 2001. The non-haemadsorbing African swine fever virus isolate ASFV/NH/P68 provides a model for defining the protective anti-virus immune response. Journal of General Virology 82: 513-523. https://doi.org/10.1099/0022-1317-82-3-513

Malmquist, W.A. and Hay, D., 1960. Hemadsorption and cytopathic effect produced by African Swine Fever virus in swine bone marrow and buffy coat cultures. American Journal of Veterinary Research 21: 104-108.

Manso Ribeiro, J. and Rosa Azevedo, J.A., 1961. Réapparition de la Peste Porcine Africaine (P.P.A) au Portugal. Bulletin de l'Office International Des Épizooties 55: 88-106.

Manso Ribeiro, J., Rosa Azevedo, J.A., Teixeira, M.J.O., Braço-Forte, M.C., Rodrigues Ribeiro, A.M. Oliveira e Noronha, F. Grave Pereira, C. and Dias Vigário, J., 1958. Peste porcine provoquée par une souche différente (Souche L) de la souche classique. Bulletin de l'Office International Des Épizooties 50: 516-534.

Martins, C., Mebus, C., Scholl, T., Lawman, M. and Lunney, J., 1988. Virus-specific CTL in SLA-inbred swine recovered from experimental African swine fever virus (ASFV) infection. Annals of the New York Academy of Sciences 532: 462-464. https://doi.org/10.1111/j.1749-6632.1988.tb36376.x

Martins, C.L.V, Lawman, M.J.P., Scholl, T., Mebus, C.A. and Lunney, J.K., 1993. African swine fever virus specific porcine cytotoxic T cell activity. Archives of Virology 129: 211-225. https://doi.org/10.1007/BF01316896

Martins, C.L.V, Scholl, T., Mebus, C.A., Fisch, H. and Lawman, M.J.P., 1987. Modulation of porcine peripheral blood-derived macrophage functions by *in vitro* infection with African swine fever virus (ASFV) isolates of different virulence. Viral Immunology 1: 177-190. https://doi.org/10.1089/vim.1987.1.177

Mellor, P.S., Kitching, R.P. and Wilkinson, P.J., 1987. Mechanical transmission of capripox virus and African swine fever virus by Stomoxys calcitrans. Research in Veterinary Science 43: 109-112. https://doi.org/10.1016/s0034-5288(18)30753-7

Michaud, V., Randriamparany, T. and Albina, E., 2013. Comprehensive phylogenetic reconstructions of African swine fever virus: proposal for a new classification and molecular dating of the virus. PLoS ONE 8: e69662. https://doi.org/10.1371/journal.pone.0069662

Montgomery, R.E., 1921. On a form of swine fever occurring in British East Africa (Kenya colony). Journal of Comparative Pathology and Therapeutics 34: 159-191 (part I), 243-269 (part II).

Olesen, A.S., Lohse, L., Hansen, M.F., Boklund, A., Halasa, T., Belsham, G.J., Rasmussen, T.B., Bøtner, A. and Bødker, R., 2018. Infection of pigs with African swine fever virus via ingestion of stable flies (*Stomoxys calcitrans*). Transboundary and Emerging Diseases 65: 1152-1157. https://doi.org/10.1111/tbed.12918

Onisk, D.V, Borca, M.V, Kutish, S., Kramer, E., Irusta, P. and Rock, D.L., 1994. Passively transferred African swine fever virus antibodies protect swine against lethal infection. Virology 198|: 350-354. https://doi.org/10.1006/viro.1994.1040

Ordas, A., Sanchez-Botija, C., Bruyel, V. and Olias, J., 1983. African swine fever. The current situation in Spain. In: Wilkinson, P.J. (ed.), African Swine Fever. CEC, Luxemburg, Luxemburg, pp. 67-73.

Oura, C.A.L., Denyer, M.S., Takamatsu, H. and Parkhouse, R.M.E., 2005. *In vivo* depletion of CD8+ T lymphocytes abrogates protective immunity to African swine fever virus. Journal of General Virology 86: 2445-2450. https://doi.org/10.1099/vir.0.81038-0

Pan, I.C. and Hess, W.R., 1984. Virulence in African swine fever: its measurement and implications. American Journal of Veterinary Research 45: 361-366.

Pan, I.C. and Hess, W.R., 1985. Diversity of African swine fever virus. American Journal of Veterinary Research 46: 314-320.

Penrith, M.L., Thomson, G.R., Bastos, A.D.S., Etter, E.M., Coetzer, J.A.W. and Tustin, R.C., 2004. African swine fever. In: Coetzer, J.A.W. and Tust (eds.), Infectious diseases of livestock. Oxford University Press, Cape Town, South Africa, pp. 1088-1119.

Perestrelo-Vieira, R., 1993. Evolution of African swine fever in Portugal. In: Galo, A. (ed.), African swine fever. Coordination of Agricultural Research. Proceedings of a Research Seminar in Lisbon, 1991. Commission of the European Communities, Luxemburg, Luxembourg, pp. 67-73.

Pini, A. and Wagenaar, G., 1974. Isolation of a non-haemadsorbing strain of African swine fever (ASF) virus from a natural outbreak of the disease. Veterinary Record 94: 2. https://doi.org/10.1136/vr.94.1.2

Plowright, W., Parker, J. and Peirce, M.A., 1969. African swine fever virus in ticks (*Ornithodoros moubata*, Murray) collected from animal burrows in Tanzania., Nature. https://doi.org/10.1038/2211071a0

Plowright, W., Thomson, G.R. and Neser, J.A., 1994. African swine fever. In: Coetzer, J.A.W., Thomson, G.R. and Tustin, R.C. (eds.), Infectious diseases of livestock with special reference to Southern Africa. Oxford University Press, Cape Town, South Africa, pp. 567-599.

Polo Jover, F. and Sanchez-Botija, C., 1961. La peste porcine Africaine en Espagne. Bulletin de l'Office International Des Épizooties 55: 148-175.

Portugal, R., Coelho, J., Höper, D., Little, N.S., Smithson, C., Upton, C., Martins, C., Leitão, A. and Keil, G.M., 2015. Related strains of African swine fever virus with different virulence: Genome comparison and analysis. Journal of General Virology 96: 408-419. https://doi.org/10.1099/vir.0.070508-0

Portugal, R., Leitão, A. and Martins, C., 2018. Modulation of type I interferon signaling by African swine fever virus (ASFV) of different virulence L60 and NHV in macrophage host cells. Veterinary Microbiology 216: 132-141. https://doi.org/10.1016/j.vetmic.2018.02.008

Reis, A.L., Abrams, C.C., Goatley, L.C., Netherton, C., Chapman, D.G., Sanchez-Cordon, P. and Dixon, L.K., 2016. Deletion of African swine fever virus interferon inhibitors from the genome of a virulent isolate reduces virulence in domestic pigs and induces a protective response. Vaccine 34: 4698-4705. https://doi.org/10.1016/j.vaccine.2016.08.011

Sanchez-Botija, C., 1963. Reservorios del virus de la P.P.A-Investigacion del virus de la P.P.A en los artropodos mediante la prueba de la hemoadsorcion. Bulletin de l'Office International Des Épizooties 60, 895-899.

Takamatsu, H., Martins, C., Escribano, J.M., Alonso, C., Dixon, L.K., Salas, M.L. and Revilla, Y., 2011. Family *Asfarviridae*. In: King, A.M.Q., Adams, M.J., Carstens, E.B. and Lefkowitz, E.J.B.T.-V.T. (eds.), Virus taxonomy. Elsevier, San Diego, CA, USA, pp. 153-162. https://doi.org/10.1016/B978-0-12-384684-6.00012-4

Takamatsu, H.H., Denyer, M.S., Lacasta, A., Stirling, C.M.A., Argilaguet, J.M., Netherton, C.L., Oura, C.A.L., Martins, C. and Rodríguez, F., 2013. Cellular immunity in ASFV responses. Virus Research 173: 110-121. https://doi.org/10.1016/j.virusres.2012.11.009

Thomson, G.R., Gainaru, M.D. and Van Dellen, A.F., 1979. African swine fever: pathogenicity and immunogenicity of two non-haemadsorbing viruses. Onderstepoort Journal of Veterinary Research 46: 149-154.

Van Furth, R. and Cohn, Z.A., 1968. The origin and kinetics of mononuclear phagocytes. Journal of Experimental Medicine 128: 415-435. https://doi.org/10.1084/jem.128.3.415

Vigário, J.D., Terrinha, A.M. and Nunes, J.F.M., 1974. Antigenic relationships among strains of African swine fever virus. Archiv für die Gesamte Virusforschung 45: 272-277. https://doi.org/10.1007/BF01249690

Wardley, R.C., Norley, S.G., Wilkinson, P.J. and Williams, S., 1985. The role of antibody in protection against African swine fever virus. Veterinary Immunology and Immunopathology 9: 201-212. https://doi.org/10.1016/0165-2427(85)90071-6

Wilson, A.J., Ribeiro, R. and Boinas, F., 2013. Use of a Bayesian network model to identify factors associated with the presence of the tick *Ornithodoros erraticus* on pig farms in southern Portugal. Preventive Veterinary Medicine 110: 45-53. https://doi.org/10.1016/j.prevetmed.2013.02.006

Yáñez, R.J., Rodríguez, J.M., Nogal, M.L., Yuste, L., Enríquez, C., Rodriguez, J.F. and Viñuela, E., 1995. Analysis of the complete nucleotide sequence of African swine fever virus. Virology 208: 249-278. https://doi.org/10.1006/viro.1995.1149

2. African swine fever virus: cellular and molecular aspects

A. Urbano[1], J.H. Forth[2], A.S. Olesen[3], L. Dixon[4], T.B. Rasmussen[5], G. Cackett[6], F. Werner[6], A. Karger[7], G. Andrés[8], X. Wang[9], D. Perez-Nuñez[8], I. Galindo[10], A. Malogolovkin[11], Y. Revilla[8], C. Alonso[10], C. Gallardo[12], S. Blome[2], E. Arabyan[13], H. Zakaryan[13] and F. Ferreira[1*]

[1]Centre for Interdisciplinary Research in Animal Health, Faculty of Veterinary Medicine, University of Lisbon, 1300-477 Lisbon, Portugal; [2]Friedrich-Loeffler-Institut, Institute of Diagnostic Virology, Federal Research Institute for Animal Health, Südufer 10, 17493 Greifswald, Germany; [3]Department of Veterinary and Animal Sciences, University of Copenhagen, Grønnegårdsvej 15, 1870 Frederiksberg C, Denmark; [4]The Pirbright Institute, Ash Road, Pirbright, Surrey, GU24 0NF, United Kingdom; [5]Department of Virus & Microbiological Special Diagnostics, Statens Serum Institut, Artillerivej 5, 2300 Copenhagen S, Denmark; [6]RNAP lab, Institute for Structural and Molecular Biology, Division of Biosciences, University College London, Gower Street, London, WC1E 6BT, United Kingdom; [7]Institute of Molecular Virology and Cell Biology, Friedrich-Loeffler-Institut, Federal Research Institute for Animal Health, Südufer 10, 17498 Greifswald, Germany; [8]Centro de Biología Molecular Severo Ochoa, Consejo Superior de Investigaciones Científicas, Calle Nicolás Cabrera, 1, 28049 Madrid, Spain; [9]Laboratory of Infection and Immunity, National Laboratory of Macromolecules, Institute of Biophysics, Chinese Academy of Sciences, Beijing 100101, China; [10]Dpt. Biotechnology, Instituto Nacional de Investigación y Tecnología Agraria y Alimentaria (INIA), Ctra. de la Coruña km 7.5, 28040 Madrid, Spain; [11]Federal Research Center for Virology and Microbiology, Pokrov, Russia; [12]European Union Reference Laboratory for African Swine Fever (EURL), Centro de Investigación en Sanidad Animal, INIA-CISA, Valdeolmos, CP: 28130 Madrid, Spain; [13]Institute of Molecular Biology of NAS, 0014, Yerevan, Armenia; fernandof@fmv.ulisboa.pt

Abstract

Over the last years, African swine fever virus (ASFV) has spread to several European and Asian countries, presently showing an unprecedented geographic distribution. The present chapter focuses on current knowledge and advances in the cellular/molecular features of ASFV, highlighting the gaps and future perspectives. The first half of the chapter addresses the general features of ASFV, its phylogeny and evolution, together with an overview of the viral transcription mechanisms and transcriptomics and the proteomics of ASFV-infected cells. The second half of the chapter summarises the structure and composition of the infectious ASFV particle, the mechanisms that lead to the infection and replication of the virus at the cellular level, and the viral-pig interactions. The last part of the chapter presents an overview of the currently described antiviral agents against ASFV.

Keywords: African swine fever virus, transcriptomics, proteome, virus-host interactions, antivirals

This publication is based upon work from COST Action CA15116, ASF-STOP, supported by COST (European Cooperation in Science and Technology).
www.cost.eu

EUROPEAN COOPERATION IN SCIENCE & TECHNOLOGY

Funded by the Horizon 2020 Framework Programme of the European Union

DOI 10.3920/978-90-8686-910-7_2,

2.1 General features of African swine fever virus

African swine fever virus (ASFV) is a large double-stranded DNA (dsDNA) virus, the sole member of the *Asfaviridae* family, genus *Asfivirus*, although a tentative new member of the family, provisionally designated as Abalone asfa-like virus (AbALV), has recently been described (Matsuyama *et al.*, 2020). ASFV is also the only DNA arbovirus, a unique group of animal viruses that depends on arthropod vectors for transmission. Virions are enveloped, with icosahedral morphology and an average diameter of around 250 nm. Virion structure consists of a nucleoid of 80 nm of diameter, which is coated by a thick protein core shell; an internal lipid membrane, surrounding the core; a capsid exhibiting icosahedral symmetry that is composed of 2,772 capsomers; and a dispensable external lipid-containing envelope that is obtained when the virus buds out through the plasma membrane (Andrés *et al.*, 2020; Liu *et al.*, 2019; Wang *et al.*, 2019). ASFV particles encode about 70 proteins, including a number of enzymes and factors needed for early mRNA transcription and processing, and enzymes involved in nucleotide metabolism, DNA replication and repair or transcription (Alejo *et al.*, 2018; Alonso *et al.*, 2018; Freitas *et al.*, 2019).

The viral genome consists of a single molecule of linear, covalently linked dsDNA. It contains terminal inverted repeats of 2.1 kbp units at both ends and complementary terminal loops present inverted flip-flop forms. The inverted repeat sequences are further characterised by numerous tandem repeat arrays. It ranges in length between 165 and 194 kbp and encodes between 151 and 167 open reading frames (Alonso *et al.*, 2018; Malogolovkin *et al.*, 2020). These differences in genome size are primarily due to variable copy numbers of several multigene families (MGFs), located in the left- and right-hand variable regions of the genome. ASFV contains five MGFs, named after the average number of codons present in each gene. Promotion of homologous recombination or unequal crossover during DNA replication within infected cells have been proposed as likely drivers of the loss and exchange of genetic material observed in virus isolates (Netherton *et al.*, 2019).

Partial sequencing of the *p72* (*B646L*) gene, which encodes the ASFV major capsid, has identified 24 different genotypes; out of the 30 complete genome sequences that are currently available, however, more than two-thirds are of three related genotypes, limiting opportunities to infer evolutionary relationships (Netherton *et al.*, 2019). Nevertheless, genome sequencing has shown a distant relationship between ASFV and other members of the nucleocytoplasmic large DNA virus (NCLDV) group, which includes *Iridoviridae*, *Poxviridae*, *Mimiviridae* and *Phycodnaviridae*, and other still unclassified giant viruses that infect lower eukaryotes, including the faustovirus, pacmanvirus and kamoebaviruses (Netherton *et al.*, 2019). The ASFV genome encodes for enzymes required for early RNA synthesis and processing, and shows DNA replication mechanisms similar to poxvirus (Alonso *et al.*, 2018). In addition, a mechanism of rapid amplification of individual genes by gene duplication under selection pressure observed in poxviruses was recently proposed as a likely contributor to the differences in copy number of individual MGFs in ASFV (Netherton *et al.*, 2019). Haemadsorption inhibition assays and cross-protection *in vivo* experiments have allowed the definition of eight different ASFV serotypes and, recently, genetic signatures of serotype specificity have been identified, which opens new insights that may fill the knowledge gap between genetic and antigenic diversity (Malogolovkin *et al.*, 2020).

ASFV replicates in cells of the mononuclear-phagocytic system, resident macrophages and specific reticular cells of natural hosts. *In vitro*, several isolates have been adapted to replicate in endothelial cell lines, a process that may lead to an attenuated phenotype (Alonso *et al.*, 2018; Malogolovkin *et al.*, 2020). Replication occurs mainly in the cytoplasm; although during early stages the nucleus is also a site of viral DNA synthesis (Galindo and Alonso, 2017; Simões *et al.*, 2015), its precise role in viral replication remains unclear. The infectious cycle starts with viral adsorption and entry into the host cell, an event which has been characterised as a low pH- and temperature-dependent process consistent with receptor-mediated dynamin-, and clathrin-dependent endocytosis; however, the receptor or receptors for the virus remain unknown and there is evidence that other mechanisms, such as phagocytosis and macropinocytosis may also be involved (Galindo and Alonso, 2017). Several ASFV proteins have been identified to be involved in the entry mechanism, such as p30, important for viral internalisation, and p12 and p54, potentially involved in viral attachment (Galindo and Alonso, 2017). Following entry, the virus enters the endosomal pathway, where it progresses through different endosomal compartments. Viral uncoating and fusion with the endosomal membrane occur when the virus particle reaches the mature endosomal compartment; host factors are then utilised for viral uncoating, with the acidic pH of the endosomal lumen facilitating disassembly (Hernáez *et al.*, 2016). Newly synthesised virions are then assembled in viral factories, where the main late phase of DNA replication also occurs, and exit the cell either by exocytosis or through the formation of apoptotic bodies (Alonso *et al.*, 2018; Galindo and Alonso, 2017). The factories are aggresome-like structures surrounded by endoplasmic reticulum (ER) membranes and vimentin cages that develop near the nucleus and the microtubule organisation centre of infected cells. They contain great amounts of viral proteins, DNA, and materials used to produce viral envelopes, but exclude host proteins, suggesting that their formation is induced by the virus. In addition to ER membranes viral factories are also surrounded by large clusters of mitochondria.

Infectivity is stable over a wide pH range, with some virions surviving treatment at pH4 or pH13. Virions are sensitive to irradiation and some disinfectants; paraphenylphenolic disinfectants, in particular, are very effective (see Chapter 11). While virions are inactivated at 60 °C within 30 minutes, they survive for years at -70 or 4 °C (Alonso *et al.*, 2018).

2.2 Genome, phylogeny and evolution of African swine fever virus

African swine fever virus contains terminal inverted repeats that consist of arrays of different tandemly repeated sequences interspersed with short open reading frames. This genome structure is shared with members of the *Poxviridae* and reflects the mechanism of cytoplasmic genome replication via head-to-head concatemers, which are resolved into unit length genomes. The genome has an A+T content of 61-62%. Genomic variation between ASFVs is mainly due to loss or gain of the genome termini among members of several MGFs. These MGFs have evolved in the ASFV genome by a process of gene duplication, presumably under strong selection pressure during the virus evolution, for example adaptation to replicate and survive in mammals. Transposition of sequences of some MGFs from one genome end to the other has occurred. The genome includes ~150-170 open reading frames (ORFs) (Cackett *et al.*, 2020; Dixon *et al.*, 2013) encoding structural proteins and non-structural proteins involved in morphogenesis, replication, and evasion of the host defence (Dixon *et al.*, 2013). Sixty-eight virus-encoded proteins have been identified within the ASFV particle (Alejo *et al.*, 2018). The ORFs are closely spaced with short upstream promoter sequences and are read from both DNA strands. As predicted for a cytoplasmic virus, no introns have been identified in ORFs.

The molecular epidemiology of ASFV has conventionally been investigated using partial sequencing of the gene encoding the major capsid protein, *p72*, (Bastos *et al.*, 2003) resulting in differentiation into 24 genotypes (Figure 2.1). Partial sequence analysis of the *p72*-gene has placed ASFVs from Georgia within genotype II along with viruses from Madagascar, Mozambique and Zambia (Rowlands *et al.*, 2008). Using *p72*-sequencing, ASFVs from Russia, Siberia, Eastern Europe, Belgium and China have also been placed within genotype II. When extending *p72*-sequencing with sequencing of additional regions, most of the analysed sequences from the currently circulating ASFVs are 100% identical (Malogolovkin *et al.*, 2012; Rowlands *et al.*, 2008). Some variation in number or sequence of tandem repeats within ORFs (e.g. B602L) or in intergenic regions has been observed and used to define a limited number of variants. Even though partial sequencing of some of these regions within the ASFV genome can be used as a tool to study the transmission of closely related ASFVs, complete genome sequences can provide more information about the phylogeny. This can include identifying useful additional markers for epidemiological tracing and providing a link between genetic changes and observed phenotypes.

Using full-length genome sequencing, complete genomes (or near complete) from genotype II ASFVs from Georgia, Russia, Europe and Asia have been described (Forth *et al.*, 2020a). Following comparisons between these full-length ASFV genotype II genomes, only low levels of variation have been reported (Figure 2.2).

The evolution of ASFV is one of the greatest mysteries in modern virology. Despite the occurrence on four continents and intensive research for almost 100 years, no closely related virus has been described yet (Alonso *et al.*, 2018). As classified by the International Committee on Taxonomy of Viruses (ICTV), ASFV is currently the only member of the family *Asfarviridae* – where 'ASFAR' stands for 'African Swine Fever and Related Viruses' – and the genus *Asfivirus* (Alonso *et al.*, 2018). Together with families of other large dsDNA viruses sharing biological features, such as replicating in the cytoplasm of their host cell, including *Poxviridae* and *Mimiviridae*, the *Asfarviridae* have only recently been classified into the Phylum 'Nucleocytoviricota' (ICTV Master Species List 2019.v1) (Figure 2.3). Although still only distantly related, the closest known relatives have been classified into the same order as ASFV, the *Asfuvirales*. These viruses belong to the family of the *Faustoviridae* or the genus *Kaumoebavirus*, both giant dsDNA viruses with genomes of over 400 kbp that were discovered in amoebae. However, only about 30 genes are recognised to be conserved between different families in the Asfuvirales, and these encode enzymes and factors required for virus replication or assembly. Interestingly, the gene for the major capsid protein (MCP) is conserved, but in *Faustoviridae* is spread across 13 exons covering a 17 kbp genome region. Both self-splicing and splicesome-like introns are present in this MCP gene. More recently, an Asfarvirus-like virus was identified as likely cause of mass die off abalone (Matsuyama *et al.*, 2020). This abalone-asfar-like virus (AbALV) shares 48 (from 159 identified) similar proteins with ASFV and shows gene-order conservation. Phylogenetically this virus is so far the closest to ASFV, although still distantly related.

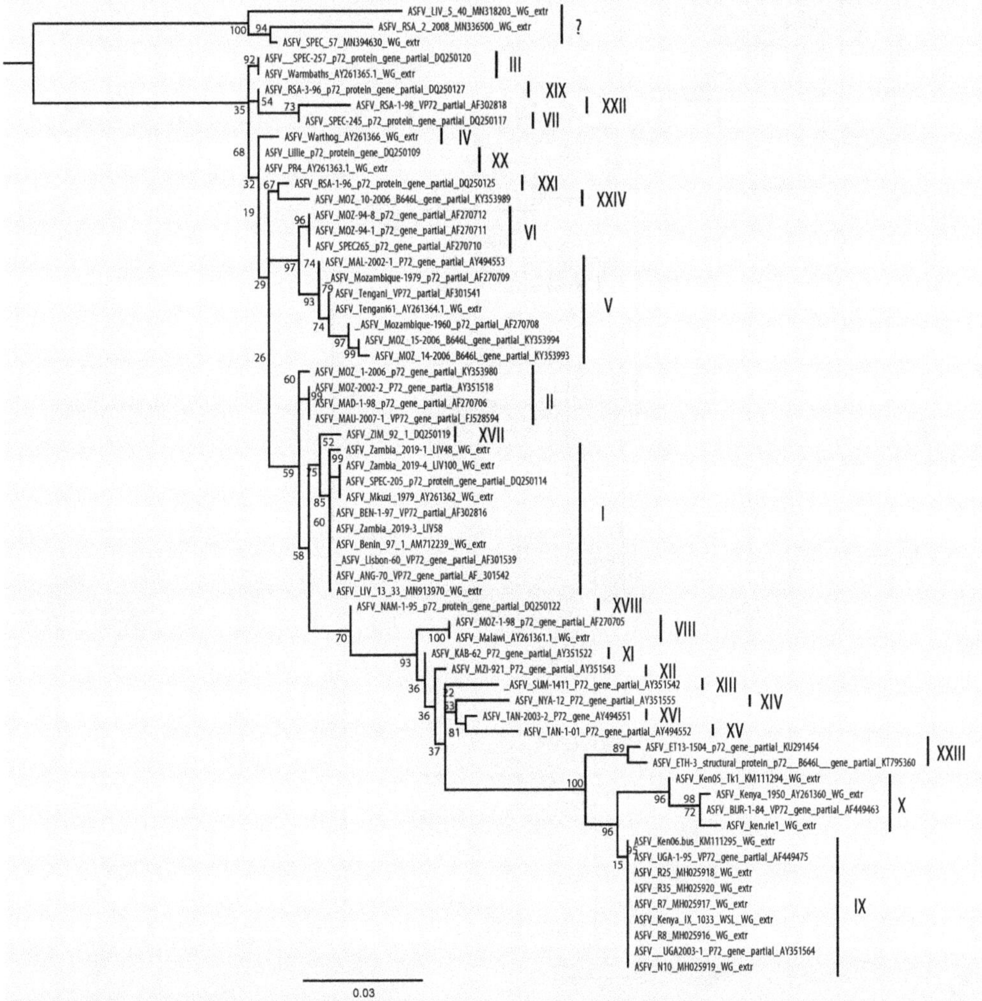

Figure 2.1. Phylogenetic tree of nucleic acid sequences of ASFV-B646L genes. The maximum-likelihood (ML) tree was constructed using IQ-TREE v1.6.5 based on MAFFT v7.388 aligned ASFV B646L gene downloaded from the International Nucleotide Sequence Database Collaboration (INSDC) (status 05.05.2020). Standard model selection was used, resulting in the best-fit model TVMe + R2 (Transversion model, AG = CT and equal base freq. + FreeRate model with 2 categories). Statistical support of 1000 ultrafast bootstraps using the ultrafast bootstrap approximation (UFBoot) (percentage values) is indicated at the nodes. Taxon names include, where available, ASFV-designation and INSDC accession number and P72 genotype. The scale bar represents number of substitutions per site.

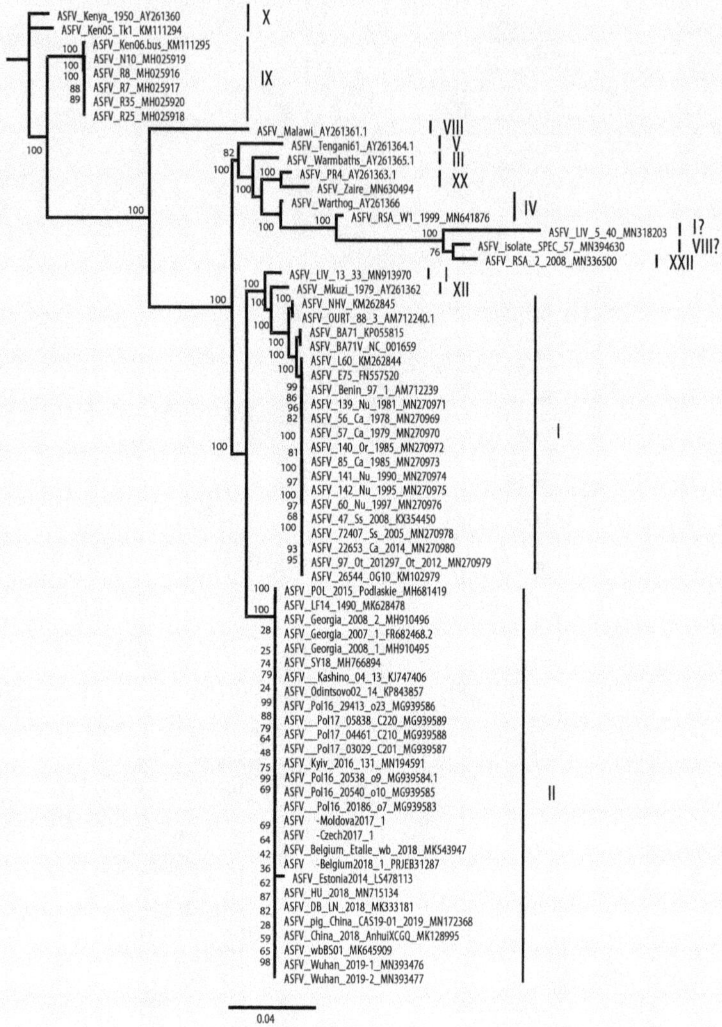

Figure 2.2. Phylogenetic tree of nucleic acid sequences of ASFV-whole genome sequences. The maximum-likelihood (ML) tree was constructed using IQ-TREE v1.6.5 based on whole-genome sequences downloaded from the International Nucleotide Sequence Database Collaboration (INSDC) (status 05.05.2020). Standard model selection was used, resulting in the best-fit model GTR+F+R4 (General time reversible model with unequal rates and unequal base freq + empirical base frequencies + FreeRate model with 4 categories). Statistical support of 1000 ultrafast bootstraps using the ultrafast bootstrap approximation (UFBoot) (percentage values) is indicated at the nodes. Taxon names include, where available, ASFV-designation and INSDC accession number and P72 genotype. The scale bar represents number of substitutions per site.

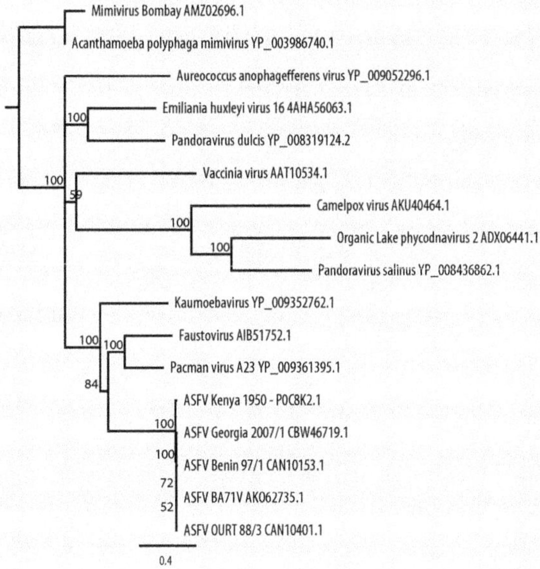

Figure 2.3. Phylogenetic tree of nucleic acid sequences of EP1242L amino acid sequences of member of the Nucleocytoviricota. The neighbour joining tree was constructed from MAFFT 7.388 aligned ASFV EP1242L amino acid sequences using the Geneious tree builder in Geneious employing the Jukes-Cantor model and the UPGMA tree build method and 1000 bootstraps. Taxon names include, where available, ASFV-designation and INSDC accession number and P72 genotype. The scale bar represents number of substitutions per site.

While studies on long-term evolution have been hindered by the absence of related viruses and the lack of old samples, studies on ASFV short-term evolution have been performed during outbreaks in swine (*Sus scrofa*) populations. However, all reported or were complicated by the limited genetic variability, caused by the very low mutation rate of the ASFV genome (Forth *et al.*, 2019) (Figure 2.2).

The highest genetic variability of ASFV strains is reported from the sylvatic cycle in Africa, where ASFV is transmitted between soft ticks of the genus *Ornithodoros* and an indigenous African suid, the warthog (*Phacochoerus africanus*; Bastos *et al.*, 2003). Due to the high degree of adaption of ASFV and its natural hosts, which show no symptoms, it has been hypothesised that they have co-evolved for a long time. This hypothesis was further strengthened by the recent discovery of endogenous viral elements in the genomes of *Ornithodoros moubata* and *Ornithodoros porcinus* ticks. These elements, representing almost 10% of the ASFV genome, probably integrated around 1.46 million years ago allowing for a glimpse into viral evolution and tick-virus coevolution (Forth *et al.*, 2020b).

2.3 African swine fever virus transcription and transcriptomics

A major limitation for developing effective antiviral treatments for ASFV has been our limited knowledge of the molecular mechanisms underlying viral gene expression. A fundamental first step in gene expression is the transcription of its sequence into mRNA by the RNA polymerase (RNAP), before translation into its subsequent polypeptide sequence – these two steps making up the 'central dogma' of biology. Unsurprisingly, the domains of life (bacteria, archaea, and eukaryotes) show conservation in RNAP structure and mechanisms for the stages of transcription initiation, elongation and termination, including transcription factors that modulate these steps. This conservation extends to viral transcription machinery encoded by nucleocytoplasmic large DNA viruses – NCLDVs (Mirzakhanyan and Gershon, 2017), where our understanding is extrapolated from well-studied poxviruses like vaccinia virus (VACV), or from the eukaryotic Pol II transcription system (Broyles, 2003; Rodríguez and Salas, 2013).

Both VACV and ASFV have 170-200 kb dsDNA genomes and replicate in the host cytoplasm. Like VACV, ASFV encodes and carries in virions all factors required to express and modify transcripts during early infection, with solubilised ASFV particles being transcriptionally competent *in vitro* (Alejo *et al.*, 2018; Rodríguez and Salas, 2013). Table 2.1 summarises components of the ASFV-encoded transcription machinery, including a multisubunit RNAP and transcription initiation factors with well-characterised homologues in either VACV or eukaryotes. NCLDVs, like ASFV, are transcriptionally self-sufficient, which may enable ASFV to propagate both in pig species and its evolutionary-distant soft tick vectors.

2.3.1 Discrete stages of viral gene expression during infection

Following the infection of the cell, viral genes are expressed in a precise and regulated fashion. It has been proposed that ASFV follows four stages of viral gene expression akin to the VACV system: immediate early, early, intermediate, and late. To some extent, these stages can be distinguished by treating cells with combinations of replication- and/or translation inhibitors (cytosine arabinoside and cycloheximide, respectively). Expression of some ASFV-BA71V genes has been studied using primer extension and S1 nuclease mapping, providing evidence for gene expression stages prior to viral proteins synthesis (immediate early), pre-DNA-replication (early) and post-replication (late). However, several transcripts could also be detected at multiple time-points (Figure 2.4A), and the intermediate stage has not yet been clearly distinguished from the late (Cackett *et al.*, 2020; Rodríguez and Salas, 2013). In summary, the experimental support for four distinct stages of gene expression is tenuous, while it is persuasive that there are at least two distinct phases to ASFV transcription during early and late infection, divided by viral genome replication.

2.3.2 A genome-wide view of transcription

The first genome-wide view of ASFV transcription (Cackett *et al.*, 2020) revealed that almost all previously identified ASFV-BA71V genes are actively transcribed. This is particularly true during late infection when the total genome-wide level of transcripts is higher, possibly due to a larger number of viral genome equivalents and thus more DNA template being available post-replication. Highly transcribed genes also agree with highly abundant proteins identified

Table 2.1. Evolutionary conservation of RNA polymerase subunits and general transcription factors.[1]

Eukaryotic Pol II subunit and factors	ASFV gene	VACV protein	VACV gene	Function
RPB1	NP1450L	Rpo147	J6R	catalytic subunit
RPB2	EP1242L	Rpo132	A24R	catalytic subunit
RPB3-11 fusion	H359L	Rpo35	A29R	assembly platform
RPB5	D205R	Rpo22	J4R	auxiliary subunit, DNA interaction
RPB6	C147L	Rpo19	A5R	auxiliary subunit
RPB7 fusion	D339L	Rpo18	D7R	stalk, RNA interaction
RPB9	C105R	–	–	auxiliary subunit, TSS selection
RPB10	CP80R	Rpo7	G5.5R	assembly platform
-	–	Rap94	H4L	early transcription
-	G1340L	VETF-L	A7L	initiation
-	D1133L	VETF-S	D6R	initiation, ATPase
TBP	B263R		A23L	initiation, TATA-binding
TFIIB	C315R		–	initiation, BRE-binding, RNAP recruitment
TFIIS	I243L		E4L (Rpo30 subunit)	elongation, release of arrested RNAP
-	Q706L	NPH I	D11L	termination, chromatin
-	NP868R	VTF/CE	D1R & D12L	capping enzyme

[1] The relationship between eukaryotic and NCDLV transcription systems from ASFV and VACV is apparent from conservation of Pol II subunit homologues, as well as core general transcription factors that guide Pol II function. The putative predicted functions of transcription factors are indicated. Note that the mRNA capping enzyme (VTF/CE) and termination factor (NPH I) is stably integrated in the VACV-RNAP.

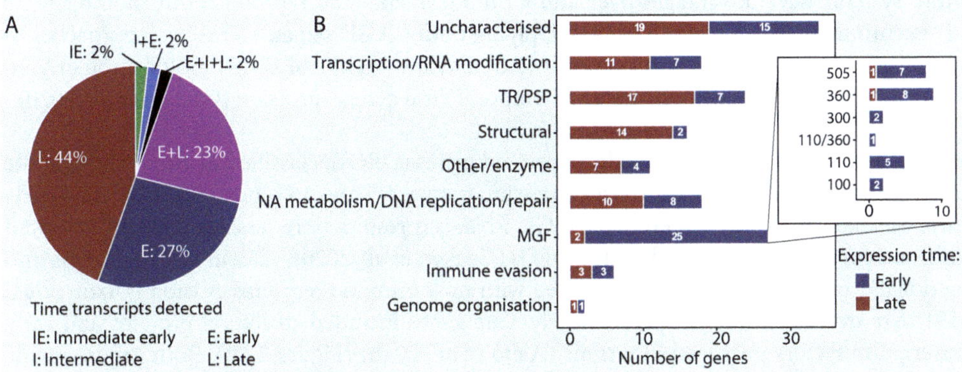

Figure 2.4. Gene expression stages of ASFV genes. (A) Summary of individual gene transcripts which have been experimentally investigated in the literature (Broyles, 2003; Rodríguez and Salas, 2013). (B) Differentially expressed genes during early and late infection of Vero cells with ASFV strain BA71V (Cackett *et al.*, 2020) are categorised according to function: early (downregulated) and late (upregulated) genes are indicated in blue and red, respectively).

in proteomic studies of ASFV-infected cells and virus particles (Alejo *et al.*, 2018), which will be discussed further in this chapter. Discrete gene classification can, however, be hindered by not achieving complete synchronicity. Furthermore, early mRNAs with long half-lives may be detected in late infection, since RNA-seq analyses represent mRNA levels, which are the sum of RNA synthesis and decay.

2.3.3 Mapping ASFV transcripts

ASFV transcriptomics employ next generation sequencing (NGS) techniques to investigate transcription from initiation to termination. Classical RNA-seq can give relative mRNA abundance but is poor at discerning the 5'- and 3'-termini of mRNAs, especially in a compact viral genome, where genes are tightly packed and RNAPs may not terminate with 100% efficiency. The resulting 'read through' of RNAP from one gene into the downstream gene can make it difficult to assign a given mRNA to its promoter. ASFV mRNAs are modified with 5'-Cap and 3'-poly(A) tails, facilitating use of specialised NGS techniques. These include CAGE-seq and 3'-RNA-seq, that are able to identify the 5'- and 3'-nucleotides of transcripts, respectively (Cackett *et al.*, 2020). Nanopore sequencing enabling the reading of very long sequences has also recently been applied to the ASFV transcriptome (Olasz *et al.*, 2020). This method can identify mRNA isoforms and, like CAGE-seq, circumvents the problem of promoter misannotation due to read-through, since complete, or near-complete, mRNAs can be read.

2.3.4 mRNAs expressed during early and late infection

Both CAGE-seq and RNA-seq have been utilised to quantify ASFV gene transcripts during infection. Classical RNA-seq has been used on blood of attenuated OURT 88/3-infected- and highly virulent Georgia 2007/1 (GRG)-infected pigs, showing viral gene expression levels varied substantially between infected animals, as well as due to differences in isolate virulence. Tissue culture systems have advantages over animal models, enabling reproducibility, homogeneity and synchronicity of infection. Analyses applying both CAGE-seq and RNA-seq approaches to quantify gene expression levels during early (5 h) and late (16 h) ASFV-BA71V infection of Vero cells, provided more in-depth insight into temporal viral gene expression (Cackett *et al.*, 2020).

CAGE-seq identified 149 differentially expressed genes (DEGs), classified as either early or late genes based on significant differential expression between 5 and 16 h post infection: 42% early genes, 54% late, and only 7 were not DEGs. RNA-seq results were less sensitive, but showed significant correlation between detected DEGs (Cackett *et al.*, 2020). Classification of DEGs into functional groups showed MGFs associated with host immune response evasion (Dixon *et al.*, 2019) were overwhelmingly expressed early. Late genes included structural proteins and early transcription factors packaged in virions (Alejo *et al.*, 2018) (Figure 2.4B). Both genome-wide ASFV expression profiles of pigs and Vero cells identified many highly expressed viral genes that remain functionally uncharacterised, highlighting genome-wide studies as a useful tool for directing further research.

2.3.5 ASFV transcription start sites and promoter motifs

CAGE-seq has been further used to map 5' ends of ASFV transcripts, locating transcription start sites (TSSs) for 151 of 153 annotated ASFV-BA71V genes, and 7 novel genes. ASFV can also utilise alternative (intragenic) TSSs, potentially encoding time-point-dependent N-terminally truncated open reading frames, though these remain to be investigated at the protein level. Information on which of the ORFs are expressed and the exact start site complements data available from genome sequencing comparisons described in Chapter 6 of this book. Mapping TSSs facilitates identification of promoter motifs upstream, potentially recognised by transcription initiation factors. Promoters for early- and late gene TSSs have been identified (EPM and LPM, respectively) (Figure 2.5A-B). A conserved transcription initiator (Inr) element overlaps TSSs: TA* for early genes, and TA*TA for late, where A* is the TSS (Figure 2.5C-D) (Cackett *et al.*, 2020). The EPM resembles the Upstream Control Element of early VACV genes, which binds the viral heterodimeric D6-A7 transcription initiation factor (Broyles, 2003), conserved in ASFV (Rodríguez and Salas, 2013). Possible LPM-interacting initiation factor candidates are ASFV homologues for TATA-binding protein (TBP) and transcription factor TFIIB, which in Pol II and archaeal systems bind the TATA-box and B-recognition elements upstream of the TSS, respectively. RNAP is then recruited, forming the preinitiation complex (Werner and Grohmann, 2011). However, there is no significant similarity between the LPM and TATA-box consensus. ASFV also encodes further putative viral initiation factors, which might play a role (Rodríguez and Salas, 2013).

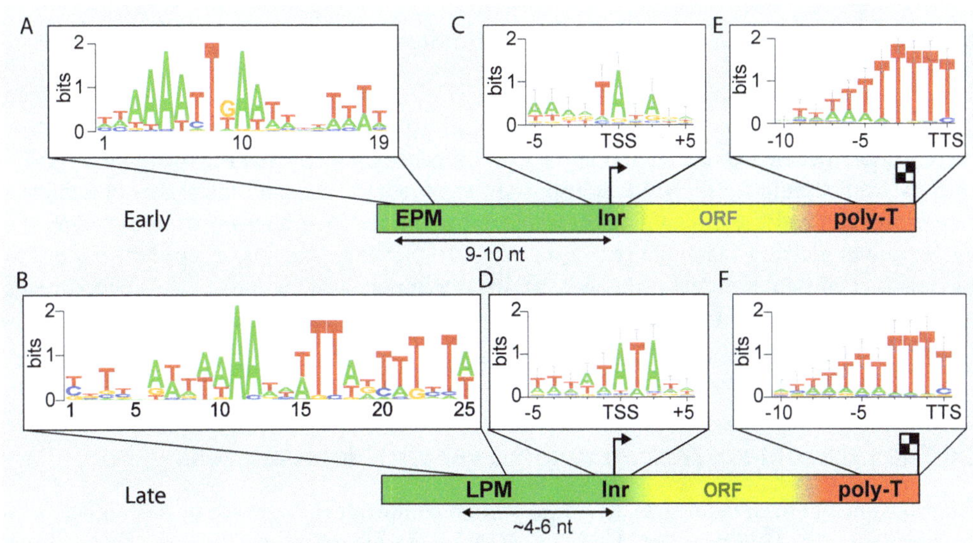

Figure 2.5. DNA consensus motifs for ASFV promoters, initiators and terminators (Cackett *et al.*, 2020). The sequence features of early and late gene promoter regions, including the EPM (A), LPM (B) and Inr elements (C-D). Note that spacing between the promoter motif and Inr varies in early and late gene promoters. Transcription termination is associated with a polyT signature (E-F), though a significant number of ASFV genes do not contain this termination motif hinting at an alternative, possibly factor-dependent, mode of termination.

2.3.6 Transcription termination signature motifs

The RNA 3'-seq analyses have identified transcription termination sites for more than two thirds of ASFV genes including a ~6-7 residue polyT (i.e. polyU in the mRNA) termination motif among both early and late genes (Figure 2.5E-F) (Cackett *et al.*, 2020). This resembles 'intrinsic' termination motifs for archaeal RNAP (Werner and Grohmann, 2011) and eukaryotic Pol III (Nielsen *et al.*, 2013). However, a third of ASFV genes showed no conserved termination motifs, indicating an alternative and possibly factor-dependent termination mechanism (Cackett *et al.*, 2020), likely by VACV-like RNA helicases, which facilitate mRNA release and termination (Freitas *et al.*, 2019; Rodríguez and Salas, 2013).

2.3.7 Untranslated regions and RNA-5' tailing

The mapping of TSS in conjunction with well-annotated ORFs (i.e. start and stop codons) enables the precise characterisation of untranslated regions. Surprisingly, many ASFV mRNAs included short extensions at the RNA-5' end comprised of 1-2 copies of an 'AU' dinucleotide. The AU and AUAU-leaders are template-encoded since all mRNAs with this feature were transcribed from genes starting with the sequence A*TA corresponding to the Inr promoter motif. These extensions are thus likely generated by transcript slippage of the promoter-associated ASFV-RNAP (Cackett *et al.*, 2020). In eukaryotes and VACV, 5'-RNA poly(A) leaders are associated with improved translation efficiency, by circumventing the need for translation initiation factors (Dhungel *et al.*, 2017).

2.3.8 The importance of transcriptomics for fundamental and applied ASFV research

Transcriptomics are invaluable for the study of ASFV transcription and the interplay between virus- and host gene expression, and for the identification of highly expressed viral genes that warrant more detailed characterisation. NGS approaches have enabled the mapping of ASFV mRNAs 'from start to finish', which is important to rationalise the molecular basis of temporal gene expression and the mechanisms of transcription initiation and termination including the core promoter and terminator consensus motifs. This knowledge can make key contributions to the design of efficient antiviral drugs and vaccine development in the future. Thus, investments in fundamental as well as applied research are equally important, because the discoveries of today can save lives tomorrow!

2.4 The proteome of African swine fever virus-infected cells

The proteome has been defined as the entire protein complement expressed by a genome, or by a cell or tissue type (Wilkins *et al.*, 1996a,b). When the term was coined by Marc Wilkins, large scale protein analysis was strongly linked to high resolution two-dimensional gel electrophoresis (2DE), which was combined with matrix-assisted laser desorption/ionisation time-of-flight (MALDI-TOF) mass spectrometry (MS) (Tanaka, 1988) as this technology evolved. With the rapid progress of bioanalytical technologies in general and of genome sequencing, mass spectrometry and bioinformatics in particular, the performance of proteomic workflows has improved dramatically in the last three decades. The combination of 2DE with MALDI-TOF

MS, a proteomic workhorse for years, was at least partially replaced by 'shotgun' approaches based on mass spectrometers coupled to liquid chromatography devices (LC-MS). These gel-free platforms have gained popularity as they require less pre-analytic sample preparation and allow large scale MS-based protein identification and quantitation at the same time. With the first scientific publication applying 2DE to ASFV infected cells appearing in 1986 (Esteves *et al.*, 1986), the analysis of the intracellular proteome of ASFV to date has a history of almost 35 years.

In virology, MS-based proteomics can be applied to pursue a variety of goals. A few examples would be the analysis of the composition of viral particles, the characterisation of the host cell response to infection, the establishment of protein-protein interaction networks, or the expression analysis of viral proteins. Although the dominant analytical platform in proteomics is mass spectrometry, other techniques may serve the purpose as well. For example, antibody-based assays like western blotting can be used to describe the expression of viral proteins in the course of an infection, if immunological reagents are available. However, the number of samples that can be analysed may be limited by the throughput of the assay format and quantitative analysis may be difficult, especially if different strains of viruses are compared. Therefore, MS-based proteomic workflows seem straightforward to analyse proteomes of viruses with large coding capacities like ASFV, especially if highly diverse genes like the members of the ASFV multi gene family (MGF) are present. Fortunately, the proteome of the domestic pig is available and the sequences of a number of ASFV strains are provided in public repositories. Recently, ASFV sequence data have been combined and made accessible in the ASFVdb database (http://asfvdb.popgenetics.net). For host organisms with less well characterised proteomes, like ticks of the *Ornithodoros* genus serving as vectors for ASFV, useful sequence databases for MS-based proteomics can be constructed from protein sequence data of higher taxonomic levels or on the basis of transcriptomic sequence databases (Oleaga *et al.*, 2017).

2.4.1 Two-dimensional gel electrophoretic studies

Already before mass-spectrometric analysis of proteins and peptides became possible or widely available, a number of 2D electrophoretic studies addressing the characterisation of the ASFV proteome in infected cells had been published. Although these studies did not establish links between the protein spots observed in 2DE and the identity of the protein, they provided valuable information. Especially the combination of 2DE with metabolic radioactive labelling in cell culture was very successful to identify newly synthesised viral proteins and to dissect the early and late phase of infection using inhibitors of viral DNA replication.

Esteves *et al.* (1986) combined 2DE, radioactive protein labelling and cytosine arabinoside treatment to dissect proteins expressed in the early and late phase of infection. A total of 106 proteins (or rather protein spots) were induced in extracts of Vero cells upon infection with ASFV strain L60. Comparison to the mobility of spots in preparations of viral particles allowed separation of 54 structural from non-structural proteins. Thirty-five proteins were classified as early, while the remaining 71 were expressed in the late phase. Of the 35 early proteins, 24 were also synthesised in the late phase. Although the proteins behind the 2DE spots were not identified, the power of 2DE for the analysis of viral proteomes, certainly those of large viruses like ASFV, was demonstrated.

In 1987, Urzainqui and co-workers demonstrated that 2DE is not only a powerful tool to analyse the kinetic phases of ASFV protein expression but also for the detection of post-translational modifications (Urzainqui *et al.*, 1987). Using a radioactive pulse labelling scheme, including pre-treatment with cycloheximide to enhance the expression of early genes and blocking of late genes with phosphonoacetic acid, they classified the expression of 34 proteins as immediate early, 13 as delayed early and 64 as late. In this study, all early proteins were expressed in the late phase as well. Moreover, 15 protein spots were labelled using $^{32}PO_4$, showing that they had been post-translationally phosphorylated.

A few years later, Alcaraz and co-workers (1992) compared the 2D protein patterns representing the proteomes of the four highly virulent ASFV field isolates E70, E75, 608, and 646, and analysed changes in the proteomes of the E70 isolate in the course of serial passages in cell culture. While most of the 86 ASFV-specific spots remained stable over the passages, it was observed that, beginning with passage 44, the p54 protein appeared in variants with modified molecular weights, representing different virus subpopulations. A concomitant decrease of viral titres in swine macrophages was observed (Alcaraz *et al.*, 1992). In follow-up studies, it was shown that the size differences in the p54 proteins were caused by differing copy numbers of a 12-nucleotide repeat introduced by serial passage (Rodriguez *et al.*, 1994).

In a 2DE study by Rodriguez and colleagues, modulations of the host cell proteome, rather than the ASFV proteome, were analysed after infection of alveolar macrophages and Vero cells with the virulent E70 strain and the Vero adapted strain BA71 (Rodriguez *et al.*, 2001). A radioactive pulse labelling scheme was implemented to assess protein synthesis at different times after infection. By quantitation of radioactive label incorporated into the protein spots over time, it was demonstrated that host cell protein synthesis was efficiently shut off by ASFV infection for the majority of host proteins. While over 900 protein spots detected in 2DE of porcine alveolar macrophages disappeared or became weaker in the course of the infection, 29 persisted with similar expression rates and 28 host proteins were stimulated two-fold, or more, in a time-dependent manner upon infection.

2.4.2 Studies including mass spectrometry

After introduction of MALDI-TOF MS and the implementation of the 2DE-MALDI-TOF MS workflow, the first study using this approach to analyse the host proteome of ASFV-infected Vero cells was published (Alfonso *et al.*, 2004). Of the 90 protein spots with differential expression that were resolved by 2DE, twelve overexpressed host proteins were identified by MALDI-TOF MS. Functionally, these proteins related to redox homeostasis, heat shock response, coagulation, and other processes that could play a role in ASFV pathogenesis. Notably, the cell culture supernatants of the infection experiments were also analysed. Two host proteins, thioredoxin and the β-galactoside soluble lectin, were reproducibly identified in the supernatants of Vero cells infected with any of four ASFV strains representing different virulence characteristics.

The studies discussed so far all used infected cultured cells as starting material. In contrast, Herrera-Uribe and co-workers analysed the proteome of lymph nodes from infected pigs. *Ex vivo* samples of organs are quite challenging for proteome analysis because of their higher heterogeneity, caused by the presence of different tissues (Herrera-Uribe *et al.*, 2018). In this study the course of the infection *in vivo* was compared between two homologous strains, virulent E75 and the attenuated strain E75CV1. In the samples from animals infected with the virulent strain a progressive loss of proteins was observed in the organ preparations, probably reflecting tissue destruction. Bioinformatic analysis showed that some cellular pathways were affected after infection with either strain, but there were also strain-specific differences. For example, cytoskeletal remodelling, clathrin-mediated endocytosis, and 14-3-3 protein mediated signalling were affected by both isolates, while the attenuated strain had a stronger influence on the expression of proteins involved in inflammatory and immunological pathways, reflecting the lack of E75 recognition by the immune system in the early times after infection. Many of the observed protein modulations fit well to known antiviral reactions of the cell and pathogenicity mechanisms, underlining the potential of proteomics to unravel the viral strategies to reprogramme cells and to monitor the cellular reaction to viral infection.

A shotgun proteomics approach was used by Kessler and co-workers to define the repertoire of ASFV proteins expressed in mammalian cell culture (Kessler *et al.*, 2018). Infection experiments were performed with a recombinant derivative of the attenuated ASFV OURT88/3 strain in three different cell lines. Vero cells were chosen as they have been extensively used for infection experiments in the past, especially together with a Vero adapted strain of BA71, whereas the wild boar WSL-HP cells and human HEK293 cells were previously infected *in vitro* and were derived from susceptible and non-susceptible host species, respectively. In the OURT88/3 strain, 157 ORFs have been annotated and for 94 of them the corresponding protein was detected by normal-phase liquid chromatography-MALDI-TOF MS/MS in one or more of the cell lines. In 23 cases, the proteins corresponding to the predicted ORFs (or to homologues in other ASFV isolates) could be experimentally confirmed in this study for the first time. The panel of expressed proteins varied with the cells. Only 54 were present in all three cell lines, indicating that these may be strictly required for the infection of cultured mammalian cells. In Vero and WSL-HP cells, the number of detected proteins was higher (88 and 83 respectively) than in HEK 293 cells (54). With 78 proteins overlapping between Vero and WSL-HP, there is some indication that the course of the infection in these cells may be similar but distinct from HEK 293 cells. This assumption was further corroborated by MS-based quantitation, which showed that expression levels of individual ASFV proteins correlated well between Vero and WSL-HP while larger variances were observed in comparison to HEK 293 cells. Surprisingly, some of the most abundantly expressed ASFV proteins, like pK145R and pI73R, are so far uncharacterised and may therefore be rewarding candidates for further functional characterisation. Screening of the identified viral proteins for post-translational modifications showed that the N-termini of four identified MFG 110 proteins (1L, 2L, 4L, and 5L) and of pI329L had been processed by cleavage of a signal peptide as predicted. Most of the 44 identified N-terminal peptides were present in the acetylated form.

Together with the shotgun proteomics study focusing on the structural proteins of strain BA71V (Alejo *et al.*, 2018), the study by Kessler and co-workers (2018) has shed some light on the dark part of the ASFV proteome. However, there still are a number of predicted ASFV ORFs for which no evidence for the expression of the corresponding protein has been presented to date (Karger *et al.*, 2019).

2.5 Structure and composition of the infectious African swine fever virus particle

Recent reports on the architecture and the protein composition of the infectious ASFV particle have brought out its huge structural and biological complexity. This subchapter summarises the most relevant features of the virion architecture, as well as the role of the different viral polypeptides characterised so far.

The extracellular ASFV particle, with an overall icosahedral morphology and a diameter of about 250 nm, displays a multi-layered structure that, according to conventional electron microscopy (EM) approaches, consists of an internal genome-containing nucleoid enclosed by a protein core shell, an inner lipid envelope, a protein outer capsid and an outer lipid envelope (Figure 2.6A) (Andrés *et al.*, 1997).

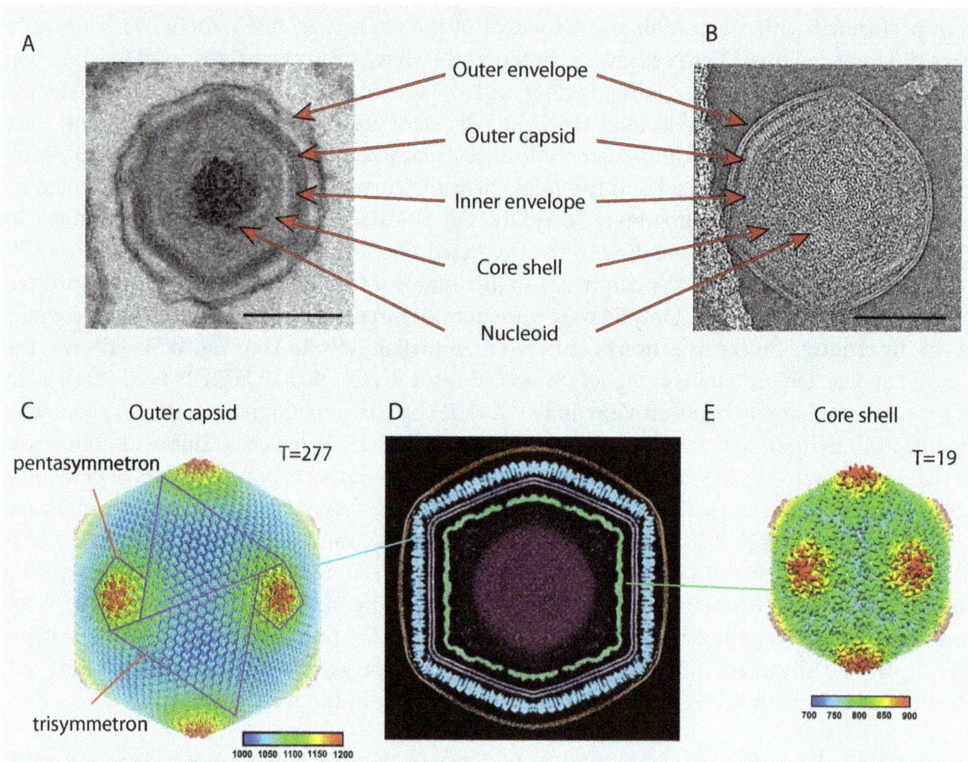

Figure 2.6. African swine fever virus (ASFV) architecture. Extracellular ASFV particles visualised by conventional EM after aldehyde fixation (A) or by cryo-EM after vitrification (B). The five concentric structural domains of the ASFV virion are indicated. Bars, 100 nm. Cryo-EM map of ASFV (D) and cryo-EM reconstructions of the outer capsid (C) and the core shell (E). The colour code indicates the radial distance from the virus centre. The triangulation number (T), the trisymmetrons and pentasymmetrons are indicated.

Like most of NCLDVs, ASFV replicates in specialised cytoplasmic areas close to the nucleus. In these viral factories, ASFV particles assemble from endoplasmic reticulum-derived membranes, which give rise to the inner viral envelope (Salas and Andrés, 2013). Icosahedral particles are formed by the progressive building of the outer capsid and the internal core shell on the opposite faces of the lipid bilayer. Subsequent packaging of the viral DNA along with viral nucleoproteins results in the assembly of the dense central nucleoid. Viral morphogenesis is followed by maturational events involving proteolytic processing of core shell components and microtubule-mediated transport of the intracellular particles to the cell surface. ASFV particles are eventually released by budding at the plasma membrane, where they acquire their outer envelope. Interestingly, both intracellular and extracellular viral forms are infectious.

2.5.1 ASFV architecture

The three-dimensional (3D) structure of the extracellular ASFV particle has been partly elucidated by using cryo-electron microscopy (cryo-EM) and single particle analysis (Andrés *et al.*, 2020; Liu *et al.*, 2019; Wang *et al.*, 2019). The cryo-EM reconstruction has unravelled a unique viral architecture consisting of two icosahedral protein capsids (the outer capsid and the core shell) and two lipoprotein membranes (the outer and inner envelopes), enclosing the central nucleoid (Figure 2.6B-E). The outer capsid is built by 2,760 pseudo-hexameric capsomers and 12 distinct pentameric capsomers arranged in an icosahedral lattice with a triangulation number T=277. The outer capsid is organised in 20 trisymmetrons (each one containing 120 pseudo-hexameric capsomers) and 12 pentasymmetrons (containing 30 pseudo-hexameric capsomers and a pentameric capsomer), that form the virion facets and vertex areas, respectively (Figure 2.6C-D).

On the other hand, the core shell displays a T=19 icosahedral lattice containing 180 six-blade propeller-like capsomers with a central channel and 12 starfish-like pentons at the apices (Figure 2.6D-E). Both outer and inner capsids are overall aligned along their relative icosahedral axis through direct interactions with the also icosahedral inner envelope.

2.5.2 ASFV proteome

In addition to these structural studies, the proteomic analysis of purified extracellular ASFV particles has identified nearly 70 viral proteins, one third with an unknown function (Alejo *et al.*, 2018). The remaining two thirds are mostly involved in virus assembly (25%) and transcription (20%), but also in the maintenance of genome integrity (6%), virus entry (4%) and the evasion of the host defences (3%). Figure 2.7 provides an atlas of the ASFV particle that summarises our current knowledge on the structure of the virion and the demonstrated or inferred subviral distribution and biological role of more than 40 viral proteins.

2.5.3 ASFV nucleoid

According to this ASFV atlas, more than 20 different viral polypeptides compose the genome-containing nucleoid. It contains two major DNA-binding proteins, p10 (pK78R) and pA104R (Alejo *et al.*, 2018), the latter showing sequence similarity with bacterial histone-like proteins of HU/IHF family. Protein A104R is essential for viral replication and, according to recent

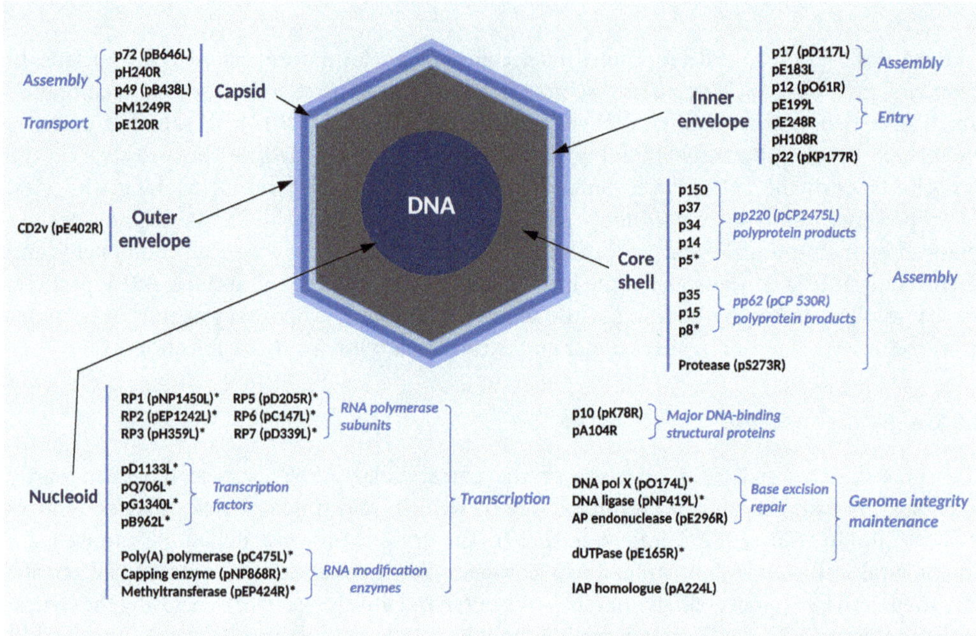

Figure 2.7. African swine fever virus (ASFV) atlas. The subviral distribution of 42 virion proteins among the five structural layers of the extracellular ASFV particle is shown. The localisation of proteins marked with an asterisk was inferred from the predicted or known role, while that of the remaining proteins was determined by immune-EM or cryo-EM reconstruction.

biochemical and structural approaches, might play a key role in DNA condensation (Liu *et al.*, 2020). Importantly, the nucleoid also contains a full set of enzymes involved in transcription and mRNA modification, which provides a high level of independence from the host transcriptional machinery. This virus-encoded transcriptional machinery includes basically all the predicted RNA polymerase subunits (RPBs), namely, pNP1450L (RPB1), pEP1242L (RPB2), pH359L (RPB3-11), pD205R (RPB5), pC147L (RPB6), and pD339L (RPB7) as well as the proposed capping enzyme (pNP868R), methyltransferase (pEP424R), and polyadenylation enzyme (pC475L). Additionally, several proteins (namely pG1340L, pD1133L, B962L and Q706L) displaying sequence similarity with VACV transcription factors, may be involved in early viral transcription.

Another interesting feature is the presence of several enzymes potentially involved in the maintenance of the genome integrity, including a dUTPase (pE165R) and a putative base excision repair (BER) system consisting of an A/P endonuclease (pE296R), an ATP-dependent DNA ligase (pNP419L) and a type X DNA polymerase (pO174L). The presence of a BER system within the nucleoid could serve to protect the incoming viral genome from the genotoxic agents present in the highly oxidative environment of the host macrophage. Finally, another nucleoid-packaged protein, pA224L, is a non-essential homologue of the cellular inhibitor of the apoptosis protein (IAP) that could prevent apoptosis in the infected cell.

2.5.4 Core shell

The core shell linking the central nucleoid to the inner envelope is primarily made from the mature products derived from the proteolytic processing of the viral polyproteins pp220 (p5, p34, p14, p37 and p150) and pp62 (p15, p35 and p8) (Andrés *et al.*, 1997, 2002a). Its membrane association depends on the N-myristoylation of pp220 precursor (Andrés *et al.*, 2002b) and, according to recent biochemical and structural studies, could be further sustained by proteins p35 and p15. Thus, patches of positively charged amino acids present on the surface of these proteins can bind phospholipids (Fu *et al.*, 2020; Li *et al.*, 2020). Interestingly, p15 also binds dsDNA, supporting the dual interactions of the core shell within the ASFV particle (Fu *et al.*, 2020). Polyprotein processing by the viral protease pS273R (which is also present in the virus particle) likely induces a deep reorganisation of the immature core shell (Alejo *et al.*, 2003), which gives rise to the capsid-like icosahedral structure identified in the mature virions (Andrés *et al.*, 2020; Wang *et al.*, 2019). A detailed description of the organisation of this matrix-like domain requires higher resolution analyses.

2.5.5 Inner envelope

The internal lipoprotein membrane contains most of the viral transmembrane proteins characterised so far (p17/pD117L, pE183L, p12/pO61R, p22/pKP177L, pEP152R, pE248R and pE199L) (Alejo *et al.*, 2018). The inner envelope plays a fundamental role in the assembly of the outer capsid and the core shell (Salas and Andrés, 2013), but it is also the fusion membrane allowing core penetration into the cell´s cytoplasm after virus entry (Hernáez *et al.*, 2010). In line with this dual role, different studies using inducible ASFV recombinants have shown that some inner envelope polypeptides (i.e. pE183L and p17) are essential for virus morphogenesis, while others are required for virus entry (pE248R and pE199L) (Hernáez *et al.*, 2016, Matamoros *et al.*, 2020). Thus, repression of p17 synthesis precludes the assembly of the outer capsid on the inner envelope (Suárez *et al.*, 2010). Consistent with this, cryo-EM studies indicate that p72 capsomers are tightly bound to the inner envelope through p17 protein (Liu *et al.*, 2019; Wang *et al.*, 2019). On the other hand, the knockdown of proteins pE248R and pE199L apparently does not interfere with the production of extracellular virus particles, which, however, does not undergo membrane fusion and core penetration after virus endocytosis (Matamoros *et al.*, 2020).

2.5.6 Outer capsid

Cryo-EM studies indicate that this protein shell is essentially composed by 8,280 copies of p72, which arranges in trimers displaying pseudo-hexameric morphology and 60 copies of pH240R, which arranges in pentamers at the 12 vertices (Andrés *et al.*, 2020; Liu *et al.*, 2019; Wang *et al.*, 2019). Each pseudo-hexameric capsomer is a homotrimer of the major capsid protein (MCP) p72/pB646L, which adopts the 'double jelly roll' fold found in many dsDNA icosahedral viruses.

In addition, protein p49/pB438L, which is involved in the assembly of the capsid vertices, mediates the docking of pH240R complex to the inner envelope. In addition, the 100-nm long tape-measure protein pM1249L interconnects neighbour pentasymmetrons, possibly determining the capsid size (Liu *et al.*, 2019; Wang *et al.*, 2019). Finally, the capsid protein pE120R has been involved in the microtubule-dependent transport of the intracellular particles to the cell surface (Salas and Andrés, 2013).

2.5.7 Outer envelope

The outer envelope is the least known of the viral layers in both composition and structural organisation. At present, the only viral protein that has been detected in this lipoprotein membrane is the protein pEP402R (Alejo *et al.*, 2018), a homologue of the T-lymphocyte surface antigen CD2 that mediates the adhesion of erythrocytes to the infected cell surface and the binding of ASFV particles to erythrocytes. Given the important role of this layer in both the exit and entry of the virus particle, the identification of new outer envelope proteins is to be expected.

2.5.8 Host proteins

In addition to the virus-encoded proteins, more than 20 cellular proteins have been consistently detected in the ASFV virion (Alejo *et al.*, 2018). Remarkably, a great proportion of them localise at the cell surface and interact with the cortical actin cytoskeleton (i.e. annexins ANXA1 and ANXA2, integrins ITGAB1 and ITGA3, tetraspanin CD9 or Rho GTPases CDC42 and RHOA). Thus, it seems plausible that they are recruited by the outcoming virus particles during their exit by budding through actin-dependent plasma membrane protrusions. A possible role of these host proteins during the virus replication cycle awaits further research.

2.6 Infection and replication of African swine fever virus at cellular level

2.6.1 ASFV infectious entry

African swine fever virus (ASFV) infects swine macrophages by receptor-mediated endocytosis engaging unknown cell receptor(s). The mechanism of entry is a very rapid dynamin- and clathrin-mediated process of endocytosis both in primary pig macrophages (Galindo *et al.*, 2015) and Vero cell lines (Hernáez and Alonso, 2010). In fact, the virus colocalises with early endosomes just a few minutes after post-adsorption warming to 37 °C. Since ASFV is a large virus (200 nm diameter), macropinocytosis and actin-dependent mechanisms can facilitate virus entry, directing the virion to the endocytic pathway (Sanchez *et al.*, 2012). ASFV endosomal passage and molecular determinants of infection found in endosomes are necessary for a successful infection. ASFV is a pH-dependent virus and requires the acidic pH of the endosomal lumen for viral decapsidation, which is maintained by the ATP-dependent proton pump (Cuesta-Geijo *et al.*, 2012; Hernáez *et al.*, 2016). Acidification occurs at the late endosome, which is the relevant compartment for infection that includes multivesicular bodies and endolysosomes, to start virion disassembly. This is a very dynamic process involving movement of virus and endosomes along microtubules, which allows transposition of the incoming virus from the plasma membrane to the perinuclear area by binding to microtubular motor dynein (Hernáez *et al.*, 2010). Then, replication sites are built at the proximity of the nucleus, at the microtubule organising centre (MTOC). Once decapsidated, virions expose their internal membrane for fusion, allowing naked viral cores to exit endosomes to start replication (Hernáez *et al.*, 2016). The viral fusion requires a number of characteristics of late endosomes, as for example, the small GTPase Rab7, which is essential for virus entry. It also requires determined lipids and receptors found at the late endosomal membrane, such as the phosphatidyl-inositol (3,5) biphosphate, related PIKfyve kinases (Cuesta-Geijo *et al.*, 2012), and cholesterol transporter protein Niemann-Pick type C type 1 (NPC1) (Cuesta-Geijo *et al.*, 2016). Cholesterol entry and transport favours ASFV infectivity at several levels. In addition, virion interplay with membranes seems very important in the process of being transferred across the endosomal membrane to the endoplasmic reticulum that will ultimately support replication. This transfer is facilitated by an oxysterol-binding protein, a cholesterol transporter protein and phosphatidyl-inositol kinases (Galindo *et al.*, 2019).

The relevance of cellular membranes for ASFV replication is emphasised, since the viral replication is followed by a marked recruitment of endosomes to the viral factory (Cuesta-Geijo *et al.*, 2017). Another fundamental element of the viral factories is the cellular cytoskeleton, which concentrates the necessary elements for the production of the viral progeny within a vimentin cage. Early recruitment of intermediate filaments to the MTOC plays a structural role in the formation of the virus assembly site (Stefanovic *et al.*, 2005). Viral DNA replication activates the calcium calmodulin-dependent protein kinase II (CaM kinase II) and the phosphorylation of the N-terminal domain of vimentin on serine 82. New virions are synthesised within these viral factories and, after maturation, they are transported by microtubules bound to kinesin, to the plasma membrane where viral exit occurs by budding or through the formation of apoptotic bodies. Intracellular and extracellular viruses are structurally and antigenically different, and this could have important implications in the host immune response against ASFV (Andrés *et al.*, 2001).

2.6.2 Innate immune response against ASFV

The innate immune system is the main first defence barrier against infections. Pathogen associated molecular patterns (PAMPs) are recognised by pattern recognition receptors (PRRs) involving (1) membrane PPRs, associated to the plasmatic membrane such as TLR7, TLR8 and TLR13, or TLR3 and TLR9, associated to endoplasmic reticulum membranes (Takeuchi and Akira, 2010); and (2) cytoplasmic sensors, mainly the retinoic acid-inducible gene I/melanoma differentiation-associated gene 5 (RIG-I/MDA5) and cyclic GMP-AMP synthase (cGAS), sensing respectively exogenous RNA or DNA. All of them are able to activate signal transduction pathways to induce innate immune responses, such as type I interferon (IFN I). In fact, the antiviral innate immunity is directed to avoid the above-described viral entry at the fusion step through IFN-induced membrane proteins named interferon-induced proteins with tetratricopeptide repeats (IFITs), which are abundant at the late endosomal membrane and have the ability to impair fusion (Munoz-Moreno *et al.*, 2016).

Cytoplasmic dsDNA acts like a potent pathogen-associated molecular pattern (PAMP), sensed by cGAS. cGAS detects cytoplasmic dsDNA and catalyses the synthesis of GMP-AMP cyclic dinucleotide (cGAMP). cGAMP acts as a second messenger and binds to the stimulator of interferon genes protein (STING), which traffics from the endoplasmic reticulum to the trans-Golgi network, where TANK-binding kinase 1 (TBK1) is recruited and phosphorylated. This event allows the recruitment of IRF3, subsequently phosphorylated by TBK1 and translocated to the nucleus, acting as a transcription factor for IFN-β gene expression (Granja *et al.*, 2008; Perez-Nunez *et al.*, 2015). Recently, García-Belmonte and co-workers demonstrated that whereas virulent Arm/07 inhibits the synthesis of IFN-β, and consequently IFIT-1 and CXCL10, the naturally attenuated strain NH/P68 modulates this response through IRF3 activation and STING phosphorylation (García-Belmonte *et al.*, 2019). Indeed, the infection by the ASFV NH/P68 strain results in (1) the induction of significant levels of IFN-β compared to those found in Arm/07-infected macrophages; (2) phosphorylation of STING; (3) traffic of STING through the Golgi to perinuclear punctuated structures; (4) localisation of IRF3 to the nuclear soluble fraction and binding to chromatin. Furthermore, inhibition of cGAS by the specific inhibitor Ru521 impairs NH/P68-induced STING phosphorylation, suggesting that cGAS is likely the main DNA sensor activating the STING cascade by the attenuated ASFV.

In contrast, cGAMP-induced phosphorylation of STING was prevented by infection of cells with Arm/07. STING trafficking to perinuclear punctuated structures was also severely impaired in Arm/07-infected cells, indicating that the virulent ASFV strain prevents the activation and trafficking of STING. These results agree with work of Portugal *et al.* (2018) describing the inhibition of IFN-β production by both virulent and attenuated ASFV strains.

2.6.3 ASFV genome replication

The ASFV replication starts soon after the virus envelope removal and breach of virus core wall, allowing virus DNA release into the cytoplasm in the perinuclear site close to microtubule organising centre (Alonso *et al.*, 2001). The ASFV genome encodes an impressive arsenal of proteins essential for DNA replication and nucleic acid metabolism (thymidine kinase, ribonucleotide reductase, dUTPase). The ASFV encapsidates the proteins to initiate genome replication independently from the host proteins. In addition, ASFV shamelessly orchestrates cellular proteins in order to prevent the sensing of viral DNA and components and promote flawless virus replication. Subsequent formation of viral factories facilitates the accumulation of viral DNA and proteins resulting in virion assembly (Stefanovic *et al.*, 2005).

Following early and late virus gene expression, ASFV genome replication occurs in the cytoplasm already after 6 hours post infection, and may gain additional protection by forming specialised replication compartments consisting of viral-encoded proteins (Dixon *et al.*, 2013). The intermediate nuclear phase in the virus replication cycle has been previously determined by *in situ* and autoradiography; however, the role of virus DNA translocation into the nucleus during the replication cycle remains unclear (Alonso *et al.*, 2001).

The ASFV genome has covalently closed terminal hairpin structures which, after precise nicking, act as a primer for DNA replication using a strain displacement mechanism. The terminal repeats (TR) elements are required for viral genome stability, transcription regulation, integration, transposition, segregation, and virion packaging. The TR elements resemble telomeres and may bind to host and viral proteins, and these protein-DNA interactions are important for viral replication and genome maintenance.

ASFV DNA replicates through intermediate forms consisting of head-to-head concatemers which later resolve into unit length genomes. ASFV DNA primase (C962R) may play an important role in replication initiation and DNA synthesis. Newly synthesised virus DNA is packaged in virus particles and assembled in viral factories. Nevertheless, the precise mechanism of virus DNA condensation and packaging is still poorly understood (Dixon *et al.*, 2013).

Despite the striking similarities between poxviruses and ASFV, ASFV has a robust model of replication that allows its efficient replication in different cell types and hosts.

2.7 African swine fever virus-pig interactions

2.7.1 African swine fever in the natural host

African swine fever (ASF) is a moderately contagious, febrile disease of pigs associated with high lethality. Besides high fever, the disease is characterised by a broad spectrum of clinical signs. The individual outcome depends on the host, route of infection and the virulence of the virus and can range from fatal to subclinical. African wild swine like bushpigs (*Potamochoerus larvatus*) and warthogs (*Phacochoerus africanus*) can be infected by ASFV showing no clinical signs of disease and acting as natural reservoir hosts. Adult warthogs are typically non-viraemic, but seropositive, although virus can usually be isolated only from lymph nodes (Penrith *et al.*, 2019). It has been reported on good evidence that neonatal warthogs become infected by *Ornithodoros* ticks in warthog burrows in the first 4-6 weeks of life. The virus replicates in the warthog and produces a transient viraemia, which is sufficient for transmission to uninfected ticks that feed on the young warthogs, thus maintaining the cycle (Netherton *et al.*, 2019; Penrith *et al.* 2019). In the case of another African wild suid, the nocturnal bushpig, which is also asymptomatic following experimental infection, ASFV has also been isolated from tissues without causing histological lesions and moderate viraemia can be observed.

In the domestic pigs and Eurasian wild boar (*Sus scrofa*) the course of ASF is comparable and there is no clear difference among ages and breeds following infection with highly virulent ASFV isolates. However, there are also reports indicating differences in disease outcome in some populations or age classes of domestic pigs, especially when moderately virulent strains are involved. Post *et al.* (2017) describe that the percentage of older pigs surviving infection with a moderately virulent isolate was shown to be higher than for younger pigs. In some regions of Africa apparently healthy pigs have tested positive for virus or had ASFV specific antibodies without showing clinical signs of the disease. One explanation may reside in the immunogenetics of the indigenous pig populations in certain African countries that present genetic composition distinct from commercial breeds and share genetic characteristic with the local wild ancestries (Mujibi *et al.*, 2018). However, as these resistance phenomena were usually not heritable (Penrith *et al.*, 2019), other explanations such as the impact of partial protection through maternally derived antibodies have to be discussed in endemically infected populations.

2.7.2 African swine fever clinical presentations

In general, ASF has an incubation period, defined as the time point of likely infection to onset of clinical symptoms, of 2-19 days during natural infection, with an accelerated incubation time (2-5 days) in experimentally infected pigs. There are no signs uniquely characteristic for ASF. Depending on the virulence of the strain involved, ASF occurs in a graded series of forms varying from peracute through acute, subacute, and chronic to unapparent (Table 2.2). Concordantly, ASFV isolates can be classified as highly virulent, moderately virulent and low virulent (Salguero, 2020). There are a number of other infections causing similar diseases, for example, classical swine fever, porcine dermatitis and nephropathy syndrome, porcine reproductive and respiratory syndrome, Aujeszky's disease, leptospirosis and erysipelas infections. However, the ASF clinical signs can not only be confused with other diseases, they may even pass unnoticed. Farmers and hunters are usually expecting much worse manifestations and may miss the unspecific disease, especially when moderate contagiousness leads to a scattered appearance of infected animals. Therefore, exclusion diagnostics should be carried out, even if ASF is not the primary suspicion. Thus, early detection depends on swift and reliable laboratory diagnosis (For more detailed information on pathological lesions, see Chapter 4). Highly virulent viruses are usually responsible for the peracute form, with high lethality rates that may reach 100% within 4-10 days post infection. In peracute ASF, affected animals present a very rapid clinical course, with high fever (up to 42 °C), anorexia, lethargy and can die suddenly within the first 4 days, sometimes without obvious signs of disease or gross lesions in pathology. The main route of infection is via the oro- or nasopharynx, and initial virus replication occurs within 24 hours of infection in the pharyngeal tonsil and in the lymphoid tissue draining the nasal mucosa (Salguero, 2020). Studies on the pathogenesis of highly virulent ASFVs show that primary viraemia can appear as early as 8 hours post-infection and secondary viraemia between 15 and 24 hours post infection. From a diagnostic point of view, viraemia is usually detectable between the 2nd and 3rd day post infection. Spleen, lymph nodes, liver, and lungs are the sites of secondary viral replication and, after 30 hours, all tissues contain the virus, reaching the maximum titres at 72 hours post infection. The ASFV is easily detected in blood and organ samples and no antibodies are developed in this disease course.

In acute infections, caused by virulent strains, clinical signs can start from 4 to 19 days post infection, depending on the dose and the route of infection (e.g. through contact), and are characterised mainly by a febrile syndrome (fever, anorexia, lethargy, weakness and recumbence). Other related ASF clinical signs are bluish-purple areas and haemorrhages (ecchymoses, petechiae) on the ears, distal limbs, and/or abdomen, ocular discharges, general reddening of the skin and bloody diarrhoea. Pigs with acute ASF usually die within 4-7 days after the first clinical sign and the mortality is 90-100% (Salguero, 2020). Internal lesions (Figure 2.8) are characterised by hyperaemic splenomegaly and haemorrhages in organs, particularly in the visceral lymph nodes, petechial haemorrhages in the kidneys, bladder mucosa, pharynx and larynx, pleura and heart, and an excess of natural fluids in body cavities and spaces. Enlarged and haemorrhagic mesenteric lymph nodes are sometimes the only obvious sign if animals die early. Haemorrhagic discharge from the anus (melaena) is sometimes observed. The distribution and frequency of these lesions are variable, and most are also seen in other swine diseases such as classical swine fever. Once the animals have been infected with virulent strains, viraemia is usually detected on average 3.75±1.4 days, two days before the onset of the clinical signs and virus excretion is mostly through the oropharyngeal route during the clinical disease phase. However, the viral load in excretions is rather low as long as no bloody admixtures are involved. The ASFV is present in all porcine samples with a high viral load, but mainly in the bone marrow, spleen, liver, lymph nodes, lung and tonsils. An initial humoral immune response can be detected from 7-8 days post infection (Gallardo *et al.*, 2019). Animals surviving the acute infection with highly virulent strains show quickly rising antibody titres and seem to recover completely over time. In these

Table 2.2. Main clinical signs and post-mortem findings observed in the different forms of African swine fever (ASF).

ASF clinical form	Mortality [days post infection]	External clinical signs	Lesions
Peracute	100% [1-4 days]	• Fever (41-42 °C), anorexia, and inactivity hyperpnoea, and cutaneous hyperaemia. Animals usually die suddenly without further clinical signs.	• No lesions are evident in organs.
Acute	90-100% [6-9 days, highly virulent ASFV]; [11-14 days, moderately virulent ASFV]	• Fever (40-42 °C), anorexia, recumbence, lethargy, weakness and show increased respiratory rate. • Bluish-purple areas and haemorrhages (spot-like or extended) on the ears, abdomen, and/or hind legs. • Ocular and nasal discharge. • Reddening of the skin of the chest, abdomen, perineum, tail, and legs. • Constipation or diarrhoea, which may progress from mucoid to bloody (melaena). • Abortion of pregnant sows at all stages of pregnancy. • Bloody froth from the nose/mouth and a discharge from the eyes. • The area around the tail may be soiled with bloody faeces.	• Haemorrhages under the skin. • Enlarged, oedematous, and completely haemorrhagic lymph nodes similar to blood clots (particularly gastrohepatic and renal). • Enlarged, friable and dark red to black spleen with rounded edges. • Petechiae (spot-like haemorrhages) on the capsule of the kidneys. • Hydro pericardium with yellowish fluid. • Hydrothorax and ascites. • Petechiae on the heart's surface (epicardium), urinary bladder, and kidneys (on the cortical and renal pelvis). • The lungs may present congestion and petechiae, with froth in the trachea and bronchus, and severe alveolar and interstitial pulmonary oedema. • Petechiae, ecchymoses (larger haemorrhages), and excess clotted blood in the stomach and small and large intestines. • Hepatic congestion and haemorrhages in the gall bladder.
Subacute	30-70% [7-22 days, moderately virulent ASFV]	• Similar (although generally less intense) to those observed in the acute form, except for vascular changes that are more intense, mainly haemorrhages and oedema. • Fluctuating fever, accompanied by depression and loss of appetite, are also common. • Painful walking; joints are often swollen with accumulated fluid and fibrin. • Laboured respiration and pneumonia. • Abortion.	• Ascites and hydro pericardium. • Oedema of the wall of the gallbladder and bile duct as well as in the surrounding area of kidneys (perineal oedema). • Partial hyperaemic splenomegaly with focal infarction. • Haemorrhagic, oedematous and friable lymph nodes (they often look like dark red haematomas). • Renal haemorrhages are more intense (petechiae and ecchymosis) and more extensive (cortex, medulla and pelvis) than in acute forms.
Chronic	<30% [>1 month, moderately virulent ASFV; Attenuated ASFV]	• Slight fever (40-40.5 °C) followed by mild respiratory distress and moderate-to-severe joint swelling. • Reddened areas of skin that become raised and necrotic. • Lymphadenopathy.	• Pneumonia with caseous necrosis (sometimes with focal mineralisation) in lungs. • Fibrinous pericarditis. • Oedematous lymph nodes, which can be partially haemorrhagic (mainly mediastinal lymph nodes).

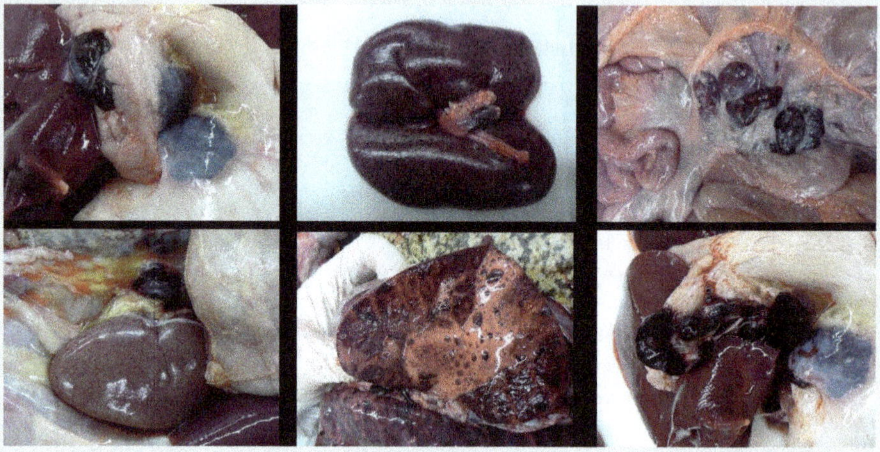

Figure 2.8. Pathomorphological lesions upon infection with highly virulent African swine fever virus strains. Upper row from left to right: ebony coloured, haemorrhagic lymph nodes in the gastro-hepatic area; petechiae and ecchymoses in the renal cortex and haemorrhagic renal lymph nodes; haemorrhagic intestinal lymph nodes. Lower row from left to right: renal petechiae; severe lung oedema and haemorrhages; haemorrhagic gastro-hepatic lymph nodes (incision).

cases, viral genome is detectable over long periods in blood and lymphatic tissues. However, the impact on the epidemiology is controversially discussed. Experimental commingling with sentinel pigs did not lead to transmission after 50 days (Nurmoja *et al.*, 2017).

Animals infected with moderately virulent isolates show a variety of clinical symptoms depending on the course of the disease that range from acute, subacute to chronic or unapparent infections. The reasons why ASF exhibits this significant variety of clinical signs are not precisely clear, but surely the immune mechanisms of defence developed by the hosts play a role. Some pigs may stay in good condition, while others show similar but less severe clinical symptoms than those observed in the acute process. In the pigs that die due to the moderately virulent viruses the disease course is usually 11-22 days and the mortality rates range 30-70%, the lowest rate displayed in adults (Gallardo *et al.*, 2019). The high fever that marks the onset of the disease may persist for up to 20 days or fluctuate irregularly throughout the course of the disease. As with the acute disease caused by virulent viruses, the reticuloendothelial system is involved, and death appears to result from vascular damage and lung oedema. Haemorrhages are most pronounced in the lymph nodes and kidneys. If the spleen is enlarged, it is due to hyperplasia of cellular elements rather than engorgement with blood. Areas of lobular consolidation are often seen in the anterior and cardiac lobes of the lung. In some cases there may be diffuse interstitial pneumonia; the entire lung may be white and does not collapse when the chest cavity is opened. Mucosal haemorrhages and bloody contents are often seen in the large intestine. Viraemia can be detected as early as 3 days after contagion if it develops into acute infections and at 8.5±3.6 days in domestic pigs with subacute infections. The ASFV is easily detected in all porcine tissues obtained from animals that succumbed within the first month to the infection. As mentioned above, animals surviving the acute course of infection usually showed increasing antibody titres and decreasing viral loads over time. Under experimental conditions (Petrov *et al.*, 2018), the viral genome could be detected for up to 100 days in blood. Replication-competent virus was isolated for up to 49 days. Sentinels commingled after 99 days remained completely negative for the remaining study period (termination at 165 days).

Chronic ASF is especially difficult to recognise, for it is extremely variable in its appearance. It may persist for several months with no particular signs other than stunting or emaciation, or it may mimic a variety of illnesses. To further complicate matters, distinction should perhaps be made between the chronic disease that is occasionally encountered in the animals that survive acute-subacute outbreaks, and the chronic disease that occurs in areas where ASF is endemic and is caused by attenuated strains. In pigs that survive acute-subacute infections caused by moderately virulent strains, the disease may last up to 70 days (Gallardo *et al.*, 2019). Pigs usually show clinical signs similar to acute cases, albeit less intense, for approximately 6-7 days, however, by day 11-14 clinical signs can subside and animals exhibit mild and unspecific clinical signs associated with a chronic ASF form for approximately one month on average after being infected. The respiratory distress, generalised lymphadenopathy, and swollen joints are the most noticeable clinical signs, whereas body temperatures can be within normal range. After a period of apparent recovery, clinical signs can reappear around two months later (50-60 days) (Gallardo *et al.*, 2018). The presence of this second excretion peak might be associated with a new cycle of virus replication in persistently infected tissues and death may occur after one of these episodes, in which case the lesions seen at necropsy are usually similar to those of the subacute disease. Haemorrhages may be prominent, but any enlargement of the spleen or lymph nodes is due to hyperplasia rather than engorgement. Haemorrhages in the lymph nodes are usually mild, sub-capsular, and occur particularly in the visceral lymph nodes. A few petechial haemorrhages may be seen in the kidneys (Salguero, 2020). The hyper-gammaglobulinemia and the high precipitating and complement-fixing antibody levels often observed are probably indicative of active synthesis of immunoglobulins in these tissues in response to the persistent viral infection. Chronic fibrinous pericarditis and pleuritis are frequently seen. One to several lobules of the lungs often contain scattered firm foci that may coalesce to form a hard, white mass involving an entire lobe. The lesion may then advance to caseous necrosis and calcification. Pneumonia accounts for more than half of the deaths in chronic ASF. Moreover, arthritis and skin ulcers have been associated with the chronic disease. Viraemia can be detected to a variable extent from the second week after the infection and can last for over two months, although viable virus is mainly obtained within the first month with occasional isolation up to day 66 in blood (Gallardo *et al.*, 2018, 2019).

The introduction of ASF in a free region is associated with highly virulent viruses and the acute-lethal disease is prevalent. Over time, the mortality can be reduced due to the natural evolution of the ASFV strains and the changing host population. This fluctuating pattern could also be observed in affected European Union Member States, with observed reduced virulence virus strains in affected wild boar populations (Gallardo *et al.*, 2018, Zani *et al.*, 2018). These attenuated strains are especially difficult to recognise, because they induce chronic or subclinical infections extremely variable in their appearance. Clinical signs and lesions are not specific, even at the beginning of the infection and may persist for several months with no particular signs, other than stunting or emaciation, or it may mimic a variety of illnesses. In addition to stunted growth and emaciation, some clinical signs may include skin ulcers, arthritis, and lameness due to swollen joints, respiratory symptoms and abortion. The low virulent ASFV isolates or strains are usually non-haemadsorbing (non-HAD) viruses. There are descriptions of attenuated and non-HAD strains that have been occasionally encountered in Portugal and in Spain in the period between 1965 and 1974 (Sanchez Botija *et al.*, 1977), in Africa and, more recently, in Latvia (Gallardo *et al.*, 2019). These non-HAD strains are more difficult to isolate than HAD viruses, since they cause a sporadic viraemia and viral particles can be isolated for organs. In most instances, the disease they cause is indistinguishable in the field. The comparatively low mortality rate and the apparent recovery of many of the infected pigs are certainly the most noteworthy features of low virulence strains.

2.8 Antiviral agents against African swine fever virus

Viral disease outbreaks occur regularly in the pig industry. While some diseases can be effectively controlled, others become endemic and cause sporadic outbreaks in affected countries. African swine fever is spreading throughout Africa, Eastern Europe and recently Asia and causing severe economic losses in pig production in involved countries. In August 2018, the first case of ASFV was reported in China (Wang *et al.*, 2018), which is one of the main centres of global pig industry. As of March 2020, ASFV outbreaks have been reported in 32 Chinese provinces and about 1.2 million pigs have been culled (FAO, Situation in Asia update, March 5, 2020), highlighting the need for development of improved control strategies including vaccines and antiviral drugs. Although all existing antiviral drugs have been developed to treat human infectious diseases, such as hepatitis B virus and HIV infections, the development of antiviral drugs for mass application in the veterinary field can be an attractive approach . For ASFV, antiviral therapy can prolong the host survival, allowing the infected pigs to generate a productive immune response against the virus. Therefore, anti-ASFV therapies can be applied in affected farms in order to isolate the epidemic area as well as to prevent further spread. In this section, a general overview of antiviral agents with identified molecular targets/mechanisms against ASFV is presented.

2.8.1 Nucleoside analogues

Since nucleosides play an essential role in the replication and transcription of genetic material, nucleoside analogues mimicking the structure of natural nucleosides can be incorporated into the DNA and RNA replication cycle, thereby terminating the nucleic acid synthesis. This approach has been widely used in anticancer, antibacterial and antiviral therapeutics for many decades.

Various nucleoside analogues were tested for their antiviral activity against ASFV infection *in vitro*. For example, Gil-Fernández and De Clercq (1987) showed that iododeoxyuridine at 100 μg/ml concentration totally inhibited ASFV infection in Vero cells. However, the authors also noted that cells exposed to the above-mentioned concentration became rounded and smaller in size, suggesting that 100 μg/ml could be toxic for cells. However, at low concentrations iododeoxyuridine may lead to the development of ASFV persistence in Vero cells. Another nucleoside analogue, (S)-9-(3-hydroxy-2-phosphonylmethoxypropyl)adenine, was found to exhibit the highest anti-ASFV activity compared to other nucleoside analogues like ribavirin and pyrazofurin. This compound decreased the ASFV titre by 5 log at 50 μg/ml concentration with selectivity index (SI) of 15,000 (Gil-Fernández and De Clercq, 1987). The antiviral activities of 3-deazaneplanocin A (SI=3,000), c3DHCaA (SI=2,500), 4'β-vinyl-DHCaA (SI=2,000) and 6'β-fluoroaristeromycin (SI=1,250) have been also reported (Villalón *et al.*, 1993).

Recently, it was reported that two nucleoside analogues, 5-(Perylen3-ylethynyl)-arabino-uridine (aUY11) and 5-(Perylen-3-ylethynyl)uracil-1-acetic acid (cm1UY11), possess a potent, dose-dependent inhibitory effect on ASFV infection in Vero cells and porcine macrophages (Hakobyan *et al.*, 2018). The major antiviral effect (3.5 log reduction) was observed when aUY11 and cm1UY11 were added at the internalisation stage of ASFV infection. These compounds are known for their ability to target the envelope lipids and prevent the curvature changes required for the fusion of viral and cellular membranes during viral entry. Thus, it is possible that aUY11 and cm1UY11 can serve as antiviral agents by interfering with ASFV infection.

2.8.2 Flavonoids: genistein and genkwanin

Because of low side effects and high availabilities, natural compounds have been the centre of attention among researchers working in antiviral drug discovery. Today numerous plant-derived compounds with potent antiviral activity have been reported, although none of them are currently used as an antiviral drug.

Genistein, a soy-derived isoflavone, has been shown to inhibit the infectivity of several RNA and DNA viruses. Recently, it was reported that genistein inhibited ASFV infection at non-cytotoxic concentrations in Vero cells and porcine macrophages by targeting the viral topoisomerase II (Arabyan *et al.*, 2018). The most significant anti-ASFV activity was seen when genistein was added to cells at the middle phase of infection (8 hours post infection), disrupting viral DNA replication. The single cell electrophoresis analysis revealed the presence of fragmented ASFV genomes in cells exposed to genistein, indicating that the compound interfered with viral type II topoisomerase. Molecular docking studies demonstrated that genistein may interact with four residues of the ATP binding site of viral topoisomerase, Asn-144, Val-146, Gly-147 and Leu148, showing stronger binding affinity (-4.62 kcal/mol) than ATP4- (-3.02 kcal/mol), suggesting that genistein may act as an ATP-competitive inhibitor in ASFV infection.

In addition, several commercially available apigenin derivatives were screened for their ability to inhibit ASFV infection *in vitro*. Among screened compounds, genkwanin, an O-methylated flavone abundant in the seeds of *Alnus glutinosa*, displayed strong inhibitory effect on ASFV infection, reducing viral titre from 6.5±0.1 to 4.75±0.25 log TCID/ml in a dose-dependent manner (IC50=2.9 μM and SI=205.2) (Hakobyan *et al.*, 2019). Further experiments indicated that genkwanin was able to inhibit ASFV infection at entry and egress stages. Since both processes depend on microtubules, and since apigenin is a highly potent inhibitor of tubulin polymerisation through the colchicine-binding site, this indicated that genkwanin may also affect the tubulin assembly, thereby suppressing the ASFV entry and egress. Indeed, *in silico* experiments revealed that genkwanin bound to the colchicine binding site of tubulin (-31.6 kcal/mol). Based on these results, it was postulated that genkwanin inhibited the ASFV infection by disrupting the virus movement along microtubules.

2.8.3 Interferons

All mammals encode three groups of interferons (IFNs) classified as type I, type II and type III according to their structural and functional characteristics. In response to viral infections, IFNs activate the JAK-STAT signalling cascade, leading to the transcriptional induction of many IFN-stimulated genes (ISGs) such as directly acting antiviral proteins OAS, PKR and Mx.

Early studies have found that the replication of both virulent and non-virulent ASFV isolates was significantly disrupted by porcine IFN-gamma. Furthermore, continuous treatment with IFN-alpha cured Vero cells from lytic and persistent infections with ASFV. Although no precise inhibitory mechanisms have been proposed for ISGs, Netherton *et al.* (2009) reported that the expression of human Myxovirus resistance protein 1 (MxA) (but not MxB) inhibited ASFV infection about 100-fold. This inhibition of ASFV was linked to the recruitment of MxA protein to perinuclear viral assembly sites, where it interacted with ASFV particles similarly to those seen between MxA and different RNA viruses. Among other ISGs, the interferon-induced transmembrane proteins (IFITMs) were also shown to reduce ASFV infectivity in Vero cells (Munoz-Moreno *et al.*, 2016). Particularly, IFITM2 and IFITM3 affected the viral entry/uncoating process through altering the endocytosis-mediated viral entry and cholesterol efflux. Recently, Fan *et al.* (2020) showed that low-dose (10^5 U/kg) recombinant porcine IFNs (PoIFN-α and PoIFN-γ) significantly reduced viral load *in vivo*, suggesting that IFN therapy has potential to be used as an emergency preventive treatment for ASFV.

2.8.4 Antibiotics

Although antibiotics are known as antimicrobial agents active against bacterial infections, some antibiotics have been reported to inhibit viral infections, including ASFV. For instance, thirty fluoroquinolones known for their inhibitory effect on topoisomerases were screened against ASFV (Mottola *et al.*, 2013). Six of them reduced the cytopathic effect in ASFV-infected Vero cells. Moreover, after continuous (7 days) treatment with fluoroquinolones the culture supernatants were unable to infect new cells and no ASFV genomes were detected by conventional PCR. Since ASFV has type II topoisomerase (Coelho *et al.*, 2015, 2016; Freitas *et al.*, 2016), other antibiotics targeting this enzyme could be studied in the future.

Acknowledgements and funding

This publication is based on work from 'Understanding and combating African swine fever in Europe (ASF-STOP COST action 15116)' supported by COST (European Cooperation in Science and Technology). Research at the Faculty of Veterinary Medicine – University of Lisbon was funded through national funds (FCT – Fundação para a Ciência e a Tecnologia, IP – project CIISA-UIDB/00276/2020). The RNAP laboratory at UCL is funded by a Wellcome Investigator Award in Science to FW (WT 207446/Z/17/Z). GC is funded by the Wellcome Trust ISMB 4-year PhD programme (WT 108877/B/15/Z). Research in the Dixon laboratory is supported by BBSRC grants BBS/E/I/0007031 and BBS/E/I/0007030. JHF was funded intramuraly trough the FLI'S ASFV Research Network. TBR and ASO were funded by Statens Serum Institute and University of Copenhagen. CA lab is supported by grants from the European Commission Horizon 2020 Framework Program VACDIVA-SFS-12-2019-1-862874 and the Spanish Ministry of Science and Innovation RTI2018-097305-R-I00. Funding from 'la Caixa' Banking Foundation under the project code HR18-00469 is acknowledged. GA was supported by grants from the Spanish Ministerio de Ciencia, Innovación y Universidades (PGC2018-098701-B-I00) and the Amarouto

Program for senior scientists from the Comunidad de Madrid. XW was supported by the Ten Thousand Talent Program and the NSFS Innovative Research Group (no. 81921005).

References

Alcaraz, C., Brun, A., Ruiz-Gonzalvo, F. and Escribano, J.M., 1992. Cell culture propagation modifies the African swine fever virus replication phenotype in macrophages and generates viral subpopulations differing in protein p54. Virus Research 23: 173-182. https://doi.org/10.1016/0168-1702(92)90076-L

Alejo, A., Andrés, G. and Salas, M.L., 2003. African swine fever virus proteinase is essential for core maturation and infectivity. Journal of Virology 77: 5571-5577. https://doi.org/10.1128/jvi.77.10.5571-5577.2003

Alejo, A., Matamoros, T., Guerra, M. and Andrés, G., 2018. A proteomic atlas of the African swine fever virus particle. Journal of Virology 92: e01293-18. https://doi.org/10.1128/JVI.01293-18

Alfonso, P., Rivera, J., Hernaez, B., Alonso, C. and Escribano, J.M., 2004. Identification of cellular proteins modified in response to African swine fever virus infection by proteomics. Proteomics 4: 2037-2046. https://doi.org/10.1002/pmic.200300742

Alonso, C., Borca, M., Dixon, L., Revilla, Y., Rodriguez, F. and Escribano, J.M., 2018. ICTV virus taxonomy profile: Asfarviridae. Journal of General Virology 99: 613-614. https://doi.org/10.1099/jgv.0.001049

Alonso, C., Miskin, J., Hernaez, B., Fernandez-Zapatero, P., Soto, L., Canto, C., Rodriguez-Crespo, I., Dixon, L. and Escribano, J.M., 2001. African swine fever virus protein p54 interacts with the microtubular motor complex through direct binding to light-chain dynein. Journal of Virology 75: 9819-9827. https://doi.org/10.1128/JVI.75.20.9819-9827.2001

Andrés, G., Alejo, A., Salas, J. and Salas, M.L., 2002a. African swine fever virus polyproteins pp220 and pp62 assemble into the core shell. Journal of Virology 76: 12473-12482. https://doi.org/10.1128/jvi.76.24.12473-12482.2002

Andrés, G., Charro, D., Matamoros, T., Dillard, R.S. and Abrescia, N.G.A., 2020. The cryo-EM structure of African swine fever virus unravels a unique architecture comprising two icosahedral protein capsids and two lipoprotein membranes. Journal of Biological Chemistry 295: 1-12. https://doi.org/10.1074/jbc.AC119.011196

Andrés, G., García-Escudero, R., Salas, M.L. and Rodríguez, J.M., 2002b. Repression of African swine fever virus polyprotein pp220-encoding gene leads to the assembly of icosahedral core-less particles. Journal of Virology 76: 2654-2666. https://doi.org/10.1128/jvi.76.6.2654-2666.2002

Andrés, G., Garcia-Escudero, R., Vinuela, E., Salas, M.L. and Rodriguez, J.M., 2001. African swine fever virus structural protein pE120R is essential for virus transport from assembly sites to plasma membrane but not for infectivity. Journal of Virology 75: 6758-6768. https://doi.org/10.1128/JVI.75.15.6758-6768.2001

Andrés, G., Simón-Mateo, C. and Viñuela, E., 1997. Assembly of African swine fever virus: role of polyprotein pp220. Journal of Virology 71: 2331-2341. https://doi.org/10.1128/JVI.71.3.2331-2341.1997

Arabyan, E., Hakobyan, A., Kotsinyan, A., Karalyan, Z., Arakelov, V., Arakelov, G., Nazaryan, N., Simonyan, A., Aroutiouniane, R., Ferreira, F. and Zakaryan, H., 2018. Genistein inhibits African swine fever virus replication *in vitro* by disrupting viral DNA synthesis. Antiviral Research 156: 128-137. https://doi.org/10.1016/j.antiviral.2018.06.014

Bastos, A.D.S., Penrith, M.-L., Cruciére, C., Edrich, J.L., Hutching, G., Roger, F., Couacy-Hymann, E. and Thomson, G.R., 2003. Genotyping field strains of African swine fever virus by partial p72 gene characterization. Archives of Virology 148: 693-706. https://doi.org/10.1007/s00705-002-0946-8

Broyles, S.S., 2003. Vaccinia virus transcription. Journal of General Virology 84: 2293-2303. https://doi.org/10.1099/vir.0.18942-0

Cackett, G., Matelska, D., Sýkora, M., Portugal, R., Malecki, M., Bähler, J., Dixon, L. and Werner, F., 2020. The African swine fever virus transcriptome. Journal of Virology 94: e00119-20. https://doi.org/10.1128/JVI.00119-20

Coelho, J., Ferreira, F., Martins, C. and Leitão, A., 2016. Functional characterization and inhibition of the type II DNA topoisomerase coded by African swine fever virus. Virology 493: 209-216. https://doi.org/10.1016/j.virol.2016.03.023

Coelho, J., Martins, C., Ferreira, F. and Leitão, A., 2015. African swine fever virus ORF P1192R codes for a functional type II DNA topoisomerase. Virology 474: 82-93. https://doi.org/ 10.1016/j.virol.2014.10.034

Cuesta-Geijo, M.A., Barrado-Gil, L., Galindo, I., Munoz-Moreno, R. and Alonso, C., 2017. Redistribution of endosomal membranes to the African swine fever virus replication site. Viruses 9: 133. https://doi.org/10.3390/v9060133

Cuesta-Geijo, M.A., Chiappi, M., Galindo, I., Barrado-Gil, L., Munoz-Moreno, R., Carrascosa, J.L. and Alonso, C., 2016. Cholesterol flux is required for endosomal progression of African swine fever virions during the initial establishment of infection. Journal of Virology 90: 1534-1543. https://doi.org/10.1128/JVI.02694-15

Cuesta-Geijo, M.A., Galindo, I., Hernaez, B., Quetglas, J.I., Dalmau-Mena, I. and Alonso, C., 2012. Endosomal maturation, Rab7 GTPase and phosphoinositides in African swine fever virus entry. PLoS One 7: e48853. https://doi.org/10.1371/journal.pone.0048853

Dhungel, P., Cao, S. and Yang, Z., 2017. The 5'-poly(A) leader of poxvirus mRNA confers a translational advantage that can be achieved in cells with impaired cap-dependent translation. PLoS Pathogens 13: e1006602. https://doi.org/10.1371/journal.ppat.1006602

Dixon, L.K., Chapman, D.A.G., Netherton, C.L. and Upton, C., 2013. African swine fever virus replication and genomics. Virus Research 173: 3-14. https://doi.org/10.1016/j.virusres.2012.10.020

Dixon, L.K., Islam, M., Nash, R. and Reis, A.L., 2019. African swine fever virus evasion of host defences. Virus Research 266: 25-33. https://doi.org/10.1016/j.virusres.2019.04.002

Esteves, A., Marques, M.I. and Costa, J.V., 1986. Two-dimensional analysis of African swine fever virus proteins and proteins induced in infected cells. Virology 152: 192-206. https://doi.org/10.1016/0042-6822(86)90384-3

Fan, W., Jiao, P., Zhang, H., Chen, T., Zhou, X., Qi, Y., Sun, L., Shang, Y., Zhu, H., Hu, R., Liu, W. and Li, J., 2020. Inhibition of African swine fever virus replication by porcine type I and type ii interferons. Frontiers in Microbiology 11: 1203. https://doi.org/10.3389/fmicb.2020.01203

Forth, J.H., Forth, L.F., Blome, S., Höper, D. and Beer, M., 2020a. African swine fever whole-genome sequencing – Quantity wanted but quality needed. PLoS Pathogens 16: e1008779. https://doi.org/10.1371/journal.ppat.1008779

Forth, J.H., Forth, L.F., King, J., Groza, O., Hübner, A., Olesen, A.S., Höper, D., Dixon, L.K., Netherton, C.L., Rasmussen, T.B., Blome, S., Pohlmann, A. and Beer, M., 2019. A deep-sequencing workflow for the fast and efficient generation of high-quality African swine fever virus whole-genome sequences. Viruses 11: 846. https://doi.org/10.3390/v11090846

Forth, J.H., Forth, L.F., Lycett, S., Bell-Sakyi, L., Keil, G., M., Blome, S., Calvignac-Spencer, S., Wissgott, A., Krause, J., Höper, D., Kampen, H. and Beer, M., 2020b. Identification of African swine fever virus-like elements in the soft tick genome provides insights into the virus' evolution. BMC Biology 18: 136. https://doi.org/10.1186/s12915-020-00865-6

Freitas, F.B., Frouco, G., Martins, C. and Ferreira, F., 2019. The QP509L and Q706L superfamily II RNA helicases of African swine fever virus are required for viral replication, having non-redundant activities. Emerging Microbes and Infections 8: 291-302. https://doi.org/10.1080/22221751.2019.1578624

Freitas, F.B., Frouco, G., Martins, C., Leitão, A. and Ferreira, F., 2016. *In vitro* inhibition of African swine fever virus-topoisomerase II disrupts viral replication. Antiviral Research 134: 34-41. https://doi.org/10.1016/j.antiviral.2016.08.021

Fu, D., Zhao, D., Zhang, W., Zhang, G., Li, M., Zhang, Z., Wang, Y., Sun, D., Jiao, P., Chen, C., Guo, Y. and Rao, Z., 2020. Structure of African swine fever virus p15 reveals its dual role for membrane-association and DNA binding. Protein Cell 11: 606-612. https://doi.org/10.1007/s13238-020-00731-9

Galindo, I. and Alonso, C. 2017. African swine fever virus: A review. Viruses 9: 103. https://doi.org/10.3390/v9050103

Galindo, I., Cuesta-Geijo, M.A., Del Puerto, A., Soriano, E. and Alonso, C., 2019. Lipid exchange factors at membrane contact sites in African swine fever virus infection. Viruses 11: 199. https://doi.org/10.3390/v11030199

Galindo, I., Cuesta-Geijo, M.A., Hlavova, K., Munoz-Moreno, R., Barrado-Gil, L., Dominguez, J. and Alonso, C., 2015. African swine fever virus infects macrophages, the natural host cells, via clathrin- and cholesterol-dependent endocytosis. Virus Research 200: 45-55. https://doi.org/10.1016/j.virusres.2015.01.022

Gallardo, C., Nurmoja, I., Soler, A., Delicado, V., Simón, A., Martin, E., Perez, C., Nieto, R. and Arias, M., 2018. Evolution in Europe of African swine fever genotype II viruses from highly to moderately virulent. Veterinary Microbiology 219: 70-79. https://doi.org/10.1016/j.vetmic.2018.04.001

Gallardo, C., Soler, A., Rodze, I., Nieto, R., Cano-Gómez, C., Fernandez-Pinero, J. and Arias M., 2019. Attenuated and non-haemadsorbing (non-HAD) genotype II African swine fever virus (ASFV) isolated in Europe, Latvia 2017. Transboundary Emerging Diseases 66: 1399-1404. https://doi.org/10.1111/tbed.13132

García-Belmonte, R., Perez-Nunez, D., Pittau, M., Richt, J.A. and Revilla, Y., 2019. African swine fever virus armenia/07 virulent strain controls interferon beta production through the cGAS-STING pathway. Journal of Virology 93: e02298-18. https://doi.org/10.1128/JVI.02298-18.

Gil-Fernández, C. and De Clercq, E., 1987. Comparative efficacy of broad-spectrum antiviral agents as inhibitors of African swine fever virus replication *in vitro*. Antiviral Research 7: 151-160. https://doi.org/10.1016/0166-3542(87)90003-9

Granja, A.G., Perkins, N.D. and Revilla, Y., 2008. A238L inhibits NF-ATc2, NF-kappa B, and c-Jun activation through a novel mechanism involving protein kinase C-theta-mediated up-regulation of the amino-terminal transactivation domain of p300. Journal of Immunology 180: 2429-2442. https://doi.org/10.4049/jimmunol.180.4.2429

Hakobyan, A., Arabyan, E., Kotsinyan, A., Karalyan, Z., Sahakyan, H., Arakelov, V., Nazaryan, K., Ferreira, F. and Zakaryan, H. 2019. Inhibition of African swine fever virus infection by genkwanin. Antiviral Research 167: 78-82. https://doi.org/10.1016/j.antiviral.2019.04.008

Hakobyan, A., Galindo, I., Nañez, A., Arabyan, E., Karalyan, Z., Chistov, A.A., Streshnev, P.P., Korshun, V.A., Alonso, C. and Zakaryan, H., 2018. Rigid amphipathic fusion inhibitors demonstrate antiviral activity against African swine fever virus. Journal of General Virology 99: 148-156. ttps://doi.org/10.1099/jgv.0.000991

Hernáez, B. and Alonso, C., 2010. Dynamin- and clathrin-dependent endocytosis in African swine fever virus entry. Journal of Virology 84: 2100-2109. https://doi.org/10.1128/JVI.01557-09

Hernáez, B., Guerra, M., Salas, M.L. and Andrés, G., 2016. African swine fever virus undergoes outer envelope disruption, capsid disassembly and inner envelope fusion before core release from multivesicular endosomes. PLoS Pathogens 12: e1005595. https://doi.org/10.1371/journal.ppat.1005595

Hernáez, B., Tarragó, T., Giralt, E., Escribano, J.M. and, Alonso, C. 2010. Small peptide inhibitors disrupt a high-affinity interaction between cytoplasmic dynein and a viral cargo protein. Journal of Virology 84: 10792-10801. https://doi.org/10.1128/JVI.01168-10

Herrera-Uribe, J., Jimenez-Marin, A., Lacasta, A., Monteagudo, P.L., Pina-Pedrero, S., Rodriguez, F., Moreno, A. and Garrido, J.J., 2018. Comparative proteomic analysis reveals different responses in porcine lymph nodes to virulent and attenuated homologous African swine fever virus strains. Veterinary Research 49: 90. https://doi.org/10.1186/s13567-018-0585-z

ICTV Master Species List 2019.v1 Available at: https://talk.ictvonline.org/files/master-species-lists/m/msl/9601

Karger, A., Perez-Nunez, D., Urquiza, J., Hinojar, P., Alonso, C., Freitas, F.B., Revilla, Y., Le Potier, M.F. and Montoya, M., 2019. An update on African swine fever virology. Viruses 11: 864. https://doi.org/10.3390/v11090864

Kessler, C., Forth, J.H., Keil, G.M., Mettenleiter, T.C., Blome, S. and Karger, A., 2018. The intracellular proteome of African swine fever virus. Scientific Reports 8: 14714. https://doi.org/10.1038/s41598-018-32985-z

Li, G., Fu, D., Zhang, G., Zhao, D., Li, M., Geng, X., Sun, D., Wang, Y., Chen, C., Jiao, P., Cao., l., Guo, Y. and Rao, Z., 2020. Crystal structure of the African swine fever virus structural protein p35 reveals its role for core shell assembly. Protein Cell 11: 600-605. https://doi.org/10.1007/s13238-020-00730-w

Liu, R., Sun, Y., Chai, Y., Li, S., Li, S., Wang, L., Su, J., Yu, S., Yan, J., Gao, F., Zhang, G., Qiu, H.-J., Gao, G.F., Qi, J. and Wang, H., 2020. The structural basis of African swine fever virus pA104R binding to DNA and its inhibition by stilbene derivatives. Proceedings of the National Academy of Sciences of the USA 117: 11000-11009. https://doi.org/10.1073/pnas.1922523117

Liu, S., Luo, Y., Wang, Y., Li, S., Zhao, Z., Bi, Y., Sun, J., Peng, R., Song, H., Zhu, D., Sun, Y., Li, S., Zhang, L., Wang, W., Sun, Y., Qi, J., Yan, J., Shi, Y., Zhang, X., Wang, P., Qiu, H.-J. and Gao, G.F., 2019. Cryo-EM structure of the African swine fever virus. Cell Host Microbe 26: 836-843.e3. https://doi.org/10.1016/j.chom.2019.11.004

Malogolovkin, A., Sereda, A. and Kolbasov, D., 2020. African swine fever virus. In: Malik, Y.S., Singh, R.K. and Yadav, M.P. (eds.), Emerging and transboundary animal viruses. Springer, Singapore, pp. 27-53. https://doi.org/10.1007/978-981-15-0402-0_2

Malogolovkin, A., Yelsukova, A., Gallardo, C., Tsybanov, S. and Kolbasov, D., 2012. Molecular characterization of African swine fever virus isolates originating from outbreaks in the Russian Federation between 2007 and 2011. Veterinary Microbiology 158: 415-419. https://doi.org/10.1016/j.vetmic.2012.03.002

Matamoros, T., Alejo, A., Rodríguez, J.M., Hernáez, B., Guerra, M., Fraile-Ramos, A. and Andrés, G., 2020. African swine fever virus protein pE199L mediates virus entry by enabling membrane fusion and core penetration. mBio 11: e00789-20. https://doi.org/10.1128/mBio.00789-20

Matsuyama, T.; Takano, T.; Nishiki, I.; Fujiwara, A.; Kiryu, I.; Inada, M.; Sakai, T.; Terashima, S.; Matsuura, Y.; Isowa, K. and Nakayasu, C. 2020. A novel *Asfarvirus*-like virus identified as a potential cause of mass mortality of abalone. Scientific Reports 10: 4620, https://doi.org/10.1038/s41598-020-61492-3

Mirzakhanyan, Y. and Gershon, P.D., 2017. Multisubunit DNA-dependent RNA polymerases from *Vaccinia* virus and other nucleocytoplasmic large-DNA viruses: impressions from the age of structure. Microbiology and Molecular Biology Reviews 81: e00010-17. https://doi.org/10.1128/MMBR.00010-17

Mottola, C., Freitas, F.B., Simões, M., Martins, C., Leitão, A. and Ferreira, F., 2013. *In vitro* antiviral activity of fluoroquinolones against African swine fever virus. Veterinary Microbiology 165: 86-94. https://doi.org/10.1016/j.vetmic.2013.01.018

Mujibi, F.D., Okoth, E., Cheruiyot, E.K., Onzere, C., Bishop, R.P., Fèvre, E.M., Thomas, L., Masembe, C., Plastow, G. and Rothschild, M., 2018. Genetic diversity, breed composition and admixture of Kenyan domestic pigs. PLoS One 22: e0190080. https://doi.org/10.1371/journal.pone.0190080

Munoz-Moreno, R., Cuesta-Geijo, M.A., Martinez-Romero, C., Barrado-Gil, L., Galindo, I., Garcia-Sastre, A. and Alonso, C., 2016. Antiviral role of IFITM proteins in African swine fever virus infection. PLoS One 11: e0154366. https://doi.org/10.1371/journal.pone.0154366

Netherton, C.L., Simpson, J., Haller, O., Wileman, T.E., Takamatsu, H.-H., Monaghan, P. and Taylor, G., 2009. Inhibition of a large double-stranded DNA virus by MxA protein. Journal of Virology 83: 2310-2320. https://doi.org/10.1128/JVI.00781-08

Netherton, C.L.; Connell, S.; Benfield, C.T.O. and Dixon, L.K., 2019. The genetics of life and death: virus-host interactions underpinning resistance to African swine fever, a viral hemorrhagic disease. Frontiers in Genetics 10: 402. https://doi.org/10.3389/fgene.2019.00402

Nielsen, S., Yuzenkova, Y. and Zenkin, N., 2013. Mechanism of eukaryotic RNA polymerase III transcription termination. Science 340: 1577-1580. https://doi.org/10.1126/science.1237934

Nurmoja, I., Petrov, A., Breidenstein, C., Zani, L., Forth, J.H., Beer, M., Kristian, M., Viltrop, A. and Blome, S., 2017. Biological characterization of African swine fever virus genotype II strains from north-eastern Estonia in European wild boar. Transboundary Emerging Diseases 64: 2034-2041. https://doi.org/10.1111/tbed.12614

Olasz, F. Tombácz, D., Torma, G., Csabai, Z., Moldován, N., Dörmő, Á., Prazsák, I., Mészáros, I., Magyar, T., Tamás, V., Zádori, Z. and Boldogkői, Z., 2020. Short and long-read sequencing survey of the dynamic transcriptomes of African swine fever virus and its host. Frontiers in Genetics 11: 758. https://doi.org/10.3389/fgene.2020.00758

Oleaga, A., Obolo-Mvoulouga, P., Manzano-Roman, R. and Perez-Sanchez, R., 2017. A proteomic insight into the midgut proteome of *Ornithodoros moubata* females reveals novel information on blood digestion in *Argasid* ticks. Parasites and Vectors 10: 366. https://doi.org/10.1186/s13071-017-2300-8

Penrith, M.-L., Thomson, G.R., Bastos, A.D.S. and Etter, E.M. 2019. African swine fever. In: J.A.W. Coetzer, G.R. Thomson, N. Maclachlan and M.-L. Penrith (eds.) Infectious disease of livestock (3rd ed.). Anipedia. https://www.anipedia.org

Perez-Nunez, D., Garcia-Urdiales, E., Martinez-Bonet, M., Nogal, M.L., Barroso, S., Revilla, Y. and Madrid, R., 2015. CD2v Interacts with adaptor protein AP-1 during African swine fever infection. PLoS One 10: e0123714. https://doi.org/10.1371/journal.pone.0123714

Petrov, A., Forth, J.H., Zani, L., Beer, M. and Blome, S., 2018. No evidence for long-term carrier status of pigs after African swine fever virus infection. Transboundary Emerging Diseases 65: 1318-1328. https://doi.org/10.1111/tbed.12881

Portugal, R., Leitao, A. and Martins, C., 2018. Modulation of type I interferon signaling by African swine fever virus (ASFV) of different virulence L60 and NHV in macrophage host cells. Veterinary Microbiology 216: 132-141. https://doi.org/10.1016/j.vetmic.2018.02.008

Post, J., Weesendorp, E., Montoya, M. and Loeffen, W.L., 2017. Influence of age and dose of African swine fever virus infections on clinical outcome and blood parameters in pigs. Viral Immunology 30: 58-69. https://doi.org/10.1089/vim.2016.0121

Rodriguez, F., Alcaraz, C., Eiras, A., Yanez, R.J., Rodriguez, J.M., Alonso, C., Rodriguez, J.F. and Escribano, J.M., 1994. Characterization and molecular basis of heterogeneity of the African swine fever virus envelope protein p54. Journal of Virology 68: 7244-7252. https://doi.org/10.1128/JVI.68.11.7244-7252.1994

Rodríguez, J.M. and Salas, M.L., 2013. African swine fever virus transcription. Virus Research 173: 15-28. https://doi.org/10.1016/j.virusres.2012.09.014

Rodriguez, J.M., Salas, M.L. and Santaren, J.F., 2001. African swine fever virus-induced polypeptides in porcine alveolar macrophages and in Vero cells: two-dimensional gel analysis. Proteomics 1: 1447-1456. https://doi.org/10.1002/1615-9861(200111)1:11<1447::aid-prot1447>3.0.co;2-y

Rowlands, R.J., Michaud, V., Heath, L., Hutchings, G., Oura, C., Vosloo, W., Dwarka, R., Onashvili, T., Albina, E. and Dixon, L.K., 2008. African swine fever virus isolate, Georgia, 2007. Emerging Infectious Diseases 14: 1870-1874. https://doi.org/10.3201/eid1412.080591

Salas, M.L. and Andrés, G., 2013. African swine fever virus morphogenesis. Virus Research 173: 29-41. https://doi.org/10.1016/j.virusres.2012.09.016

Salguero, F.J., 2020. Comparative pathology and pathogenesis of African swine fever infection in swine. Frontiers in Veterinary Science 7: 282. https://doi.org/10.3389/fvets.2020.00282

Sanchez Botija, A.C., Ordas, A., Ruiz Gonzalvo, F. and Solana, A., 1977. Procedures in use for diagnosis of ASF. Commission of the European Communities 5904EN

Sanchez, E.G. Quintas, A., Perez-Nunez, D., Nogal, M., Barroso, S., Carrascosa, A.L. and Revilla, Y., 2012. African swine fever virus uses macropinocytosis to enter host cells. PLoS Pathogens 8: e1002754. https://doi.org/10.1371/journal.ppat.1002754

Simões, M., Martins, C. and Ferreira, F., 2015. Early intranuclear replication of African swine fever virus genome modifies the landscape of the host cell nucleus. Virus Research 210: 1-7. https://doi.org/10.1016/j.virusres.2015.07.006

Stefanovic, S., Windsor, M., Nagata, K.I., Inagaki, M. and Wileman, T., 2005. Vimentin rearrangement during African swine fever virus infection involves retrograde transport along microtubules and phosphorylation of vimentin by calcium calmodulin kinase II. Journal of Virology 79: 11766-11775. https://doi.org/10.1128/JVI.79.18.11766-11775.2005

Suárez, C., Gutiérrez-Berzal, J., Andrés, G., Salas, M.L. and Rodríguez, J.M., 2010. African swine fever virus protein p17 is essential for the progression of viral membrane precursors toward icosahedral intermediates. Journal of Virology 84: 7484-7499. https://doi.org/10.1128/JVI.00600-10

Takeuchi, O. and Akira, S., 2010. Pattern recognition receptors and inflammation. Cell 140: 805-820. https://doi.org/10.1016/j.cell.2010.01.022

Tanaka, K., 1988. Protein and polymer analyses up to m/z 100,000 by laser ionization time-of-flight mass spectrometry. Rapid Communications in Mass Spectrometry 2: 151-153. https://doi.org/10.1002/rcm.1290020802

Urzainqui, A., Tabares, E. and Carrasco, L., 1987. Proteins synthesized in African swine fever virus-infected cells analyzed by two-dimensional gel electrophoresis. Virology 160: 286-291. https://doi.org/10.1016/0042-6822(87)90076-6

Villalón, M.D., Gil-Fernández, C. and De Clercq, E., 1993. Activity of several S-adenosylhomocysteine hydrolase inhibitors against African swine fever virus replication in Vero cells. Antiviral Research 20: 131-144. https://doi.org/10.1016/0166-3542(93)90003-2

Wang, N., Zhao, D., Wang, J., Zhang, Y., Wang, M., Gao, Y., Li, F., Wang, J., Bu, Z., Rao, Z. and Wang, X., 2019. Architecture of African swine fever virus and implications for viral assembly. Science 366: 640-644. https://doi.org/10.1126/science.aaz1439

Wang, T., Sun, Y. and Qiu, H.J., 2018. African swine fever: an unprecedented disaster and challenge to China. Infectious Diseases of Poverty 7: 111. https://doi.org/10.1186/s40249-018-0495-3

Werner, F. and Grohmann, D., 2011. Evolution of multisubunit RNA polymerases in the three domains of life. Nature Reviews Microbiology 9: 85-98. https://doi.org/10.1038/nrmicro2507

Wilkins, M.R., Pasquali, C., Appel, R.D., Ou, K., Golaz, O., Sanchez, J.C., Yan, J.X., Gooley, A.A., Hughes, G., Humphery-Smith, I., Williams, K.L. and Hochstrasser, D.F., 1996a. From proteins to proteomes: large scale protein identification by two-dimensional electrophoresis and amino acid analysis. Nature Biotechnology 14: 61-65. https://doi.org/10.1038/nbt0196-61

Wilkins, M.R., Sanchez, J.C., Gooley, A.A., Appel, R.D., Humphery-Smith, I., Hochstrasser, D.F. and Williams, K.L., 1996b. Progress with proteome projects: why all proteins expressed by a genome should be identified and how to do it. Biotechnology and Genetic Engineering Reviews 13: 19-50. https://doi.org/10.1080/02648725.1996.10647923

Zani, L., Forth, J.H., Forth, L., Nurmoja, I., Leidenberger, S., Henke, J., Carlson, J., Breidenstein, C., Viltrop, A., Höper, D., Sauter-Louis, C., Beer, M. and Blome, S., 2018. Deletion at the 5′-end of Estonian ASFV strains associated with an attenuated phenotype. Scientific Reports 8: 6510. https://doi.org/10.1038/s41598-018-24740-1

3. Immune responses against African swine fever virus infection

M. Montoya[1*], G. Franzoni[2], D. Pérez-Nuñez[3], Y. Revilla[3], I. Galindo[4], C. Alonso[4], C.L. Netherton[5] and U. Blohm[6]

[1]Centro de Investigaciones Biológicas Margarita Salas (CIB-CSIC), Ramiro de Maeztu 9, 20840 Madrid, Spain; [2]Department of Animal Health, Istituto Zooprofilattico Sperimentale della Sardegna, 07100 Sassari, Italy; [3]Centro de Biología Molecular Severo Ochoa (CBMSO), Calle Nicolás Cabrera, 1, 28049 Madrid, Spain; [4]Dpt. Biotechnology, Instituto Nacional de Investigación y Tecnología Agraria y Alimentaria (INIA), Ctra. de la Coruña km 7.5, 28040 Madrid, Spain; [5]The Pirbright Institute, Ash Road, Pirbright, Woking, GU24 ONF, United Kingdom; [6]Friedrich-Loeffler-Institut, Federal Research Institute for Animal Health, Institute of Immunology, Suedufer 10, 17493 Greifswald-Insel Riems, Germany; maria.montoya@cib.csic.es

Abstract

Infection with African swine fever virus (ASFV) leads to a short haemorrhagic course of disease that, depending on the virus isolate, results in up to 100% lethality in domestic and Eurasian wild pigs. Consequently, ASFV infection in swine is of considerable economic significance. This chapter explains the basics of antiviral immunity in swine, focusing on the 'knowns' and 'unknowns' of the response against ASFV. In particular, monocytes and macrophages play an essential role as the main targets of infection and are crucial in viral persistence and dissemination. Furthermore, ASFV has developed several mechanisms to influence the antiviral and cell biological activity of infected monocytes, including down-regulation of cell surface receptors (e.g. CD14 and MHC-I) and modulation of interferon and cytokine/chemokine responses. ASFV infected pigs also develop virus-specific antibodies that can be used diagnostically, and while the neutralising effect of these antibodies has led to their involvement in protective immunity being controversially discussed, they may still exhibit protective functions through complement-mediated lysis and/ or antibody dependent cell-mediated cytotoxicity. Indeed, T cells (presumably CD8+) also play a central role in the elimination of the virus, as can be seen in experiments where, after depletion of these cells, pigs previously primed with an avirulent ASFV become ill, while non-depleted animals are protected from highly virulent challenge. Nonetheless, despite these advances in our knowledge, much remains unknown about antiviral immunity generated during the course of

This publication is based upon work from COST Action CA15116, ASF-STOP, supported by COST (European Cooperation in Science and Technology).
www.cost.eu

EUROPEAN COOPERATION IN SCIENCE & TECHNOLOGY

Funded by the Horizon 2020 Framework Programme of the European Union

DOI 10.3920/978-90-8686-910-7_3,

a natural ASFV infection, or in response to attenuated virus strains or immunisation. Although such studies would undoubtedly be technically challenging, a deeper understanding of the immunity developed by the natural hosts (i.e. bushpigs and warthogs) against ASFV infection would teach us a lot about an effective protection from ASFV infection, and the involvement of both the innate and adaptive immune systems in this process.

Keywords: innate immunity, adaptive immunity, pig, anti-viral immune responses

3.1 Introduction to anti-viral immune responses

The main aim of the immune system is to detect and subsequently to eliminate harmful pathogens whilst maintaining a homeostatic relationship with beneficial microbes. Pathogens can be very different - parasites, fungi, bacteria, viruses – that must be recognised by the immune system as harmful and foreign. This chapter focuses on the arms of the immune system that are required for anti-viral immune responses in general and particularly the 'knowns' and 'unknowns' of such responses against African swine fever virus (ASFV).

When any pathogen comes into contact with a possible host, the first line of defence is provided by the innate immune system, which comprises soluble factors like type I interferons (IFNs) and cells, like macrophages or dendritic cells that detect pathogens through distinct pathogen-associated molecular patterns (PAMPs). PAMPs are found only in microbes but not in the host and are recognised by innate microbial sensors, collectively known as pattern recognition receptors (PRRs). PRRs consist of several families, each of which recognises unique PAMPs associated with infections. Noteworthy, most if not all components of replicating viruses are synthesised by the host cell using host cell machinery. Thus, viruses can interact with PAMPs either extracellularly or intracellularly. The result of stimulating PRRs is a transcriptional activation of genes involved in innate defence as well as those that activate antigen-presenting cells for successful priming of adaptive responses.

Unlike the innate immune system, which is activated only based on the identification of general threats, adaptive immunity is activated by exposure to specific pathogens, in this case viruses, and uses an immunological memory to learn about the threat to enhance the immune response accordingly. The adaptive immune response is much slower to respond to pathogens than innate immune response, but it is more specific. They respond to pathogens by recognising parts of them as antigens, which are considered harmful or foreign substances entering the body. Soluble factors such as cytokines or chemokines and effector cells such as T or B cells are the main players in adaptive immune responses. Figure 3.1 shows the main stages of an antiviral immune response. Section 2 in this chapter comprises the elements of the innate immune systems that have been characterised during ASFV infection whereas section 3 and 4 will focus on adaptive humoral and cellular responses respectively in the context of ASFV infection.

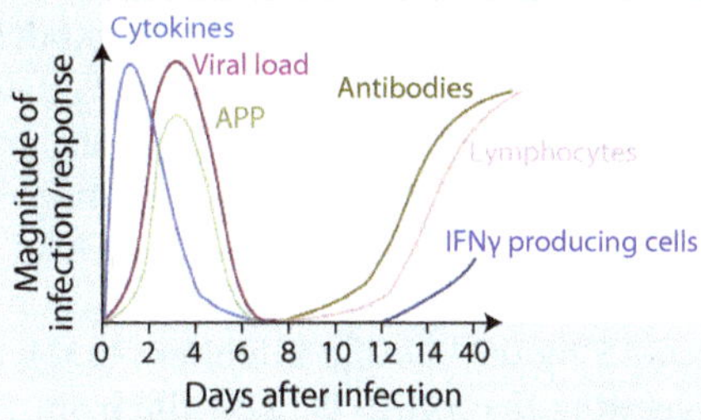

Figure 3.1. General kinetics of lung viral load, acute phase proteins (APPs), cytokines and different immune system components (lymphocytes, interferon (IFN)-γ producing cells and antibodies) during African swine fever viral infection in pigs. The magnitude of responses is related to the days after infection (0-14).

3.2 Innate immunity: the first line of defence

As mentioned, innate immune responses provide a first line of defence against pathogens and promote the development of acquired immunity (Medzhitov and Janeway, 2000). Virulent ASFV isolates have developed mechanisms to outpace these defences and to evade the innate immune system (Dixon *et al.*, 2019; Franzoni *et al.*, 2018). In this subchapter, an overview of how ASFV infection impacts two key cell types of the innate immune system, macrophages and dendritic cells, is provided. In addition, the main findings on ASFV modulation of type I IFN, a crucial element of the innate response to viral infection, are summarised.

PAMPs are recognised by PRRs involving (1) membrane PPRs, associated to plasmatic membrane such as toll-like receptors (TLR)7, TLR8 and TLR13, or TLR3 and TLR9, associated to endoplasmic reticulum membranes (Takeuchi and Akira *et al.*, 2010), and (2) cytoplasmic sensors, mainly the retinoic acid-inducible gene I/melanoma differentiation-associated gene 5 (RIG-I/MDA5) and cyclic GMP-AMP synthase (cGAS), sensing respectively exogenous RNA or DNA (Wu and Chen, 2014). All of them are able to activate signal transduction pathways to induce innate immune responses, such as type I IFN. In fact, the antiviral innate immunity is directed to avoid the viral entry at the fusion step through IFN-induced membrane proteins named interferon-induced proteins with tetratricopeptide repeats (IFITs), which are abundant at the late endosomal membrane and have the ability to impair fusion (Munoz-Moreno *et al.*, 2016).

Cytoplasmic dsDNA acts like a potent PAMP, sensed by cGAS. cGAS detects cytoplasmic dsDNA and catalyses the synthesis of GMP-AMP cyclic dinucleotide (cGAMP) (Sun *et al.*, 2013). cGAMP acts as a second messenger and binds to the stimulator of interferon genes protein (STING), which traffics from the endoplasmic reticulum (ER) to the trans-Golgi network (TGN), where tumour necrosis factor (TNF) receptor-associated factors (TRAF)-associated NF-κB activator (TANK)-binding kinase 1 (TBK1) is recruited and phosphorylated. This event allows the recruitment of interferon regulatory factor (IRF)3, subsequently phosphorylated by TBK1 and translocated to the nucleus, acting as a transcription factor for IFN-β gene expression (Granja *et al.*, 2008; Perez-Nunez *et al.*, 2015; Wu and Chen *et al.*, 2014). Recently, Garcia-Belmonte and co-workers demonstrated that whereas virulent Arm/07 inhibits the synthesis of IFN-β, and consequently

IFIT-1 and C-X-C motif chemokine ligand 10 (CXCL10), by modulating IRF3 activation and STING phosphorylation, the naturally attenuated strain NH/P68 does not (Garcia-Belmonte *et al.*, 2019). Indeed, the infection by the ASFV NH/P68 strain results in: (1) the induction of significant levels of IFN-β compared to those found in Arm/07-infected macrophages; (2) phosphorylation of STING; (3) traffic of STING through the Golgi to perinuclear punctuated structures; and (4) localisation of IRF3 to the nuclear soluble fraction and binding to chromatin. Furthermore, inhibition of cGAS by the specific inhibitor Ru521, impairs NH/P68-induced STING phosphorylation, suggesting that cGAS is likely the main DNA sensor activating the STING cascade by the attenuated ASFV.

In contrast, cGAMP-induced phosphorylation of STING was prevented by infection of cells with Arm/07. STING trafficking to perinuclear punctuated structures was also severely impaired in Arm/07-infected cells, indicating that the virulent ASFV strain prevents the activation and trafficking of STING. These results are in agreement with works describing the inhibition of IFN-β production during virulent ASFV strain infection, despite that the molecular mechanisms involved were not approached by these authors (Gil *et al.*, 2008; Reis *et al.*, 2016). In contrast, Portugal *et al.* (2018) reported recently that IFN-β inhibition was exerted by both virulent and attenuated ASFV strains.

3.2.1 Macrophages

Macrophages (Mφ) are phagocytic cells which are at the frontline of pathogen defence. They detect foreign molecules as they are equipped with an array of sensing molecules named PRRs (Hume, 2015). These cells contribute to the initiation of acquired immune responses, by processing and presenting antigens to naïve T lymphocytes (Hume, 2015). Both monocytes and Mφ are the main targets of ASFV and are thought to be crucial for viral persistence and dissemination (Sánchez-Cordón *et al.*, 2008). Virulent ASFV isolates can inhibit Mφ defences, in order to survive and efficiently replicate in these cells.

Considering the central role of Mφ in ASFV immunopathogenesis, several studies analysed the impact of this virus on the phenotype and functionality of these cells. Susceptibility of myeloid cells to ASFV seems to be linked to maturity: *in vitro* differentiation of porcine blood monocytes into Mφ, with accompanying CD163, major histocompatibility complex (MHC) class II DR, and CD203a up-regulation, increased susceptibility to ASFV infection (Basta *et al.*, 1999; McCullough *et al.*, 1999; Sánchez-Torres *et al.*, 2003). A role for CD163 as an ASFV receptor was postulated in the past (Sánchez-Torres *et al.*, 2003), but subsequent studies showed that CD163 expression was not necessary for ASFV infection of Mφ (Popescu *et al.*, 2017) whereas it was demonstrated that both virulent and attenuated ASFV down-regulated expression of CD14 and CD16 on infected Mφ (Franzoni *et al.*, 2017). Down-regulation of CD14 (lipopolysaccharide (LPS) receptor) and CD16 (a low-affinity receptor for the immunoglobulin (Ig)G Fc involved in antibody opsonisation) might impair Mφ's antimicrobial and antiviral activities (Franzoni *et al.*, 2018). Overall, ASFV infection seems to compromise Mφ ability to respond to different external stimuli: it was recently described that both attenuated NH/P68 and virulent 22653/14-infected Mφ released lower amounts of interleukin (IL)-6, IL-12 and TNF-α compared to mock-infected Mφ in response to stimulation with IFN-γ and LPS or a TLR2 agonist (Franzoni *et al.*, 2020).

ASFV strains of diverse virulence can differently modulate MHC class I expression on Mφ. It was observed that attenuated (BA71V, NH/P68) but not virulent (22653/14) ASFV strains down-regulated MHC I expression on infected Mφ (Franzoni *et al.*, 2017, 2020), and this might promote natural killer (NK) cell activation *in vivo* (Lanier., 2005). In fact, an *in vivo* study reported a correlation between NK activation and protection in pigs inoculated with NH/P68 and then challenged with the homologous virulent L60 strain (Leitão *et al.,* 2001). Through this mechanism, virulent ASFV isolates might elude the early recognition of the innate immune system; however, differences between ASFV genotypes have been observed: Mφ infected with virulent isolates belonging to genotype II (Arm07), IX (Ken06.Bus) or VIII (Malawi Lil20) resulted in MHC I expression similar to those infected with attenuated NH/P68 and lower than those infected with virulent genotype I isolates (Benin 97/1, Sardinian 22653/14) (Franzoni *et al.*, 2020; Arav, personal communication). In addition, deletion of the gene coding for CD2v, whose ORF is interrupted in NH/P68 (Portugal *et al.,* 2015), from genotype I Benin 97/1 resulted in lower MHC I expression on bone marrow derived Mφ comparable to that of NH/P68 infected Mφ (Arav, personal communication), suggesting that CD2v might represent a relevant virulence factor for genotype I ASFV isolates.

Several studies have analysed the chemokine/cytokine responses of Mφ to ASFV infection *in vitro.* It was speculated that cytokine release by infected Mφ might contribute to ASFV pathogenesis, and it was recently reported that Mφ's infection with virulent Georgia 2007 down-regulated expression of anti-inflammatory IL-10, whereas it induced up-regulation of pro-inflammatory IL-17 and cytokines of TNF superfamily, including (Fas-ligand) FASL, leukotriene (LT)A, LTB, TNF, TNFSF4, TNFSF10, TNFSF13B and TNFSF18. *In vivo,* these cytokines might promote death of bystander cells and tissue inflammation (Zhu *et al.*, 2019). Differences between strains were observed, with attenuated ASFV strains inducing enhanced expression of key cytokines (IFN-α, IFN-β, IL-1β, IL-12p40, TNF-α) and chemokines (CCL4, CXCL8, CXCL10) compared to highly virulent strains (Fishbourne *et al.*, 2013; Franzoni *et al.*, 2020; Gil *et al.*, 2003, 2008; Reis *et al.*, 2016). Virulent ASFV isolates might have evolved mechanisms to circumvent Mφ cytokine/chemokine responses, and these mechanisms might be at least partially lost in attenuated strains, with consequent enhancement of immune surveillance and thus induction of effective adaptive immune responses against the virus. Another hypothesis is that virulent ASFV isolates are so aggressive that the Mφ ability to mount cytokine responses is affected from the early steps of infection.

Despite this central role for Mφ in ASFV infection biology, little is known about responses of polarised Mφ to ASFV. Mφ are a heterogeneous population and present extraordinary plasticity, changing *in vivo* their phenotype and functions in response to the subtle and continuous changes of environmental signals (Italiani and Boraschi, 2014; Mosser, 2003). 'Classical' and 'alternative' activated Mφ, called M1 and M2 respectively, represent the two extremes of diverse functional status, with M1 Mφ providing antimicrobial and pro-inflammatory functions, and M2 being associated with mechanisms of immunosuppression and wound healing (Mosser, 2003). Nevertheless, M1 and M2 are not ontogenetically defined and the M1/M2 paradigm is just a limited attempt to define the complexity and plasticity of the mononuclear phagocyte system (Italiani and Boraschi, 2014). In pigs, as in other species, M1 polarisation can be achieved *in vitro* through IFN-γ and LPS, with consequent up-regulation of MHC and co-stimulatory molecules and release of pro-inflammatory cytokines, whereas M2 polarisation can be achieved through IL-4 (Mosser, 2003; Singleton *et al.*, 2016).

It was recently described that virulent isolates present a greater ability to infect Mφ activated with type I or type II IFNs compared to attenuated strains (Franzoni *et al.*, 2017, 2020; Golding *et al.*, 2016), whereas alternative activation did not affect susceptibility to ASFV infection. Both attenuated (NH/P68) and virulent (22653/14) ASFV strains efficiently replicated in different monocyte-derived macrophage subsets (moMφ, M1, M2, Mφ activated with IFN-α), but with delayed kinetics in moM1. The data suggested that the ability of ASFV to infect moM1 is initially impaired, but then the virus overcame cellular defence mechanisms and efficiently replicated in this subset (Franzoni *et al.*, 2020). MoM1 can mount a robust cytokine response after infection with attenuated ASFV strains (BA71V or NH/P68), with release of IL-1α, IL-1β and IL-18 (Franzoni *et al.*, 2017, 2020), whereas negligible release of all the tested cytokines (IL-1α, IL-1β, IL-6, IL-10, IL-12, IL-18, TNF-α) was detected from moMφ, moM2, IFN-α-activated moMφ (Franzoni *et al.*, 2020). As previously reviewed, release of IL-18 might enhance the development of T cell responses, whereas IL-1β release might promote apoptosis of bystander uninfected cells, reducing ASFV replication and spread *in vivo* (Franzoni *et al.*, 2018). The virulent 22653/14 ASFV seems to have developed mechanisms to counteract cytokine responses by moM1, and further studies should be performed, aimed at understanding which viral factors, lost in attenuated strains, inhibit release of these key cytokines (Franzoni *et al.*, 2017, 2020). A summary of ASFV impact on Mφ phenotype and functionality is shown in Figure 3.2.

3.2.2 Dendritic cells

Dendritic cells (DCs) are regarded as sentinels of the immune system and play a central role in activating and shaping adaptive immune responses against pathogens. They express a variety of PRRs and can detect, take up and process antigen, and then migrate into secondary lymphoid tissues to present the processed antigens to T lymphocytes (Banchereau and Steinman, 1998). These cells can orchestrate host immune responses, being able to prime naïve T lymphocytes and to modulate the type and the characteristics of T and B cell responses (Banchereau and Steinman, 1998). DCs are a heterogeneous population and can be broadly divided into three subsets: conventional DC1 (cDC1), conventional DC2 (cDC2) and plasmacytoid DC (pDC) (Collin and Bigley, 2018). pDC are specialised in secreting large amounts of type I interferon following viral stimulation (Collin and Bigley, 2018), whereas cDC1 are specialised in priming a Th1 response through IL-12 secretion and antigen presentation via MHC class I to $CD8^+$ T cells, and cDC2 can stimulate Th2 and Th17 responses (Collin and Bigley, 2018). Each species has its own peculiarities and DC phenotypes are also organ specific, as recently reviewed (Franzoni *et al.*, 2019).

To date, few studies have analysed DC responses to ASFV infection, as recently reviewed (Franzoni *et al.*, 2019). In the 1990s, researchers described that ASFV was able to infect skin-derived DCs *in vitro*, with subsequent impairment of DCs infection by foot-and-mouth disease virus (Gregg *et al.*, 1995). *In vivo*, it was reported that virulent ASFV L60 infected interdigitating DC (iDC): ASFV antigens were identified in iDC in mandibular lymph nodes at 3 days post infection and this was followed by a reduction in the number of iDC. These results suggest that early DC depletion in lymph nodes during infection with virulent ASFV isolate might contribute

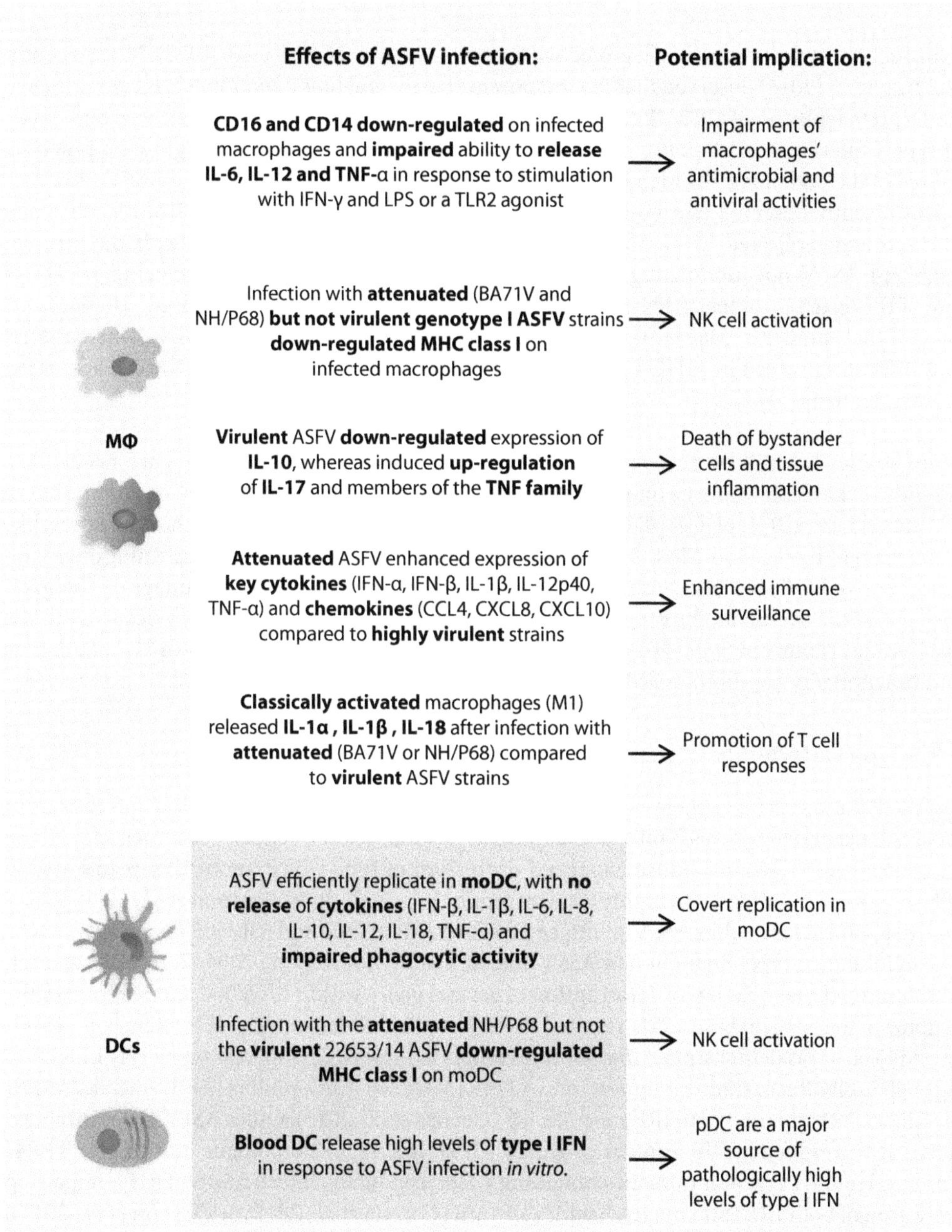

Figure 3.2. African swine fever virus modulation of macrophages and dendritic cell phenotype and functionality and possible implications. CXCL = C-X-C motif chemokine ligand; DC = dendritic cell; IFN = interferon; IL = interleukin; LPS = lipopolysaccharides; MHC = major histocompatibility complex; Mφ = macrophages; TNF = tumour necrosis factor; TRL = toll like receptor.

to unsuccessful development of a protective immune response (Gregg *et al.*, 1995). More recently, Golding *et al.* (2016) observed that *in vitro* porcine peripheral blood mononuclear cells enriched for DC (by depletion of $CD3^+$ T cells, $CD14^+$ monocytes and $CD21^+$ B cells) released high levels of type I interferon in response to ASFV infection, suggesting that pDC might be a potential source of the pathologically high levels of type I interferon in pigs during acute ASFV infection. It was recently observed that both attenuated NH/P68 and virulent 22653/14 ASFV efficiently infected and replicated in monocyte derived DC (moDC), with infection with the virulent 22653/14 ASFV not modulating any surface marker expression and not inducing release of any of the tested cytokines (IFN-β, IL-1β, IL-6, IL-8, IL-10, IL-12, IL-18, TNF-α). These results suggest that, similar to Mφ, virulent 22653/14 strain covertly replicates in moDC, whereas NH/P68 infection resulted in MHC I down-regulation, thus potentially affecting NK cell activation *in vivo* (Franzoni *et al.*, 2018).

More studies focused on the interaction of ASFV with different DC subsets are needed. As recently reviewed, it would be interesting to assess ASFV interaction with different porcine blood DC subsets, i.e. cDC1, cDC2, pDC and to characterise DC interaction with other cell types, like NK cells, γδ-T cells, and Mφs. More studies focused on the infection of ASFV not only with blood derived DC, but also with DC from tissues are required, taking advantage of the great increase of the overall toolbox for the analysis of the pig immune system which occurred in the last decades (Franzoni *et al.*, 2019). Current knowledge of ASFV modulation of DC phenotype and functionality are summarised in Figure 3.2.

3.2.3 Type I interferon

Type I IFN is a crucial element of the innate immune response to viral infection and thus most viruses have developed mechanisms to counteract its induction or effects (Summerfield, 2012). Attenuated ASFV strains induce enhanced expression of type I IFNs (both IFN-β and several IFN-α subtypes) compared to highly virulent strains (Gil *et al.*, 2008; Razzuoli *et al.*, 2020; Reis *et al.*, 2016) and genes within ASFV multigene families (MGF) 360 and 530/505 may play a central role in inhibiting type I induction in ASFV-infected Mφ (Afonso *et al.*, 2004; Reis *et al.*, 2016). It was described that deletion or interruption of several genes within MGF360 and 530/505 in the genome of the virulent Benin 97/1 resulted in enhanced induction of type I IFN (Reis *et al.*, 2016). In addition, it was demonstrated that some members of these MGF suppress type I IFN induction through inhibition of transcription factors: A276R (MGF360-15R) inhibits IRF 3, whereas A528R (MGF505-7R) impaired both IRF3 and NK-kβ (Correia *et al.*, 2013). Other ASFV genes inhibited type I IFN production: the protein coded by I329L is a TLR3 homologue and targets TLR4/toll-interleukin-1 receptor domain-containing adapter inducing interferon-β (TRIF), impairing activation of both IRF3 and nuclear factor (NF)-κB (Correia *et al.*, 2013; de Oliveira *et al.*, 2011).

ASFV can subvert not only type I IFN induction, but also that of interferon stimulated genes (ISGs). It was described that infection with attenuated NH/P68 but not virulent Armenia/07 resulted in enhanced CCLX10 and IFIT-1 expression in porcine alveolar Mφ (Garcia-Belmonte *et al.*, 2019). Portugal and colleagues (2018) reported similar findings, with virulent L60 but not attenuated NH/P68 inhibiting induction of several ISGs in Mφ following IFN-α stimulation (Portugal *et al.*, 2018). NH/P68 lacks several genes within MGF360/530 (Portugal *et al.*, 2018) and these genes seem to play a crucial role in modulating ISG responses: deletion of six MGF360 and two MGF530 genes from the virulent Pr4 strain resulted in enhanced expression of several ISGs in infected Mφ, including cytokine IP-10, IFIT-1, and ISG43 (Afonso *et al.*, 2004).

Despite ASFV possessing strategies to inhibit type I IFN production/responses by Mφ or other cell types, the virus likely promotes release of these cytokines by pDC, as described in other viruses (Summerfield, 2012). High levels of type I IFN have been observed in pigs infected with virulent ASFV (Golding *et al.*, 2016) and pDC are likely to be the source of this cytokine (Golding *et al.*, 2016).

Viruses have evolved mechanisms to become 'tolerant' to the action of type I IFN (Summerfield, 2012) and several studies suggest that ASFV has employed this strategy. It was described that high doses of type I IFN (2,000 U/ml) impaired ability of attenuated OUR T88/3 but not virulent ASFV isolates to infect porcine alveolar Mφ (Golding *et al.*, 2016). Similar findings were obtained in moDC infected with ASFV strains of diverse virulence (attenuated BA71V and NH/P68 and virulent 22653/14) (Franzoni *et al.*, 2018). It was recently reported that IFN-α inhibited ASFV ability to infect moMφ in a dose dependent manner, and only the virulent 22653/14 was tolerant to low levels of IFN-α (100 U/ml) (Franzoni *et al.*, 2020).

Ability of virulent ASFV isolates to overcome the type I IFN induced antiviral state seems to be linked to MGF360 and MGF505 genes: attenuated OUR T88/3 lacks genes within MGF360 and 505, deletion of five genes within MGF360 and two within MGF505 from virulent Pr4 ASFV resulted in enhanced sensitivity to IFN-α (Golding *et al.*, 2016). Tolerance to type I IFNs seems to be linked to virulence *in vivo*: deletion of genes within MGF360 and 505 from the virulent genotype II Georgia 2007 or genotype I Benin 91/7 resulted in attenuation *in vivo*, and deleted mutants were also able to induce protection to homologous challenge (O'Donnell *et al.*, 2015; Reis *et al.*, 2016).

Not only type I IFN, but also some ISGs were able to inhibit ASFV infection. Two interferon-induced transmembrane proteins (IFITM2 and IFITM3) affected BA71V infectivity in Vero cells, impairing viral entry and uncoating (Munoz-Moreno *et al.*, 2016), whereas MxA protein affected BA71V entry in Vero cells due to its recruitment to perinuclear viral factories (Netherton *et al.*, 2009). It would be interesting to assess whether differences exist between strains of diverse virulence, and which viral factors are involved in resistance to type I IFN/ISG mediated antiviral activities.

3.3 Humoral responses against African swine fever virus upon infection

The way viral antigens are recognised by B cells are through inmunoglobulins (Ig). These proteins are produced by B cells in a vast range of specificities, each B cell is producing Ig of a single specificity. Membrane-bound Ig on the B cell surface serves as receptor for antigen specificity and it is known as the B cell receptor (BCR). Ig of the same antigen specificity is secreted as antibody (Ab) by terminally differentiated B cells or plasma cells (Janeway *et al.*, 2017). Even though research on B cell development has received some attention in pigs, there are still several gaps into B cell development and characterisation of B cell subpopulations. On the other hand, Ab repertoires in pigs have been extensively studied (Butler *et al.*, 2017). Their studies showed that pigs resemble most mammals in using both kappa and lambda light chains of Ab. Similar to in humans, but unlike in many other mammals, the kappa-lambda expression ratio is essentially equal in adult swine, but not during development (reviewed in Butler *et al.*, 2017).

Antibodies are secreted in response to viral infections and they have been extensively used in diagnosis of infections such ASFV. This topic will be later explained in Chapter 5 in this book. Here, we will focus on the knowledge of humoral responses during ASFV infection.

Surprisingly, the role of Ab within ASFV infection remains controversial nowadays. It has been described that ASFV infection induces neutralising Ab but the existence of non-neutralising Ab, which binds to virus without diminishing infectivity, has also long been recognised (reviewed in Escribano *et al.*, 2013). Early experiments showed that antibodies obtained from pigs that had recovered from ASFV infection with one of the most used strains in experimental protection experiments, E75CV1-4 virus, a Spanish strain (E75) which was adapted to grow in CV1 cells and propagated in pig macrophages, appeared to confer protection against ASFV infection. Sera from E75CV1-4 infected convalescent pigs neutralised the infectivity of virulent ASFV isolates E75, E70, Lisbon 60, Malawi Lil 20/1 and a low-passage tissue culture-adapted variant of E75, namely E75CV/V3, by 86-97% in Vero and macrophage cell cultures (Zsak *et al.*, 1993). Antiviral antibodies' role in homologous protective immunity to E75 virulent ASFV strain was examined by passive transfer experiments in swine. In those experiments, 85% of the animals receiving anti-ASFV immunoglobulin survived challenge infection. In contrast, 100% mortality was observed in the experimental group that received control immunoglobulin sera fractions or phosphate-buffered saline. With the exception of a significantly delayed and transient fever response, the animals who received anti-ASFV antibodies remained clinically normal following challenge, whereas the control group presented clinical ASF on day 4 post-challenge. In addition, a significant 3-day delay in the onset of viraemia and a 10,000-fold reduction in both mean and maximum virus titres were observed for animals given anti-ASFV Ig. These data strongly suggest that anti-ASFV antibodies alone protect swine from lethal infection with virulent ASFV (Onisk *et al.*, 1994). Moreover, they support the view that the antibody-mediated protective effect is an early event that effectively delays disease onset (Onisk *et al.*, 1994). Following this line of thought, other reports showed that when ASFV antibodies are transferred through colostrum, they also confer a degree of protection against viral challenge in suckling piglets (Schlafer *et al.*, 1984a,b) showing a representation of *in vivo* antibody-mediated protection against ASFV. Several ASFV proteins are able to induce neutralising antibodies in immunised pigs. Among these proteins, p54 and p30 (Gómez-Puertas and Escribano, 1997; Gómez-Puertas *et al.*, 1996) were shown to be involved in various steps of virus attachment and internalisation. Nevertheless, immunisation of pigs with recombinant p54 and p30 proteins expressed in baculovirus did not protect against virulent ASFV challenge. Similarly, a vaccine based on ASFV proteins p30, p54, p22, and p72 produced in baculovirus failed to protect pigs against a virulent ASFV challenge, despite producing neutralising antibodies (Neilan *et al.*, 2004); the above results added to the controversy about the role of antibody-mediated neutralisation of ASFV in protection against ASF infection (Escribano *et al.*, 2013).

Several *in vivo* and *in vitro* studies indicate a potential protective role of Ab by mechanisms including complement mediated cell lysis or antibody dependent cell-mediated cytotoxicity (ADCC) (Rock, 2017; Takamatsu *et al.*, 2013). A remarkable correlation has been established between the presence of haemadsorption (HAD) inhibitory antibodies in serum with its capacity to inhibit the infection of ASFV *in vitro* and to partially protect against ASFV challenge *in vivo* (Burmakina *et al.*, 2019; Ruiz-Gonzalvo *et al.*, 1996).

In summary, ASFV Ab and their neutralisation ability have received little attention recently, and most of the publications only use antibodies to monitor infection without further analysis on their activity, mechanisms, localisation and isotypes. For example, virus antigens important for protection or cross-protection have not been fully characterised although antibodies against the virus CD2-like protein have been shown to be involved in cross-protection (Burmakina *et al.*, 2019). Further studies on Ab generated after infection with different isolates of ASFV are required to elucidate their role in ASFV infection.

3.4 Cellular response against African swine fever virus upon infection

T cells are defined in pigs, as in other species, by the expression of the CD3 T cell co-receptor complex on their surface. Additionally, they express either αβ or γδ T cell receptors (TCR) as antigen receptors. T cells, like all antigen-specific lymphocytes, are able to recognise antigens specifically and react differently depending on their function (helper function, cytotoxicity, cytokine production, and sometimes even antigen presentation). A special characteristic of the adaptive immune system is that after antigen contact and the immune response that follows, antigen-specific memory cells are produced, so that a subsequent reaction to an antigen can be faster and/or stronger. However, both the antigen presentation, followed by the antigen-specific reaction and even more so the generation of memory cells take time. Therefore, the adaptive immune system reacts with very high antigen-specificity but with a significant time delay compared to the innate immune system. This time factor is problematic in the case of an acute and usually lethal virus infection such as that caused by infection with ASFV. Therefore, failure to detect a specific cellular immune response after infection with virulent ASFV strains is commonly reported. However, mechanisms of both specific and non-specific cellular immune responses to infections with moderately virulent ASF viruses and after immunisations with ASFV are described.

3.4.1 αβ TCR+ T cells

When T cells contact their specific antigen using their TCR, they differentiate into effector cells. Apart from a small population of regulatory T cells, the main distinction is made between cytotoxic T cells and helper T cells.

3.4.1.1 CD8+ cytotoxic T cells

Cytotoxic T cells play an important role in protection against intracellular antigens. These cells recognise short sections of viral antigen displayed on the surface of infected cells via MHC class I (referred to as swine leukocyte antigen I (SLA I) in pigs), and then kill such cells usually by releasing perforin and granzyme. Cytotoxic T cells can also kill cells by engaging Fas/FasL, but this is only marginally described in pigs. In pigs, classical MHC I restricted T cells with lytic activity are described as CD2+CD3+CD4-CD5^{high}CD6+CD8α^{high}CD8β+ (Denyer *et al.*, 2006).

Depending on their function as effector or memory cells, they can also express CCR7 and CD27. As CD8 T cells also produce IFN-γ and TNF-α they play an important role in antimicrobial immunity. Not least because of this cytokine secretion and the production of effector molecules, such as perforin, cytotoxic T cells can also be responsible for severe immunopathology after viral infection.

T cells, especially CD8α+ T cells, play an important role in protective immunity against ASFV (Oura *et al.*, 2005). After exposure to the low virulent ASFV strain OUR T88/3, pigs were depleted of CD8α+ lymphocytes with monoclonal anti-CD8 antibodies (clones 76-2-11 and 11/295/33). A subsequent challenge with homologous, but virulent, OUR T88/1 revealed that depleted animals suffered from severe acute ASF and died, whereas non-depleted animals showed only mild clinical signs and survived. Since CD8α is expressed on different T cell populations like cytotoxic T cells, γδ T cells, NK cells, invariant T cells or memory helper T cells, it remains unclear which of these subpopulations might mediate the described protection. As cytotoxic T cells in the pig typically express a heterodimer of CD8α and CD8β, in order to define the protective CD8 expressing cell type the authors tried to specifically deplete CD8β T cells in a follow-on experiment. However, depletion with the anti CD8β antibody (clone PPT22) was not successful in all animals. Only one animal out of seven showed a significant depletion of CD8β+ cells. Interestingly, this pig did not survive the challenge infection. Overall, these depletion experiments proved the crucial role of CD8α+ T cells for protection. It remains to be clarified which CD8+ T cell type is responsible for the protective effect. Furthermore, although this experiment proves the protective effect of CD8+ ASFV-specific memory cells, it does not explain the role of cytotoxic T cells in acute infections.

An early reference to ASFV specific CD8+ T effector cells is found in Norley and Wardley (1984): the authors infected pigs with a virulent Uganda isolate and examined peripheral blood mononuclear cells (PBMCs) of these animals in a killing assay with virus-infected autologous testicular cells as target. The effector cells were able to lyse the ASFV-infected cells virus-specifically. Since unseparated PBMCs were used, it can only be assumed that CD8+ cells were involved in cytotoxicity. Furthermore, the statement is limited to early effector cells, since the animals in this experiment did not survive long enough to investigate the cytotoxic properties of ASFV-specific memory cells. Such ASFV specific CD8+ memory cells could be described in a study by Martins *et al.* (1993): the authors used SLA inbred mini-pigs that had survived infection with the non-lethal ASFV strain NH/P68. Their PBMCs were re-stimulated *in vitro* and their cytotoxic properties were tested in a killing assay with homologous infected macrophages. A specific lysis of ASFV infected macrophages against non-infected or classical swine fever virus (also called hog cholera virus) infected controls could be shown. The authors demonstrated that the observed killing is SLA I dependent, by showing that both anti-SLA I antibodies and anti-CD8 antibodies strongly reduce cytotoxicity. In addition, they showed a preferential lysis of infected macrophages with matching SLA I.

Which viral epitopes are recognised by ASFV specific CD8 cells is still largely unclear. Knowledge of CD8 epitopes is important in any viral infection, not only for the production of a possible vaccine, but also for the elucidation of CD8-dependent protective effects and immunopathogenesis. Two studies used inbred mini-pigs immunised with two different attenuated isolates that were subsequently challenged with homologous virulent strains (Alonso *et al.*, 1997; Leitão *et al.*, 1998). In the study by Alonso *et al.* (1997), these pigs served as donors for the effector cells tested in killing assays against infected alveolar macrophages. Specific lysis was achieved with isolated CD8+ cells but not with CD4+ cells. In a further experiment, the target cells were not infected with complete ASFV, but with a recombinant vaccinia virus expressing the early ASFV protein encoded by the CP204L gene (also referred to as p30 or p32). These target cells were also specifically lysed, albeit less strongly than those infected with the full virus. The result indicates that specific CD8 epitopes are present in CP204L (Alonso *et al.*, 1997). Leitão *et al.* (1998) showed that effector cells isolated from ASF recovered pigs were capable of lysing macrophages loaded with a 25 amino acid peptide derived from the B646L gene that encodes the major capsid protein p72. This lysis could be efficiently blocked with antibodies against SLA-I, again suggesting a role for CD8+ cells (Leitão *et al.*, 1998). More recent work using an attenuated strain of ASFV has shown that both B646L and CP204L, along with other 16 antigens, are capable of inducing secretion of IFNγ from cells from recovered inbred pigs (Netherton *et al.*, 2019). Individual epitopes have also been mapped within the EP153R and EP402R proteins (Argilaguet *et al.*, 2012; Burmakina *et al.*, 2019) genes; however, the relative importance of a given antigen or epitope in the protection mediated by these attenuated viruses is unclear.

An unusual facet of the swine immune system is the significant number of CD4+CD8+ double positive (DP) T cells (Saalmüller *et al.*, 1987). DP T cells play a distinct role in porcine immune responses (Okutani *et al.*, 2018) and are often described as memory or effector T cells (Saalmüller *et al.*, 2002). *In vitro* stimulation studies demonstrated CD8 upregulation on porcine CD4 T cells (Reutner *et al.*, 2013) or showed proliferation and cytokine production in DP T cells e.g. (Lefevre *et al.*, 2012). DP T cells are capable of expressing perforin and granzyme, also indicating a role as cytotoxic effector cells. Alonso *et al.* (1997) discussed in the abovementioned study the involvement of DP T cells, which account for a considerable proportion of T cells after re-stimulation, in the measurable specific cytotoxicity (Schäfer *et al.*, 2019). Antigen stimulation of PBMCs from ASF recovered pigs induced both proliferation of DP T cells that are also perforin positive (Takamatsu *et al.*, 2013), as well as secretion of IFN-γ (Netherton *et al.*, 2019). After infection with the virulent Armenia strain, an increase of DP T cell frequency with a corresponding decrease of CD4+ T cell frequency in peripheral blood and lymphoid organs of domestic pigs was detectable. In wild boar, no direct correlation between CD4+ T helper cells and DP T cells was detectable. DP T cells of wild boar showed a high proliferative activity. However, data on the role of DP T cells during acute primary infections are not available. Nevertheless, these studies indicate an activation-dependent expression of CD8 on CD4+ T cells and underline the effector functions of DP T cells (Hühr *et al.*, 2020).

A rare cell population of αβTCR+ T cells are the invariant NK T cells (iNKT). Their frequency in swine is typically between 0.01 and 1% among CD3+ T cells and decreases with age. These cells bridge and orchestrate both untargeted innate and specific adaptive responses, which are crucial for pathogen clearance and survival. In contrast to the vast heterogeneity of TCRs among conventional CD3+ T cells (cTC), iNKT cells possess a semi-invariant TCR. This TCR is restricted to the non-classical MHC class I-related CD1d, presenting lipid or glycolipid antigens. iNKT cells can be activated antigen-dependently with glycolipids derived from microbes or the host by TCR-CD1d interactions or antigen-independently via cytokines, mainly IL-12 and IL-18 or type I IFN. Effector cytokines are present as immediately available preformed mRNA transcripts in iNKT cells. Therefore, iNKT cells rapidly proliferate and secrete effector molecules like IFN-γ, IL-17 or granulocyte-macrophage colony-stimulating factor after activation. Moreover, they are able to lyse infected cells by perforin and Fas/FasL interaction. Pig iNKT cells can be stained by a cross reacting murine CD1d tetramer, loaded with PBS57. After infection with the virulent Armenia strain, an average 5-fold increase in the frequency of iNKT cells can be observed in blood. In the examined gastrohepatic lymph nodes, liver and lung a less strong increase in frequency is measurable. This first study does not yet make any statements about the activation and effector mechanisms of iNKT cells after ASFV infection, but it does provide indications of the possible involvement of this rare T cell population in the pathogenesis of ASFV infection (Schäfer *et al.*, 2019). In the same line are *in vitro* experiments by Denyer *et al.* (2006), who investigated perforin-producing lymphocytes in a co-culture experiment, in which they co-cultivated PBMCs from naïve pigs with ASFV-infected MHC I haplotype-matching cells and observed a stronger increase in perforin positive cells in culture with ASFV compared to the control. Furthermore, they also described that after stimulation with ASFV infected cells, the majority of perforin positive cells can be assigned to NK cells and MHC I independent NKT cells (CD3+CD6-CD8+CD16+), which indicates a specific proliferation of these cells (Deyner *et al.*, 2006).

3.4.1.2 CD4+ T helper cells

In contrast to the CD8+ T cells described above, CD4+ cells recognise antigens presented by MHC class II. While almost every cell can present an antigen via MHC I, only classical antigen presenting cells (APC) present phagocytosed extracellular antigen in MHC II. However, as will be discussed later, pigs also have T cell subtypes and NK cells that express MHC II on their surface and are therefore able to present antigen for CD4+ T cells. Depending on the cytokine environment, CD4+ T helper cells develop into a number of functionally different subgroups, which can be differentiated by means of different transcription factors, also in pigs. The existence of Th1, Th2, Th17 and regulatory T (Tregs) in pigs is undisputed. The effector molecules produced by the cells correspond to those of comparable helper cells in humans and mice. Of interest in connection with ASFV, is the group of perforin-producing cytolytic T-helper cells characterised by the transcription factor EOMES. These cells belong to the CD4+CD8+ phenotype and were discussed in the previous paragraph because of their cytolytic activity.

The activation of CD4+ cells after infection is measured by their proliferation and/or IL-2 and IFN-γ production. Some early studies report T cell proliferation and cytokine production in assays using T cells from pigs that had survived low virulent ASFV infection (Revilla *et al.*, 1992). However, there is usually no consistent differentiation of proliferating T cell subtypes. Regarding a specific response of CD4+ cells, Canals *et al.* (1992) presented a more detailed study: they re-stimulated PBMCs of ASFV E75 immune pigs with active virus and with ultraviolet (UV) inactivated virus. The measured proliferation after stimulation with inactivated virus (and thus extracellular antigen) could be completely inhibited with anti-CD4 antibodies, whereas anti-CD8, anti-SLA I and anti-SLA II antibodies only partially inhibited this proliferation (Canals *et al.*, 1992). ASFV seems to influence the ability of leukocytes to proliferate in general. CD4+ and CD8+ splenocytes from mini-pigs infected with the Malawi ASFV strain showed a constantly decreased proliferation performance after mitogenic stimulation with concanavalin A or pokeweed mitogen over the duration of the infection. The effect could not be reversed by rIL-2, but there was no evidence of apoptosis in T cells (Childerstone *et al.*, 1998). *In vitro* studies have also shown an effect of ASFV infection on proliferation of naïve cells and this may depend in part on the EP402R gene that is responsible for the haemadosorption phenomenon (Borca *et al.*, 1998).

A major task of the CD4+ T cells is to assist in the so-called T cell dependent B cell proliferation and maturation. In an *in vitro* antibody synthesis assay, this T cell-help could be investigated. PBMCs of BA71 immune pigs were re-stimulated *in vitro* with ASFV and their antibody production was measured *in vitro*: if the T cells are removed by rosetting or the CD4+ cells are specifically removed with anti-CD4 antibodies, ASFV specific antibody production is significantly reduced (Casal *et al.*, 1987). This result would have to be tested for other ASFV strains, but indicates a T cell dependent antibody production and thus, probably also a T cell dependent class switch.

3.4.2 γδ T cells

Unlike many other mammals, pigs have a large number of γδ T cells in the periphery and in the lymphoid organs. In young pigs up to 50% of all PBMCs are γδ T cells, the percentage decreases with age. They are defined as CD3+ γδ TCR+ CD4-. Depending on the activation status, three cell populations can be distinguished in this population: CD2-CD8- (naïve); CD2+CD8- (proliferation and cytokine secretion) and CD2+CD8+ (cytotoxicity, proliferation and cytokine secretion) (Yang and Parkhouse, 1996).

Due to their activation via TCR, γδ T cells are, by definition, assigned to the adaptive immune system. Thus, they can be activated directly or by contact with APCs and then fulfil functions of cytotoxicity and cytokine production (mainly IFN-γ, TNF-α and IL-17A) (Sedlak *et al.*, 2014). In parallel, however, they also fulfil tasks of the innate immune system, since activation is also possible via other receptors, such as TLR. The extent of this reaction is regulated by an interaction of activating and inhibiting receptors. A third function of γδ T cells in pigs is their ability to present antigen. For this purpose, the cells express MHC II on the surface. This function may be particularly relevant after ASFV infections: it is generally agreed that ASFV replicates predominantly in monocytes/macrophages and thus influences their function. Can antigen-presenting γδ T cells fill this infection-related gap in antigen presentation?

Despite the large number of γδ T cells in pigs and their ability to present antigens, their involvement in the anti ASFV response has been studied only on few occasions. After infection with the moderately virulent strain Netherlands 86, an association between circulating γδ T cells and survival of the animals was described (Post *et al.*, 2017). Takamatsu *et al.* (2006) showed that γδ T cells, incubated with ASFV, are able to stimulate ASFV specific T cells. However, the mechanism of this cell interaction has yet to be clarified (Takamatsu *et al.*, 2006). After lethal infection with the virulent Armenia strain, the frequency of γδ T cells tends to decrease. In domestic pigs, the composition of γδ T cell-subpopulation remains unchanged over the time of infection. In infected wild boar, however, an activation and thus a percentage shift towards CD2+ CD8- or CD2+ CD8+ γδ T cells is observed in blood and spleen (Hühr *et al.*, 2020).

3.4.3 Natural killer cells

NK cells, which by definition belong to the innate immune system, play a crucial role in fighting virus-infected cells. They produce a relevant amount of IFN-γ and TNF-α in an antiviral immune response. Their activation status is regulated by activating and inhibiting ligands that can bind ligands to different target cells. NK cells interact with APCs, especially DCs, and thus can induce their maturation and increase their functionality. In pigs, NK cells are detected as CD3-CD2+CD8+CD16+perforin+ cells. Based on the expression of NKp46, NK subtypes with different functionality can be described. While cytolytic activity could be measured independently from NKp46 expression, NKp46+ NK cells produce significantly higher levels of IFNγ after stimulation (Mair *et al.*, 2012).

Similarly, as described above for γδ T cells, NK cells in pigs also appear to be able to act as APCs. After cytokine stimulation (IL-2, IL-12, IL-18, IFN-α) they express MHC II on their surface (Franzoni *et al.*, 2014). Additionally, after stimulation (IL-2, IL-12, IL-18), co-stimulatory molecules CD80/CD86 are also found on the surface of NK cells (De Pelsmaeker *et al.*, 2019). The same authors also report that NK cells internalise antigen of the cells they lyse. Thus, NK cells, like γδ T cells, would theoretically be able to take over APC function in ASFV infected animals instead of the infected classical APCs. This mechanism has not yet been described after ASFV infection. However, it has been shown that the increase in cytotoxic activity of NK cells correlates with the absence of clinical signs after infection with the non-haemabsorbing NH/P68 strain. Conversely, a lower cytotoxicity of NK cells could be measured in pigs with clinical signs of chronic ASF (Lanier, 2005). Since in the described experiment the animals with NK cell activity were also protected from challenge with the virulent strain Lisbon 60, participation of NK cells in an antigen-specific memory response can be considered. After infection with the moderately virulent strain Malta 78, a decrease in NK activity was measured in 4 of 6 animals (Norley and Wardley, 1983). However, an interesting temperature phenomenon was also discussed: the NK activity measured *in vitro* disappears at 40 °C. Therefore, reduced or missing protective NK activity in ASFV infected pigs with high fever and severe courses of disease is possible. However, some *in vitro* experiments show a decrease of NK activity after infection with virulent virus strains, independent of temperature (Martins and Leitão, 1994; Mendoza *et al.*, 1991).

3.4.4 Regulatory T cells

Regulatory T (Tregs) cells are essential for maintaining peripheral tolerance, preventing autoimmune diseases and limiting chronic inflammatory diseases. Generally, they have multiple mechanisms to mediate their suppressive effects: suppression by inhibitory cytokines, suppression by cytolysis, suppression by metabolic disruption and suppression by modulation of DC maturation or function.

In pigs, Tregs defined as CD3+CD4α+CD25+FoxP3+ cells have not received much attention in the context of viral infections until recently. A study by Sanchez-Cordon *et al.* (2020) has recently shown that Tregs exhibited a progressive increase in two groups of immunised animals with OURT88/3 or BeninΔMGF ASFV strains. In fact, they peak at day 130 pi, showing significant statistically differences as compared with pre-immunisation values. They also showed that those Tregs were able to inhibit proliferative responses *in vitro*, suggesting that higher percentage of Tregs circulating in blood might represent an inhibition of specific immune responses. Therefore, Tregs might represent a viral strategy to prevent immune responses against ASFV.

3.5 Immunity unknowns

Given the relevance of ASFV infection in swine, it is surprising that immune responses against ASFV are still poorly characterised. Deeper understanding of the immunity developed by the natural hosts, including bushpig (*Potamochoerus larvatus*) and warthog (*Phacochoerus africanus*), by ASFV infection is an area that should be addressed. However, this approach is very difficult since few reagents are available for wild African pigs with little cross reactivity among them. Unfortunately, many gaps still exist regarding immune responses in the natural infection vs experimental infection. Besides, identification of the correlates of immune protection would allow applying the three Rs (Replace, Reduce, Refine) guiding principles of animal science and reducing unnecessary painful challenges with ASFV (Arias *et al.*, 2017). Despite these advances, very little is known about how immune responses are generated during the course of ASFV natural infection.

Acknowledgements

This publication is based on work from 'Understanding and combating African swine fever in Europe (ASF-STOP COST action 15116)' supported by COST (European Cooperation in Science and Technology).

References

Afonso, C.L., Piccone, M.E., Zaffuto, K.M., Neilan, J., Kutish, G.F., Lu, Z., Balinsky, C.A., Gibb, T.R., Bean, T.J., Zsak, L. and Rock, D.L., 2004. African swine fever virus multigene family 360 and 530 genes affect host interferon response. Journal of Virology 78: 1858-1864. https://doi.org/10.1128/jvi.78.4.1858-1864.2004

Alonso, F., Dominguez, J., Viñuela, E. and Revilla, Y., 1997. African swine fever virus-specific cytotoxic T lymphocytes recognize the 32 kDa immediate early protein (vp32). Virus Research 49: 123-130. https://doi.org/10.1016/s0168-1702(97)01459-7

Argilaguet J.M., Pérez-Martín E., Nofrarías M., Gallardo C., Accensi F., Lacasta A., Mora M., Ballester M., Galindo-Cardiel I., López-Soria S., Escribano J.M., Reche P.A. and Rodríguez F., 2012. DNA vaccination partially protects against African swine fever virus lethal challenge in the absence of antibodies. PLoS One 7: e40942. https://doi.org/10.1371/journal.pone.0040942

Arias, M.L., De La Torre, A., Dixon, L., Gallardo, C., Jori, F., Laddomada, A., Martins, C., Parkhouse, M., Revilla, Y., Rodríguez, F. and Sánchez-Vizcaíno, J.M., 2017. Blueprint and Roadmap on the possible development of a vaccine for African swine fever prepared by the African swine fever EU reference laboratory on Commission request. European Commission, Directorate-General for Health and Food Safety, Brussels, Belgium.

Banchereau, J. and Steinman, R.M., 1998. Dendritic cells and the control of immunity. Nature 392: 245-252. https://doi.org/10.1038/32588

Basta, S., Knoetig, S.M., Spagnuolo-Weaver, M., Allan, G. and McCullough, K.C., 1999. Modulation of monocytic cell activity and virus susceptibility during differentiation into macrophages. Journal of Immunology 162: 3961-3969.

Borca, M.V., Carrillo, C., Zsak, L., Laegreid, W.W., Kutish, G.F., Neilan, J.G., Burrage, T.G. and Rock, D.L., 1998. Deletion of a CD2-like gene, 8-DR, from African swine fever virus affects viral infection in domestic swine. Journal of Virology 72: 2881-2889.

Burmakina, G., Malogolovkin, A., Tulman, E.R., Xu, W., Delhon, G., Kolbasov, D. and Rock, D.L., 2019. Identification of T-cell epitopes in African swine fever virus CD2v and C-type lectin proteins. Journal of General Virology 100: 259-265. https://doi.org/10.1099/jgv.0.001195

Butler, J.E., Wertz, N. and Sinkora, M., 2017. Antibody repertoire development in swine. Annual Review of Animal Biosciences 5: 255-279. https://doi.org/10.1146/annurev-animal-022516-022818

Canals, A., Alonso, F., Tomillo, J. and Domínguez, J., 1992. Analysis of T lymphocyte subsets proliferating in response to infective and UV-inactivated African swine fever viruses. Veterinary Microbiology 33: 117-127. https://doi.org/10.1016/0378-1135(92)90040-z

Casal, I., Viñuela, E. and Enjuanes, L., 1987. Synthesis of African swine fever (ASF) virus-specific antibodies *in vitro* in a porcine leucocyte system. Immunology 62: 207-213.

Childerstone, A., Takamatsu, H., Yang, H., Denyer, M. and Parkhouse, R.M., 1998. Modulation of T cell and monocyte function in the spleen following infection of pigs with African swine fever virus. Veterinary Immunology and Immunopathology 62: 281-296. https://doi.org/10.1016/s0165-2427(97)00173-6

Collin, M. and Bigley, V., 2018. Human dendritic cell subsets: an update. Immunology 154: 3-20. https://doi.org/10.1111/imm.12888

Correia, S., Ventura, S. and Parkhouse, R.M., 2013. Identification and utility of innate immune system evasion mechanisms of ASFV. Virus Research 173: 87-100. https://doi.org/10.1016/j.virusres.2012.10.013

De Oliveira, V.L., Almeida, S.C., Soares, H.R., Crespo, A., Marshall-Clarke, S. and Parkhouse, R.M., 2011. A novel TLR3 inhibitor encoded by African swine fever virus (ASFV). Archives of Virology 156: 597-609. https://doi.org/10.1007/s00705-010-0894-7

De Pelsmaeker, S., Devriendt, B., Leclercq, G. and Favoreel, H.W., 2018. Porcine NK cells display features associated with antigen-presenting cells. Journal of Leukocyte Biology 103: 129-140. https://doi.org/10.1002/JLB.4A0417-163RR

Denyer, M.S., Wileman, T.E., Stirling, C.M., Zuber, B. and Takamatsu, H.H., 2006. Perforin expression can define CD8 positive lymphocyte subsets in pigs allowing phenotypic and functional analysis of natural killer, cytotoxic T, natural killer T and MHC un-restricted cytotoxic T-cells. Veterinary Immunology and Immunopathology 110: 279-292. https://doi.org/10.1016/j.vetimm.2005.10.005

Dixon, L.K., Islam, M., Nash, R. and Reis, A.L., 2019. African swine fever virus evasion of host defences. Virus Research 266: 25-33. https://doi.org/10.1016/j.virusres.2019.04.002

Escribano, J.M., Galindo, I. and Alonso, C., 2013. Antibody-mediated neutralization of African swine fever virus: myths and facts. Virus Research 173: 101-109. https://doi.org/10.1016/j.virusres.2012.10.012

Fishbourne, E., Abrams, C.C., Takamatsu, H.H. and Dixon, L.K., 2013. Modulation of chemokine and chemokine receptor expression following infection of porcine macrophages with African swine fever virus. Veterinary Microbiology 162: 937-943. https://doi.org/10.1016/j.vetmic.2012.11.027

Franzoni, G., Dei Giudici, S. and Oggiano, A., 2018. Infection, modulation and responses of antigen-presenting cells to African swine fever viruses. Virus Research 258: 73-80. https://doi.org/10.1016/j.virusres.2018.10.007

Franzoni, G., Edwards, J.C., Kurkure, N.V., Edgar, D.S., Sanchez-Cordon, P.J., Haines, F.J., Salguero, F.J., Everett, H.E., Bodman-Smith, K.B., Crooke, H.R. and Graham, S.P., 2014. Partial Activation of natural killer and gammadelta T cells by classical swine fever viruses is associated with type I interferon elicited from plasmacytoid dendritic cells. Clinical and Vaccine Immunology 21: 1410-1420. https://doi.org/10.1128/CVI.00382-14

Franzoni, G., Graham, S.P., Dei Giudici, S. and Oggiano, A., 2019. Porcine dendritic cells and viruses: an update. Viruses 11: 445. https://doi.org/10.3390/v11050445

Franzoni, G., Graham, S.P., Giudici, S.D., Bonelli, P., Pilo, G., Anfossi, A.G., Pittau, M., Nicolussi, P.S., Laddomada, A. and Oggiano, A., 2017. Characterization of the interaction of African swine fever virus with monocytes and derived macrophage subsets. Veterinary Microbiology 198: 88-98. https://doi.org/10.1016/j.vetmic.2016.12.010

Franzoni, G., Graham, S.P., Sanna, G., Angioi, P., Fiori, M.S., Anfossi, A., Amadori, M., Dei Giudici, S. and Oggiano, A., 2018. Interaction of porcine monocyte-derived dendritic cells with African swine fever viruses of diverse virulence. Veterinary Microbiology 216: 190-197. https://doi.org/10.1016/j.vetmic.2018.02.021

Franzoni, G., Razzuoli, E., Dei Giudici, S., Carta, T., Galleri, G., Zinellu, S., Ledda, M., Angioi, P., Modesto, P., Graham, S.P. and Oggiano, A., 2020. Comparison of macrophage responses to African swine fever viruses reveals that the NH/P68 strain is associated with enhanced sensitivity to type I IFN and cytokine responses from classically activated macrophages. Pathogens 9: 209. https://doi.org/10.3390/pathogens9030209

García-Belmonte, R., Pérez-Núñez, D., Pittau, M., Richt, J.A. and Revilla, Y., 2019. African swine fever virus Armenia/07 virulent strain controls interferon beta production through the cGAS-STING pathway. Journal of Virology 93. https://doi.org/10.1128/JVI.02298-18

Gil, S., Sepúlveda, N., Albina, E., Leitão, A. and Martins, C., 2008. The low-virulent African swine fever virus (ASFV/NH/P68) induces enhanced expression and production of relevant regulatory cytokines (IFNalpha, TNFalpha and IL12p40) on porcine macrophages in comparison to the highly virulent ASFV/L60. Archives of Virology 153: 1845-1854. https://doi.org/10.1007/s00705-008-0196-5

Gil, S., Spagnuolo-Weaver, M., Canals, A., Sepúlveda, N., Oliveira, J., Aleixo, A., Allan, G., Leitão, A. and Martins, C.L., 2003. Expression at mRNA level of cytokines and A238L gene in porcine blood-derived macrophages infected *in vitro* with African swine fever virus (ASFV) isolates of different virulence. Archives of Virology 148: 2077-2097. https://doi.org/10.1007/s00705-003-0182-x

Golding, J.P., Goatley, L., Goodbourn, S., Dixon, L.K., Taylor, G. and Netherton, C.L., 2016. Sensitivity of African swine fever virus to type I interferon is linked to genes within multigene families 360 and 505. Virology 493: 154-161. https://doi.org/10.1016/j.virol.2016.03.019

Gómez-Puertas, P. and Escribano, J.M., 1997. Blocking antibodies inhibit complete African swine fever virus neutralization. Virus Research 49: 115-122. https://doi.org/10.1016/s0168-1702(97)01463-9

Gómez-Puertas, P., Rodríguez, F., Oviedo, J.M., Ramiro-Ibáñez, F., Ruiz-Gonzalvo, F., Alonso, C. and Escribano, J.M., 1996. Neutralizing antibodies to different proteins of African swine fever virus inhibit both virus attachment and internalization. Journal of Virology 70: 5689-5694.

Granja, A.G., Perkins, N.D. and Revilla, Y., 2008. A238L inhibits NF-ATc2, NF-kappa B, and c-Jun activation through a novel mechanism involving protein kinase C-theta-mediated up-regulation of the amino-terminal transactivation domain of p300. Journal of Immunology 180: 2429-2442. https://doi.org/10.4049/jimmunol.180.4.2429

Gregg, D.A., Mebus, C.A. and Schlafer, D.H., 1995. Early infection of interdigitating dendritic cells in the pig lymph node with African swine fever viruses of high and low virulence: immunohistochemical and ultrastructural studies. Journal of Veterinary Diagnostic Investigation 7: 23-30. https://doi.org/10.1177/104063879500700104

Hühr, J., Schäfer, A., Schwaiger, T., Zani, L., Sehl, J., Mettenleiter, T.C., Blome, S. and Blohm, U., 2020. Impaired T-cell responses in domestic pigs and wild boar upon infection with a highly virulent African swine fever virus strain. Transboundary and Emerging Diseases 67: 3016-3032. https://doi.org/10.1111/tbed.13678

Hume, D.A., 2015. The many alternative faces of macrophage activation. Frontiers in Immunology 6: 370. https://doi.org/10.3389/fimmu.2015.00370

Italiani, P. and Boraschi, D., 2014. From monocytes to M1/M2 macrophages: phenotypical vs. functional differentiation. Frontiers in Immunology 5: 514. https://doi.org/10.3389/fimmu.2014.00514

Karalyan, Z., Zakaryan, H., Sargsyan, K., Voskanyan, H., Arzumanyan, H., Avagyan, H. and Karalova, E., 2012. Interferon status and white blood cells during infection with African swine fever virus *in vivo*. Veterinary Immunology and Immunopathology 145: 551-555. https://doi.org/10.1016/j.vetimm.2011.12.013

Lanier, L.L., 2005. NK cell recognition. Annual Review of Immunology 23: 225-274. https://doi.org/10.1146/annurev.immunol.23.021704.115526

Lefevre, E.A., Carr, B.V., Inman, C.F., Prentice, H., Brown, I.H., Brookes, S.M., Garcon, F., Hill, M.L., Iqbal, M., Elderfield, R.A., Barclay, W.S., Gubbins, S., Bailey, M., Charleston, B. and Cosi, 2012. Immune responses in pigs vaccinated with adjuvanted and non-adjuvanted A(H1N1)pdm/09 influenza vaccines used in human immunization programmes. PLoS One 7: e32400. https://doi.org/10.1371/journal.pone.0032400

Leitão, A., Cartaxeiro, C., Coelho, R., Cruz, B., Parkhouse, R.M.E., Portugal, F.C., Vigário, J.D. and Martins, C.L.V., 2001. The non-haemadsorbing African swine fever virus isolate ASFV/NH/P68 provides a model for defining the protective anti-virus immune response. Journal of General Virology 82: 513-523. https://doi.org/10.1099/0022-1317-82-3-513

Leitão, A., Malur, A., Cornelis, P. and Martins, C.L., 1998. Identification of a 25-aminoacid sequence from the major African swine fever virus structural protein VP72 recognised by porcine cytotoxic T lymphocytes using a lipoprotein-based expression system. Journal of Virology Methods 75: 113-119. https://doi.org/10.1016/s0166-0934(98)00105-0

Mair, K.H., Essler, S.E., Patzl, M., Storset, A.K., Saalmüller, A. and Gerner, W., 2012. NKp46 expression discriminates porcine NK cells with different functional properties. European Journal of Immunology 42: 1261-1271. https://doi.org/10.1002/eji.201141989

Martins, C.L. and Leitão, A.C., 1994. Porcine immune responses to African swine fever virus (ASFV) infection. Veterinary Immunology and Immunopathology 43: 99-106. https://doi.org/10.1016/0165-2427(94)90125-2

Martins, C.L., Lawman, M.J., Scholl, T., Mebus, C.A. and Lunney, J.K., 1993. African swine fever virus specific porcine cytotoxic T cell activity. Archives of Virology 129: 211-225. https://doi.org/10.1007/bf01316896

McCullough, K.C., Basta, S., Knötig, S., Gerber, H., Schaffner, R., Kim, Y.B., Saalmüller, A. and Summerfield, A., 1999. Intermediate stages in monocyte-macrophage differentiation modulate phenotype and susceptibility to virus infection. Immunology 98: 203-212. https://doi.org/10.1046/j.1365-2567.1999.00867.x

Medzhitov, R. and Janeway, C., Jr., 2000. Innate immunity. New England Journal of Medicine 343: 338-344. https://doi.org/10.1056/NEJM200008033430506

Mendoza, C., Videgain, S.P. and Alonso, F., 1991. Inhibition of natural killer activity in porcine mononuclear cells by African swine fever virus. Research in Veterinary Science 51: 317-321. https://doi.org/10.1016/0034-5288(91)90084-2

Mosser, D.M., 2003. The many faces of macrophage activation. Journal of Leukocyte Biology 73: 209-212. https://doi.org/10.1189/jlb.0602325

Muñoz-Moreno, R., Cuesta-Geijo, M.Á., Martínez-Romero, C., Barrado-Gil, L., Galindo, I., García-Sastre, A. and Alonso, C., 2016. Antiviral role of IFITM proteins in African swine fever virus infection. PLoS One 11: e0154366. https://doi.org/10.1371/journal.pone.0154366

Neilan, J.G., Zsak, L., Lu, Z., Burrage, T.G., Kutish, G.F. and Rock, D.L., 2004. Neutralizing antibodies to African swine fever virus proteins p30, p54, and p72 are not sufficient for antibody-mediated protection. Virology 319: 337-342. https://doi.org/10.1016/j.virol.2003.11.011

Netherton, C.L., Goatley, L.C., Reis, A.L., Portugal, R., Nash, R.H., Morgan, S.B., Gault, L., Nieto, R., Norlin, V., Gallardo, C., Ho, C.S., Sanchez-Cordon, P.J., Taylor, G. and Dixon, L.K., 2019. Identification and immunogenicity of African Swine fever virus antigens. Frontiers in Immunology 10: 1318. https://doi.org/10.3389/fimmu.2019.01318

Netherton, C.L., Simpson, J., Haller, O., Wileman, T.E., Takamatsu, H.H., Monaghan, P. and Taylor, G., 2009. Inhibition of a large double-stranded DNA virus by MxA protein. Journal of Virology 83: 2310-2320. https://doi.org/10.1128/JVI.00781-08

Norley, S.G. and Wardley, R.C., 1983. Investigation of porcine natural-killer cell activity with reference to African swine-fever virus infection. Immunology 49: 593-597.

Norley, S.G. and Wardley, R.C., 1984. Cytotoxic lymphocytes induced by African swine fever infection. Research in Veterinary Science 37: 255-257.

O'Donnell, V., Holinka, L.G., Gladue, D.P., Sanford, B., Krug, P.W., Lu, X., Arzt, J., Reese, B., Carrillo, C., Risatti, G.R. and Borca, M.V., 2015. African swine fever virus Georgia isolate harboring deletions of MGF360 and MGF505 genes is attenuated in swine and confers protection against challenge with virulent parental virus. Journal of Virology 89: 6048-6056. https://doi.org/10.1128/JVI.00554-15

Okutani, M., Tsukahara, T., Kato, Y., Fukuta, K. and Inoue, R., 2018. Gene expression profiles of CD4/CD8 double-positive T cells in porcine peripheral blood. Animal Science Journal 89: 979-987. https://doi.org/10.1111/asj.13021

Onisk, D.V., Borca, M.V., Kutish, G., Kramer, E., Irusta, P. and Rock, D.L., 1994. Passively transferred African swine fever virus antibodies protect swine against lethal infection. Virology 198: 350-354. https://doi.org/10.1006/viro.1994.1040

Oura, C.A.L., Denyer, M.S., Takamatsu, H. and Parkhouse, R.M.E., 2005. *In vivo* depletion of CD8+ T lymphocytes abrogates protective immunity to African swine fever virus. Journal of General Virology 86: 2445-2450. https://doi.org/10.1099/vir.0.81038-0

Pérez-Núñez, D., García-Urdiales, E., Martínez-Bonet, M., Nogal, M.L., Barroso, S., Revilla, Y. and Madrid, R., 2015. CD2v interacts with adaptor protein AP-1 during African swine fever infection. PLoS One 10: e0123714. https://doi.org/10.1371/journal.pone.0123714

Pietschmann, J., Mur, L., Blome, S., Beer, M., Pérez-Sánchez, R., Oleaga, A. and Sánchez-Vizcaíno, J.M., 2016. African swine fever virus transmission cycles in Central Europe: evaluation of wild boar-soft tick contacts through detection of antibodies against *Ornithodoros erraticus* saliva antigen. BMC Veterinary Research 12: 1. https://doi.org/10.1186/s12917-015-0629-9

Popescu, L., Gaudreault, N.N., Whitworth, K.M., Murgia, M.V., Nietfeld, J.C., Mileham, A., Samuel, M., Wells, K.D., Prather, R.S. and Rowland, R.R.R., 2017. Genetically edited pigs lacking CD163 show no resistance following infection with the African swine fever virus isolate, Georgia 2007/1. Virology 501: 102-106. https://doi.org/10.1016/j.virol.2016.11.012

Portugal, R., Coelho, J., Höper, D., Little, N.S., Smithson, C., Upton, C., Martins, C., Leitão, A. and Keil, G.M., 2015. Related strains of African swine fever virus with different virulence: genome comparison and analysis. Journal of General Virology 96: 408-419. https://doi.org/10.1099/vir.0.070508-0

Portugal, R., Leitão, A. and Martins, C., 2018. Modulation of type I interferon signaling by African swine fever virus (ASFV) of different virulence L60 and NHV in macrophage host cells. Veterinary Microbiology 216: 132-141. https://doi.org/10.1016/j.vetmic.2018.02.008

Post, J., Weesendorp, E., Montoya, M. and Loeffen, W.L., 2017. Influence of age and dose of African swine fever virus infections on clinical outcome and blood parameters in pigs. Viral Immunology 30: 58-69. https://doi.org/10.1089/vim.2016.0121

Razzuoli, E., Franzoni, G., Carta, T., Zinellu, S., Amadori, M., Modesto, P. and Oggiano, A., 2020. Modulation of Type I interferon system by African swine fever virus. Pathogens 9: 361. https://doi.org/10.3390/pathogens9050361

Reis, A.L., Abrams, C.C., Goatley, L.C., Netherton, C., Chapman, D.G., Sánchez-Cordón, P.J. and Dixon, L.K., 2016. Deletion of African swine fever virus interferon inhibitors from the genome of a virulent isolate reduces virulence in domestic pigs and induces a protective response. Vaccine 34: 4698-4705. https://doi.org/10.1016/j.vaccine.2016.08.011

Reutner, K., Leitner, J., Müllebner, A., Ladinig, A., Essler, S.E., Duvigneau, J.C., Ritzmann, M., Steinberger, P., Saalmüller, A. and Gerner, W., 2013. CD27 expression discriminates porcine T helper cells with functionally distinct properties. Veterinary Research 44: 18. https://doi.org/10.1186/1297-9716-44-18

Revilla, Y., Pena, L. and Viñuela, E., 1992. Interferon-gamma production by African swine fever virus-specific lymphocytes. Scandinavian Journal of Immunology 35: 225-230. https://doi.org/10.1111/j.1365-3083.1992.tb02854.x

Rock, D.L., 2017. Challenges for African swine fever vaccine development – ... perhaps the end of the beginning. Veterinary Microbiology 206: 52-58. https://doi.org/10.1016/j.vetmic.2016.10.003

Ruiz-Gonzalvo, F., Rodríguez, F. and Escribano, J.M., 1996. Functional and immunological properties of the baculovirus-expressed hemagglutinin of African swine fever virus. Virology 218: 285-289. https://doi.org/10.1006/viro.1996.0193

Saalmüller, A., Reddehase, M.J., Bühring, H.J., Jonjić, S. and Koszinowski, U.H., 1987. Simultaneous expression of CD4 and CD8 antigens by a substantial proportion of resting porcine T lymphocytes. European Journal of Immunology 17: 1297-1301. https://doi.org/10.1002/eji.1830170912

Saalmüller, A., Werner, T. and Fachinger, V., 2002. T-helper cells from naive to committed. Veterinary Immunology and Immunopathology 87: 137-145. https://doi.org/10.1016/s0165-2427(02)00045-4

Sánchez-Cordón, P.J., Jabbar, T., Chapman, D., Dixon, L.K. and Montoya, M., 2020. Absence of long-term protection in domestic pigs immunized with attenuated African swine fever virus isolate OURT88/3 or BeninDeltaMFG correlates with increased levels of regulatory T cells and IL-10. Journal of Virology 94: e00350-20. https://doi.org/10.1128/JVI.00350-20

Sánchez-Cordón, P.J., Romero-Trevejo, J.L., Pedrera, M., Sanchez-Vizcaino, J.M., Bautista, M.J. and Gomez-Villamandos, J.C., 2008. Role of hepatic macrophages during the viral haemorrhagic fever induced by African swine fever virus. Histolology and Histopathology 23: 683-691. https://doi.org/10.14670/HH-23.683

Sánchez-Torres, C., Gómez-Puertas, P., Gómez-del-Moral, M., Alonso, F., Escribano, J.M., Ezquerra, A. and Domínguez, J., 2003. Expression of porcine CD163 on monocytes/macrophages correlates with permissiveness to African swine fever infection. Archives of Virology 148: 2307-2323. https://doi.org/10.1007/s00705-003-0188-4

Schäfer, A., Hühr, J., Schwaiger, T., Dorhoi, A., Mettenleiter, T.C., Blome, S., Schröder, C. and Blohm, U., 2019. Porcine invariant natural killer T cells: functional profiling and dynamics in steady state and viral infections. Frontiers in Immunology 10: 1380. https://doi.org/10.3389/fimmu.2019.01380

Schlafer, D.H., McVicar, J.W. and Mebus, C.A., 1984a. African swine fever convalescent sows: subsequent pregnancy and the effect of colostral antibody on challenge inoculation of their pigs. American Journal of Veterinary Research 45: 1361-1366.

Schlafer, D.H., Mebus, C.A. and McVicar, J.W., 1984b. African swine fever in neonatal pigs: passively acquired protection from colostrum or serum of recovered pigs. American Journal of Veterinary Research 45: 1367-1372.

Sedlak, C., Patzl, M., Saalmüller, A. and Gerner, W., 2014. CD2 and CD8alpha define porcine gammadelta T cells with distinct cytokine production profiles. Developmental and Comparative Immunology 45: 97-106. https://doi.org/10.1016/j.dci.2014.02.008

Singleton, H., Graham, S.P., Bodman-Smith, K.B., Frossard, J.P. and Steinbach, F., 2016. Establishing porcine monocyte-derived macrophage and dendritic cell systems for studying the interaction with PRRSV-1. Frontiers in Microbiology 7: 832. https://doi.org/10.3389/fmicb.2016.00832

Summerfield, A., 2012. Viewpoint: factors involved in type I interferon responses during porcine virus infections. Veterinary Immunology and Immunopathology 148: 168-171. https://doi.org/10.1016/j.vetimm.2011.03.011

Sun, L., Wu, J., Du, F., Chen, X. and Chen, Z.J., 2013. Cyclic GMP-AMP synthase is a cytosolic DNA sensor that activates the type I interferon pathway. Science 339: 786-791. https://doi.org/10.1126/science.1232458

Takamatsu, H.H., Denyer, M.S., Lacasta, A., Stirling, C.M., Argilaguet, J.M., Netherton, C.L., Oura, C.A., Martins, C. and Rodriguez, F., 2013. Cellular immunity in ASFV responses. Virus Research 173: 110-121. https://doi.org/10.1016/j.virusres.2012.11.009

Takamatsu, H.H., Denyer, M.S., Stirling, C., Cox, S., Aggarwal, N., Dash, P., Wileman, T.E. and Barnett, P.V., 2006. Porcine gammadelta T cells: possible roles on the innate and adaptive immune responses following virus infection. Veterinary Immunology and Immunopathology 112: 49-61. https://doi.org/10.1016/j.vetimm.2006.03.011

Takeuchi, O. and Akira, S., 2010. Pattern recognition receptors and inflammation. Cell 140: 805-820.

Wu, J. and Chen, Z.J., 2014. Innate immune sensing and signaling of cytosolic nucleic acids. Annual Review of Immunology 32: 461-488. https://doi.org/10.1146/annurev-immunol-032713-120156

Yang, H. and Parkhouse, R.M., 1996. Phenotypic classification of porcine lymphocyte subpopulations in blood and lymphoid tissues. Immunology 89: 76-83. https://doi.org/10.1046/j.1365-2567.1996.d01-705.x

Zhu, J.J., Ramanathan, P., Bishop, E.A., O'Donnell, V., Gladue, D.P. and Borca, M.V., 2019. Mechanisms of African swine fever virus pathogenesis and immune evasion inferred from gene expression changes in infected swine macrophages. PLoS One 14: e0223955. https://doi.org/10.1371/journal.pone.0223955

Zsak, L., Onisk, D.V., Afonso, C.L. and Rock, D.L., 1993. Virulent African swine fever virus isolates are neutralized by swine immune serum and by monoclonal antibodies recognising a 72-kDa viral protein. Virology 196: 596-602. https://doi.org/10.1006/viro.1993.1515

4. Pathology of African swine fever

P.J. Sánchez-Cordón[1,2*], B. Vidaña[1,3], A. Neimanis[4], A. Núñez[1], E. Wikström[4] and D. Gavier-Widén[4,5]

[1]Pathology Department, Animal and Plant Health Agency, APHA-Weybridge, Addlestone, KT153NB, United Kingdom; [2]European Union Reference Laboratory for African Swine Fever (EURL), Centro de Investigación en Sanidad Animal, CISA-INIA, Valdeolmos, 28130 Madrid, Spain; [3]Bristol Veterinary School, Faculty of Health Science, University of Bristol, Langford, BS40 4UD, United Kingdom; [4]Department of Pathology and Wildlife Diseases, National Veterinary Institute (SVA), 751 89, Uppsala, Sweden; [5]Department of Biomedical Sciences and Veterinary Public Health, Swedish University of Agricultural Sciences (SLU), Box 7028, 750 07 Uppsala, Sweden; pedro.sanchez-cordon@apha.gov.uk; pedrojose.sanchez@inia.es

Abstract

African swine fever (ASF) is a haemorrhagic viral disease of domestic pigs and wild boar (both *Sus scrofa* species) that seriously impacts pig production worldwide. Pathology plays a key role not only in identifying and characterising macroscopic and histopathological lesions, but also in studying pathogenesis, thus complementing other disciplines to provide a broad understanding of host-virus interactions, which contributes to the development of vaccines and therapies. The aim of the present chapter is to provide a compilation of the most characteristic macroscopic and histopathological lesions associated with the different clinical forms of ASF (peracute, acute, subacute and chronic). Another aim is to present the current state of knowledge regarding ASF pathogenesis, with special attention to target cells, virus distribution and virus load in multiple organ locations and its correlation with the appearance of clinical signs and lesions from early stages to terminal disease. For that purpose, the chapter includes a complete collection of original macroscopic and microscopic photographs, the latter including illustrations of haematoxylin and eosin-stained sections and immunohistochemistry to identify the ASF-virus (ASFV). The photographs were obtained from experimental infections with ASFV isolates of different virulence carried out over the last two decades in different research centres. The purpose is to provide a reference that may be useful for the early recognition of ASF by veterinary practitioners working in the field and researchers focused on characterising new ASFV isolates or testing new vaccines and treatments.

Keywords: ASF clinical forms, macroscopic lesions, histopathological lesions, ASFV target cells, pathogenic mechanisms

This publication is based upon work from COST Action CA15116, ASF-STOP, supported by COST (European Cooperation in Science and Technology).
www.cost.eu

EUROPEAN COOPERATION IN SCIENCE & TECHNOLOGY

Funded by the Horizon 2020 Framework Programme of the European Union

DOI 10.3920/978-90-8686-910-7_4,

4.1 Introduction

African swine fever (ASF) is a haemorrhagic viral disease of domestic pigs and wild boar (all *Sus scrofa* species) that is seriously impacting pig production worldwide. Understanding disease pathogenesis (host-virus interactions) is needed to guide vaccine development and help inform disease control. Pathology plays a key role in studying pathogenesis and complements other disciplines to provide a broad understanding of host-virus interactions. Studies on vaccine development and biological characterisation of African swine fever virus (ASFV) isolates should therefore employ a transdisciplinary approach that includes research in immunology, virology and also pathology. Pathology contributes to such a broad approach by providing information on: (1) routes of virus uptake; (2) co-localisation of virus with lesions at the cellular, tissue and organism level; (3) distribution of virus and viral load in various organs during different stages of the disease; (4) routes of virus shedding at various stages of infection; (5) correlation of lesions with clinical signs, virus presence and immunological response; and (6) development of lesions over time, from early stage to terminal or late disease.

Animal experiments on ASF require clinico-pathological assessments to determine and follow the health status of the animals. To fully understand host-virus interactions, pathology data collected should include a complete macroscopic and histopathological evaluation, complemented by other pathological techniques focused on determining target organs, virus distribution within tissue compartments, target cells and quantification of infected cells if needed. For that purpose, immunohistochemistry is an optimal technique to determine the presence of ASFV, detect viral antigen on tissue sections, and assess the correlation between virus presence and severity of histopathological lesions. Other complementary techniques include *in situ* hybridisation, RNAscope, special stains, confocal microscopy, electron microscopy or laser-capture microdissection. Most of these pathological techniques are expensive and time-consuming and are therefore applied mostly to specialised ASF studies in which routine screening tools such as macroscopic lesions assessment and clinico-pathological evaluations are not sufficient.

The purpose of this chapter is to provide a synthesis of clinical and gross pathological findings associated with different forms of ASF and to present the current state of knowledge regarding ASF pathogenesis, including microscopic lesions and virus tissue distribution.

4.2 Clinical signs and forms of African swine fever

ASF displays different forms that range from peracute to subclinical or inapparent depending on different factors among which the virulence of the isolates is particularly important. The range of clinical forms is matched by an equally wide range of immune responses and suggests a complex array of virus-host interactions. Based on the molecular phylogeny, 24 distinct genotypes of ASFV have been described to date (Achenbach *et al.*, 2017). Depending on their virulence, the phenotypes of ASFV isolates have been traditionally classified within three main groups: highly, moderately and low virulent isolates (Pan and Hess, 1984). Highly virulent isolates are usually involved in peracute or acute forms of disease, moderately virulent isolates can induce acute, subacute and even chronic forms of disease, while low virulent isolates produce mild and nonspecific signs that are easily confused with other diseases, giving rise to chronic forms of ASF.

As determined by experimental studies, in addition to the virulence of the isolate and health status of the animals other factors such as age (Post *et al.*, 2017), dose and route of infection (Howey *et al.*, 2013; Sánchez-Cordón *et al.*, 2017, 2020a) have been identified as critical for the evolution of clinical forms. There are different experimental protocols that have been used to carry out the biological characterisation of ASFV isolates. Experimental infection of immunologically mature outbred pigs (older than 8 weeks) with low doses of ASFV isolates by the oronasal route (the natural route of infection) or intramuscular route, along with non-infected in-contact pigs housed in the same pen, has been demonstrated as one of the most reliable approaches to experimentally assess the real virulence of ASFV isolates (Galindo-Cardiel *et al.*, 2013; Gallardo *et al.*, 2018). Host characteristics and breed may also influence the outcome of the experimental infections. Different experimental studies have suggested that less selected domestic pig breeds or wild suids, such as wild boar or feral pigs (all *Sus scrofa*), are more susceptible to virus infection and disease development (Sanchez-Cordon *et al.*, 2019). On the other hand, in comparative experimental infections between domestic pigs with an European genetic lineage and indigenous domestic African pigs of the same age inoculated with the same dose of virus, indigenous African pigs showed a delay in the incubation period, onset of viraemia, appearance of clinical signs and death of individuals (Gallardo *et al.*, 2012).

Transcontinental spread of ASFV outside the endemic areas in Africa as in the outbreaks in Spain and Portugal in the late 1950s, and during the current epidemic (see Chapter 9), resulted in mostly acute and subacute forms of disease. However, subacute, subclinical and chronic forms have also been described after the initial outbreaks, resulting from the innate or acquired immunity of the animals or the appearance of related ASFV isolates of reduced virulence (Gallardo *et al.*, 2018; Zani *et al.*, 2018).

4.2.1 Peracute/hyperacute forms of African swine fever

Peracute disease means that the disease is very severe, of very short duration and in general quickly fatal. Peracute forms of ASF are induced by high doses of highly virulent isolates. Natural cases of peracute forms have been described as a result of the first introduction of ASFV into a free or non-endemic area. These forms are characterised by short incubation periods, sudden death of animals with either no clinical signs before death or marked pyrexia and moribund status. The affected pigs die without presenting macroscopic external or internal lesions or showing only nonspecific changes (Kleiboeker, 2002; Sánchez-Vizcaíno *et al.*, 2015).

However, the existence of peracute forms should be questioned. In experimental infections with high doses of highly virulent isolates administered by intramuscular route, peracute forms with the clinical pathological pictures mentioned above (no clinical signs or lesions) have not actually been described. Instead, after inoculation with high doses of highly virulent isolates animals show high viraemia levels, short incubation periods (1-3 days) and short clinical courses (1-3 days) with obvious clinical signs (pyrexia up to 42 °C, skin erythema, inappetence, increased respiratory rate, death, or animals reaching the pre-determined point for euthanasia, at day 3-4 post inoculation) and macroscopic lesions characteristic of ASF. These results suggest that the so called peracute forms in natural disease might actually correspond to early acute forms of ASF (P. Sánchez-Cordón, unpublished data).

4.2.2 Acute forms of African swine fever

Acute forms of ASF are induced by highly or moderately virulent isolates. It is the most common clinical presentation in natural outbreaks. Based on existing information, the incubation period for ASF under natural conditions is between 4 to 19 days. In experimental infections, incubation periods may range between 2 and 7 days and clinical courses may last around 1 week (3-8 days). Animals usually die between 3 and 9 days post inoculation (dpi) with highly virulent isolates or later after inoculation with moderately virulent isolates (usually 11-15 dpi, although some animals may die up to 18-20 dpi). This form results in mortality of up to 90-100% in immunologically naïve herds if the infection is left unnoticed for a rather long period (Kleiboeker, 2002; Sánchez-Cordón *et al.*, 2018; Sánchez-Vizcaíno *et al.*, 2015).

Animals with acute forms of ASF usually develop an increase in body temperature (40.5-42 °C). Animals within the same pen also tend to huddle together or pile on top of one another (Figure 4.1A), and express loss of appetite, apathy, weakness, recumbency and inability or unwillingness to stand up. Abortions (caused by febrile syndrome) are frequent at all stages of pregnancy, being one of the first signs observed at the early stages of ASF outbreaks. Although not very well documented, inconsistent lesions have been described in aborted foetuses such as petechial haemorrhages in skin and myocardium, mottled lung and liver or anasarca. Petechiae may also be found in the placenta. However, some authors have indicated that there is no evidence of intrauterine transmission of ASFV (Penrith and Vosloo, 2009; Schulz *et al.*, 2017). During acute forms of ASF, animals may also display breathing difficulty (dyspnoea), cough, increased pulse and respiratory rate, mucopurulent nasal discharges, nasal haemorrhages (epistaxis) (Figure 4.1B) and congested ocular mucosa with mucopurulent discharges (Figure 4.1C). Other clinical signs include vomiting, abdominal pain, constipation or diarrhoea which may be initially mucoid but that, as disease progresses, may show blood-stained mucus in the faeces; other animals may develop watery diarrhoea with fresh or blackened blood adhering to the tail and perineum (melaena). Neurological signs like head tilt and occasional tremors as well as incoordination or stiff gait have also been reported in late stages of the disease, just before death. Among the most characteristic changes in blood parameters are the presence of early leukopenia, with a decrease in the number of monocytes and lymphocytes, and moderate thrombocytopenia (48-72 hours). Animals with light-coloured skin show skin erythema and cyanotic areas mainly in the tips of ears (Figure 4.1B, 4.1D-E), snout, distal limbs, tails, perianal area, scrotal sac, ventral chest and abdomen (Figure 4.1D, 4.1F-H). Cyanotic areas usually appear in these locations 24-48 hours before death. Skin petechial haemorrhages and ecchymoses may also be observed (Figure 4.1E, 4.1G-H). Towards the end of the acute disease, rectal temperature falls rapidly, often to subnormal levels, followed by coma and death. Other animals die in circulatory shock. The presence of foam in mouth and nose due to severe pulmonary oedema is a common finding in animals found dead (Kleiboeker, 2002; Sánchez-Cordón *et al.*, 2018; Sánchez-Vizcaíno *et al.*, 2015).

4.2.3 Subacute forms of African swine fever

In subacute forms of ASF, induced by moderately virulent isolates, some pigs may not develop any clinical signs while others display only mild signs. Clinical signs are like those of acute forms, but generally milder and persisting longer (up to 4 weeks). Incubation periods may range between 2 and 7 days and clinical courses usually last between 7 and 20 days. Affected pigs usually die

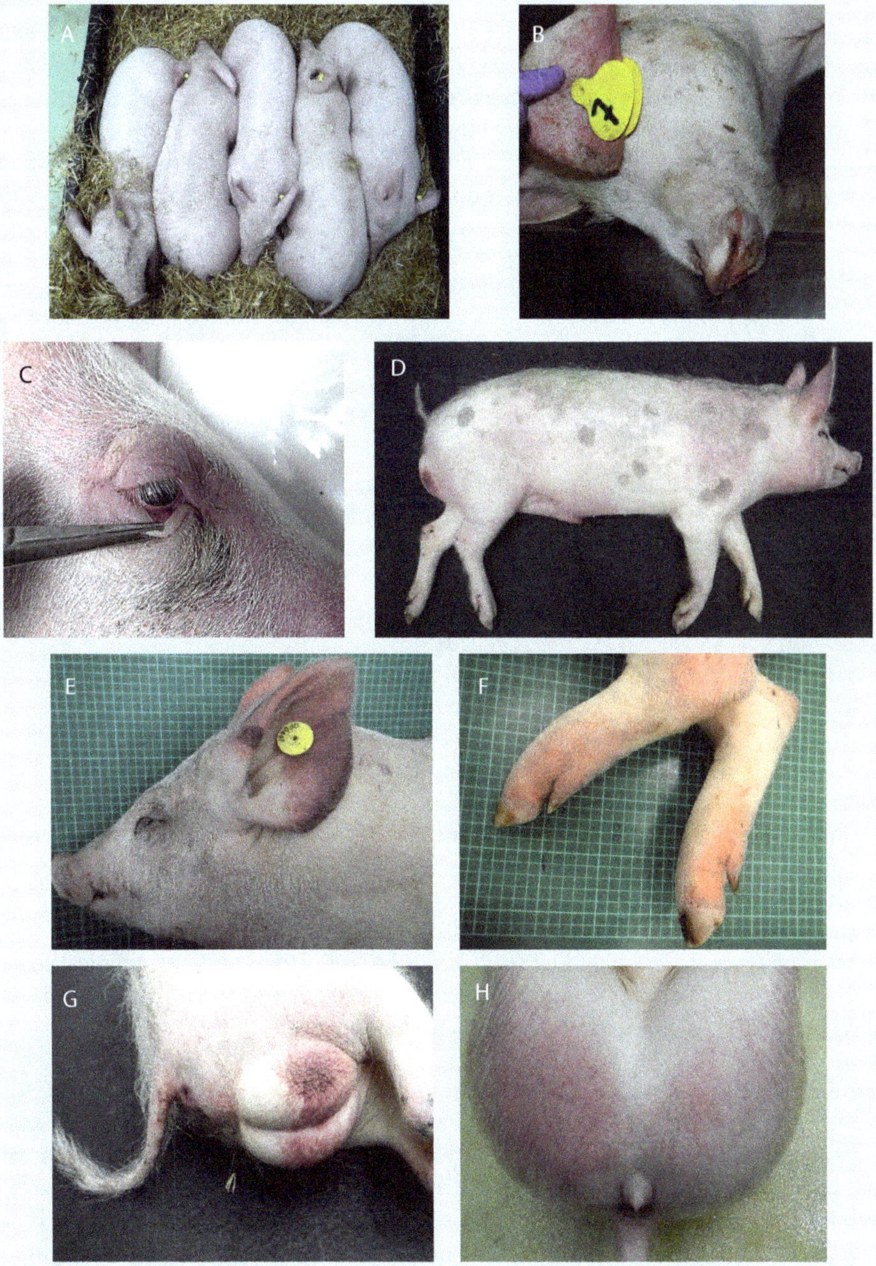

Figure 4.1. Clinical signs and external lesions in animals with acute African swine fever. (A) Lethargic pigs huddling together; (B) Nasal haemorrhages (epistaxis) and skin erythema in the ear; (C) Congested ocular mucosa and ocular discharge; (D) Skin erythema and cyanosis in the ears, neck, dorsal and lateral skin areas, abdomen and scrotal sac; (E) Erythema, severe cyanosis and haemorrhages in the ears; (F) Erythema and cyanosis in distal limbs; (G) Erythema, cyanosis and haemorrhages in the tail, perianal area and scrotal sac; (H) Erythema and petechial haemorrhages in caudal skin areas.

within 14 to 30 days after infection (although some may die up to 45 dpi) or recover within 3 to 4 weeks. In recovered pigs, viral DNA can be detected in secretions up to several weeks after infection. The mortality rate is generally lower in adult pigs (20-70%) than in young animals (70-100%), which besides the higher mortality rates also exhibit more severe clinical signs. Animals may die at two different stages: during the phase of intense thrombocytopenia/leukopenia or in the recovery phase when haemorrhages due to erythrodiapedesis from vasodilation appear, especially in young animals.

Pyrexia may fluctuate, occurring at irregular intervals and may exceed 40.5 °C (up to 42 °C). Affected pigs may huddle together as in acute forms. The body temperature may decline and pigs become more active by 3-5 dpi. Abortion may be one of the first clinical signs observed also in subacute forms. Pregnant sows may die acutely. Thrombocytopenia and leukopenia usually appear later than in acute forms and may be transient. However, thrombocytopenia is generally more intense than in acute forms and may cause bleeding from injection sites. Animals may also display laboured breathing, diarrhoea with blood-stained mucus in the faeces or even melaena. Neurological signs may be more frequent and severe than those described in acute forms. A purple colour over the pig's body near death along with petechial haemorrhages, ecchymoses (Figure 4.2A-B) and occasional small foci of necrosis in skin (Figure 4.2C-D) can be observed. Haemorrhagic skin lesions have been described in recovered animals (Kleiboeker, 2002; Sánchez-Cordón *et al.*, 2018; Sánchez-Vizcaíno *et al.*, 2015).

4.2.4 Chronic forms of African swine fever

Due to the wide range of non-specific signs and lesions, the identification of chronic forms of ASF (also known as atypical ASF) is difficult and suspicion will depend largely on clinical history and exclusion of other diseases. Chronic forms have usually been detected during serological screenings in late stages of disease eradication, and have been associated with moderately and low virulent isolates. To date, experimental infections focused on studying or inducing chronic ASF forms are scarce (Boinas *et al.*, 2004; Gallardo *et al.*, 2019; Sánchez-Cordón *et al.*, 2017, 2020a). Natural cases attributed to low virulence isolates are characterised by clinical courses of more than 1 month. Mortality rates are low (less than 30%) and affected animals may die over an extended time period lasting between 2 and 15 months after infection.

The most characteristic clinical signs include recurrent transient pyrexia without reaching high temperature, poor body condition, loss of appetite, depression, stunting, emaciation and joint swelling (Figure 4.3A-B) attributed to arthritis and periarthritis (Figure 4.3C-E) with or without lameness. Cutaneous ulcers and reddened or raised necrotic skin foci, particularly in areas that overlie bony protuberances, but also in ears, nose, neck, flanks, abdomen and limbs are generally more frequent and severe than those described occasionally in subacute forms (Figure 4.3F-I). Difficult breathing, coughing, diarrhoea, occasional vomiting and abortions may be present (Pérez *et al.*, 1994).

The existence of subclinical or unapparent infections has also been suggested in pigs surviving the subacute forms. These animals are infected but do not display clinical signs or lesions described traditionally in pigs with chronic forms. ASFV could persist for prolonged periods in tissues or blood from pigs recovered from subacute forms or following infection with low virulence

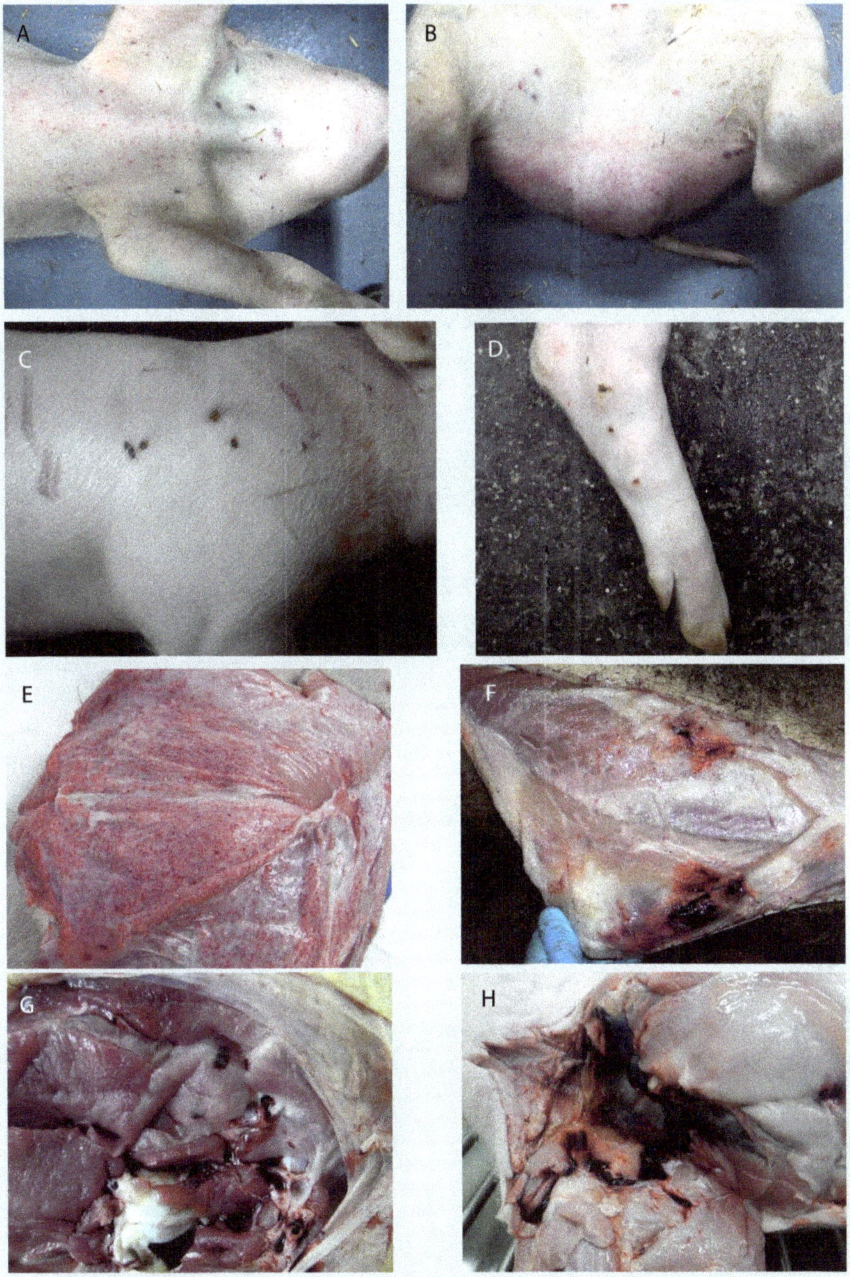

Figure 4.2. Lesions in skin and skeletal muscle in animals with subacute African swine fever. (A) Petechial haemorrhages and ecchymoses in the neck and chest; (B) Petechial haemorrhages and ecchymoses in abdomen, scrotal sac and caudal skin areas; (C-D) Small foci of necrosis in the skin of flank and limb; (E) Petechial haemorrhages in the subcutis and skeletal muscles of ventral areas; (F) Subcutaneous haematomas and haemorrhages; (G-H) Haemorrhages, haematomas and oedema in skeletal muscle.

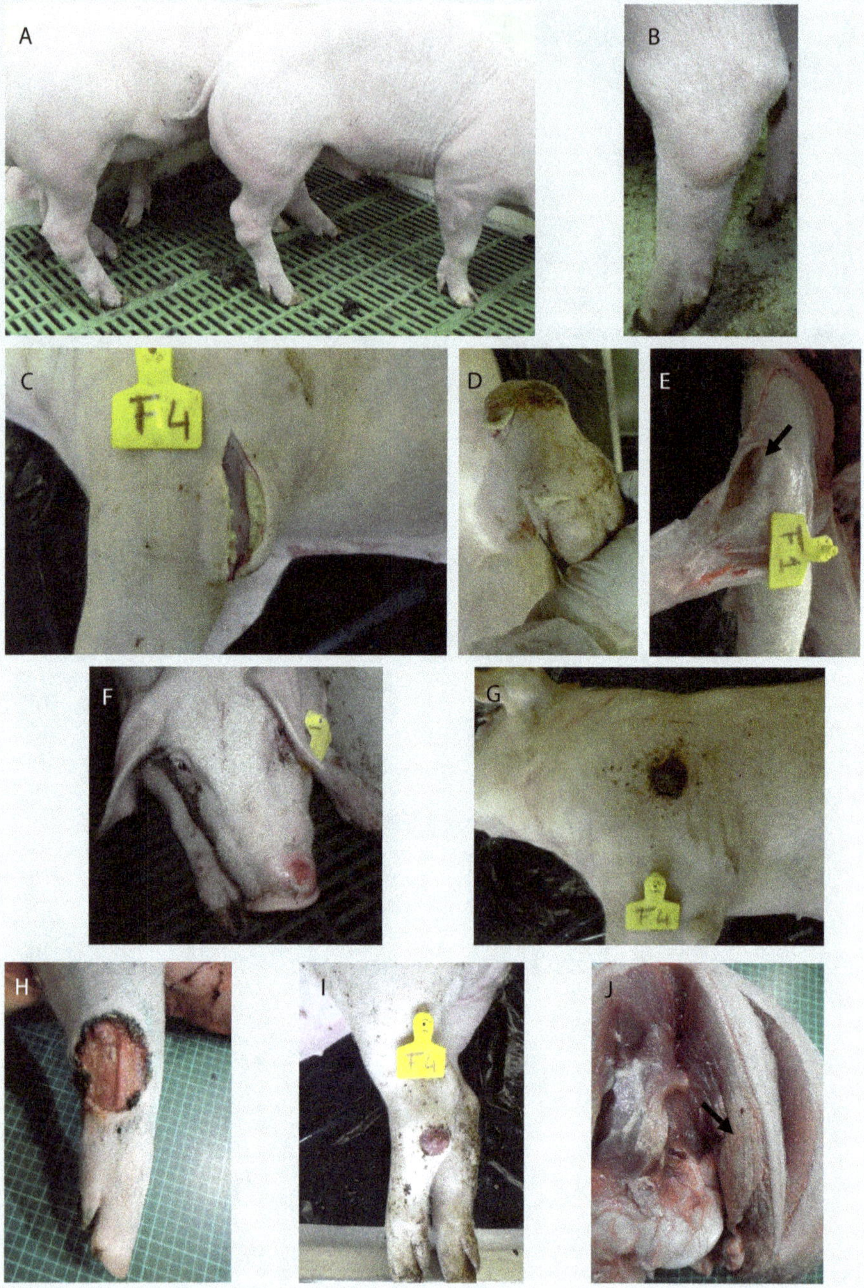

Figure 4.3. Lesions in joints skin and skeletal muscle in animals with chronic African swine fever. (A-B) Joint swelling; (C) Purulent periarthritis; (D) Purulent arthritis; (E) Serofibrinous periarthritis; (F-I) Skin ulcers in the nose (F), flank (G) and limbs (H-I); (J) Necrosis of skeletal muscle.

isolates (up to 3 months), which in turn may contribute to virus transmission, disease persistence, sporadic outbreaks and ASFV introduction into disease-free zones (Gallardo *et al.*, 2015a; Penrith and Vosloo, 2009).

In experimental infections with the low virulent ASFV isolate NH/P68 (genotype I), two clinical pictures were described. In the first group, pigs remained asymptomatic after infection and displayed low or no fever and low viraemia levels before 14 dpi. These animals were neither viraemic nor febrile after 14 dpi. The second group represented pigs that showed viraemia and fever in a late phase of infection after 14 dpi and went on to develop chronic forms of ASF. These animals displayed gradual weight loss, respiratory signs with coughing, joint swelling and raised necrotic skin areas, changes that were also correlated with late bouts of viraemia and pyrexia (Gallardo *et al.*, 2015a; Leitão *et al.*, 2001). From the second week after infection, domestic pigs infected with Lv17/WB/Rie1ASFV (genotype II) displayed low viraemia levels and similar clinical signs including a weak increase in temperature and joint swelling (Gallardo *et al.*, 2019). In other studies in which different doses of the naturally attenuated ASFV isolate OURT88/3 (genotype I) were administered by different routes to domestic pigs, some animals developed subclinical infection with only transient mild clinical signs and recurrent low viraemia levels from 7 dpi. Other animals, mainly those infected with the highest doses (10^5 median tissue culture infectious dose ($TCID_{50}$)/ml), developed clinical signs and lesions characteristic of chronic forms of ASF from 12-14 dpi. These included joint swelling, laboured breathing, skin erosion and ulcers. The changes correlated with a late and mild increase of viraemia levels and body temperature (Sánchez-Cordón *et al.*, 2017). The appearance of lesions in joints and skin from the second week after infection that are not described in experimental acute and subacute forms of ASF induced by highly/moderately virulent isolates suggest the existence of specific pathogenic mechanisms induced by low virulent ASFV isolates. The exact nature of these mechanisms remains unclear.

In subsequent experimental infections of domestic pigs with intramuscular inoculation of lower doses (10^4 $TCID_{50}$/ml) of the naturally attenuated ASFV isolate OURT88/3 and the attenuated deletion mutant BeninΔMFG (Sánchez-Cordón *et al.*, 2020b), mild joint swelling of forelimbs (knee and metacarpophalangeal joints) and hindlimbs (hock and metatarsophalangeal joints) were transiently (~2-3 weeks) observed in most animals between 15-18 weeks post infection. Joint swelling was not associated with pain, heat or lameness in any of the affected pigs.

4.3 Macroscopic lesions

As a consequence of the antigenic differences and virulence range of ASFV isolates, the presence and severity of internal macroscopic lesions described at post-mortem examinations are highly variable and may affect numerous organs. There may be a degree of skin flushing of extremities and the ventral abdomen in light-skinned pigs, general congestion of organs with fluid exudation into body cavities and possibly fibrin strands on organ surfaces along with petechial haemorrhages and ecchymoses on serosal and mucosal surfaces.

4.3.1 Acute forms of African swine fever

In acute ASF forms, animals are often in good body condition. Affected animals usually show fluid in body cavities, in particular mild hydrothorax and hydropericardium with transparent, yellowish or red-tinged fluid (Figure 4.4A, 4.4C) and mild ascites with yellow-tinged fluid (Figure 4.4E). The most significant macroscopic lesions affect the lymphoid system. The spleen displays an increase in size (up to 2 to 6 times its normal size) with rounded edges, friable consistency and dark red to black colour (Figure 4.5A-B), showing moderate to marked bleeding after sectioning and occasional presence of petechiae in the capsule.

The presence of small multifocal splenic infarcts in the spleen margins has been occasionally described. The spleen may occupy the abdominal cavity from side to side, characteristic of the so-called hyperaemic splenomegaly. In other cases, the spleen may be enlarged but not friable, and the colour closer to normal. The changes described are not always present in affected animals. The lymph nodes are often swollen, oedematous and haemorrhagic, sometimes with the appearance of blood clots. Haemorrhages can be present in both cortex and medulla. The gastrohepatic and renal lymph nodes (Figure 4.5D-E) and, to a lesser extent, the mesenteric lymph nodes are often the most severely affected. These lymph nodes are also, usually, the only ones with visible haemorrhagic lesions at early acute forms of ASF which may range from petechiae to severe haemorrhages. Superficial inguinal, parotid, pre-scapular, submandibular (Figure 4.5C), retropharyngeal, cervical, tracheobronchial and mediastinal lymph nodes, among others, may also display lymphadenopathy along with less severe haemorrhagic lesions (petechiae and ecchymoses) in medulla and cortex. Petechial haemorrhages have been occasionally described in the thymus. Some animals may also display erythema or petechiae on the surface of the palatine tonsils.

In the cardiorespiratory system, the lungs usually display lack of collapse with scarce or mild rib impressions, diffuse congestion, minimal distension of interlobular septa (interstitial oedema) with presence of yellow gelatinous infiltrate, mild to moderate alveolar oedema with the presence of exudation of fluid and froth on cut surface along with foamy material (which may be blood-stained) in the trachea (Figure 4.6A-C). Occasional subpleural petechial haemorrhages are also present (Figure 4.6F). All these lung lesions are commonly induced by highly virulent isolates of ASFV, and observed histologically. In the heart, the presence of petechial haemorrhages and ecchymoses in the epicardium and endocardium is frequent (Figure 4.6L).

In the urinary system, the kidneys characteristically show subcapsular petechiae in the renal cortex as well as petechiae in the medulla and renal pelvis (Figure 4.7A-B). Minimal perirenal oedema may be present in some cases. Petechial haemorrhages may appear in the submucosa of the urinary bladder (Figure 4.7K-L).

Regarding the gastrointestinal tract, different segments of small and large intestine may appear normal to mildly congested with petechial haemorrhages in the serosa. In addition, mucosal surfaces may be congested and haemorrhagic containing bloody faecal fluid or hard, blood-stained formed faeces. Hyperaemia and inflamed mucosal surfaces in stomachs containing food have been described. In some animals the liver may display mild hepatomegaly and congestion (Figure 4.8A). Minimal oedema may affect the bile duct and gall bladder wall in some cases, occasionally with petechiae appearing on the serosal and mucosal surfaces.

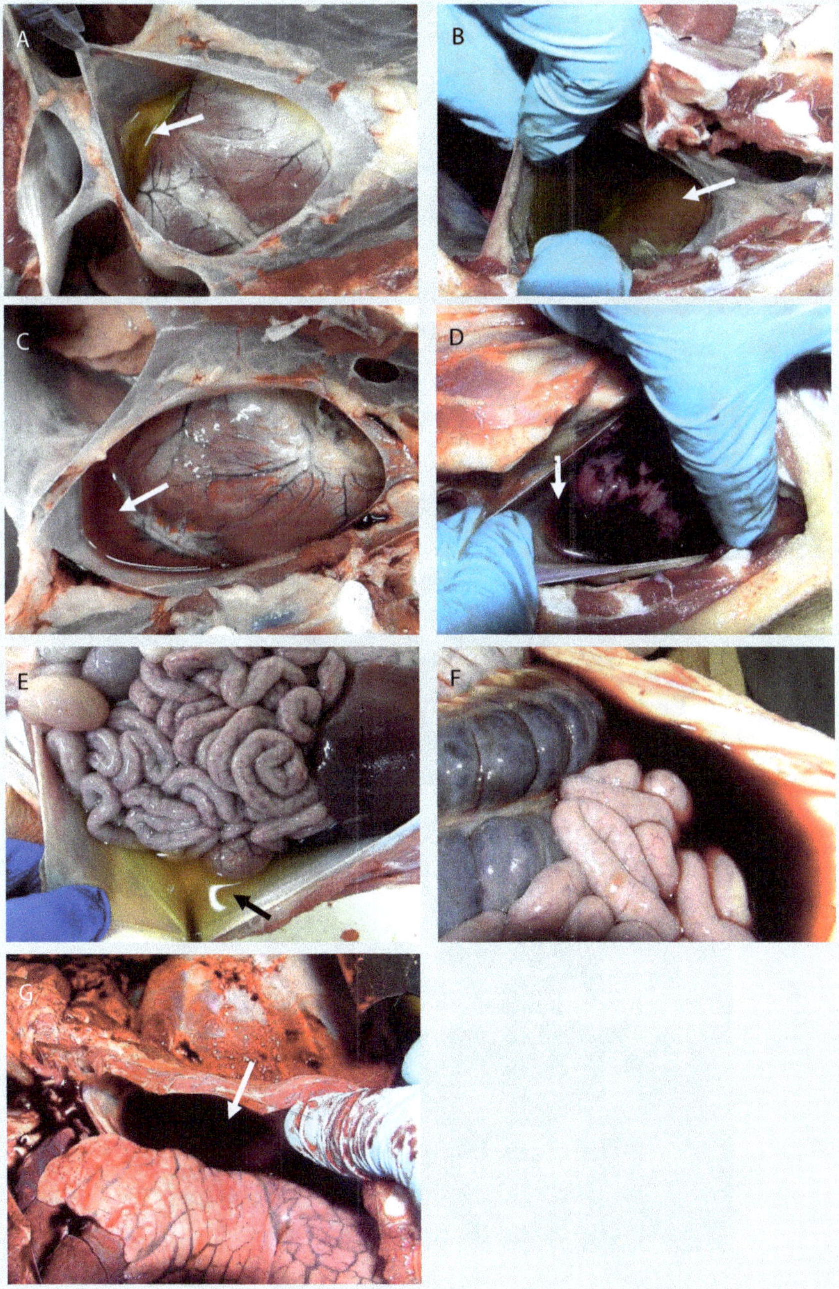

Figure 4.4. Fluids in body cavities in pigs with acute and subacute forms of African swine fever (ASF). (A) Moderate hydropericardium with yellowish fluid (Acute ASF); (B) Severe hydropericardium with yellowish fluid (Subacute ASF); (C) Moderate hydropericardium with red-tinged fluid (Acute ASF); (D) Severe hydropericardium with red-tinged fluid (Subacute ASF); (E) Moderate ascites (Acute ASF); (F) Severe ascites with red-tinged fluid (Subacute ASF); (G) Severe hydrothorax with reddish fluid (Subacute ASF).

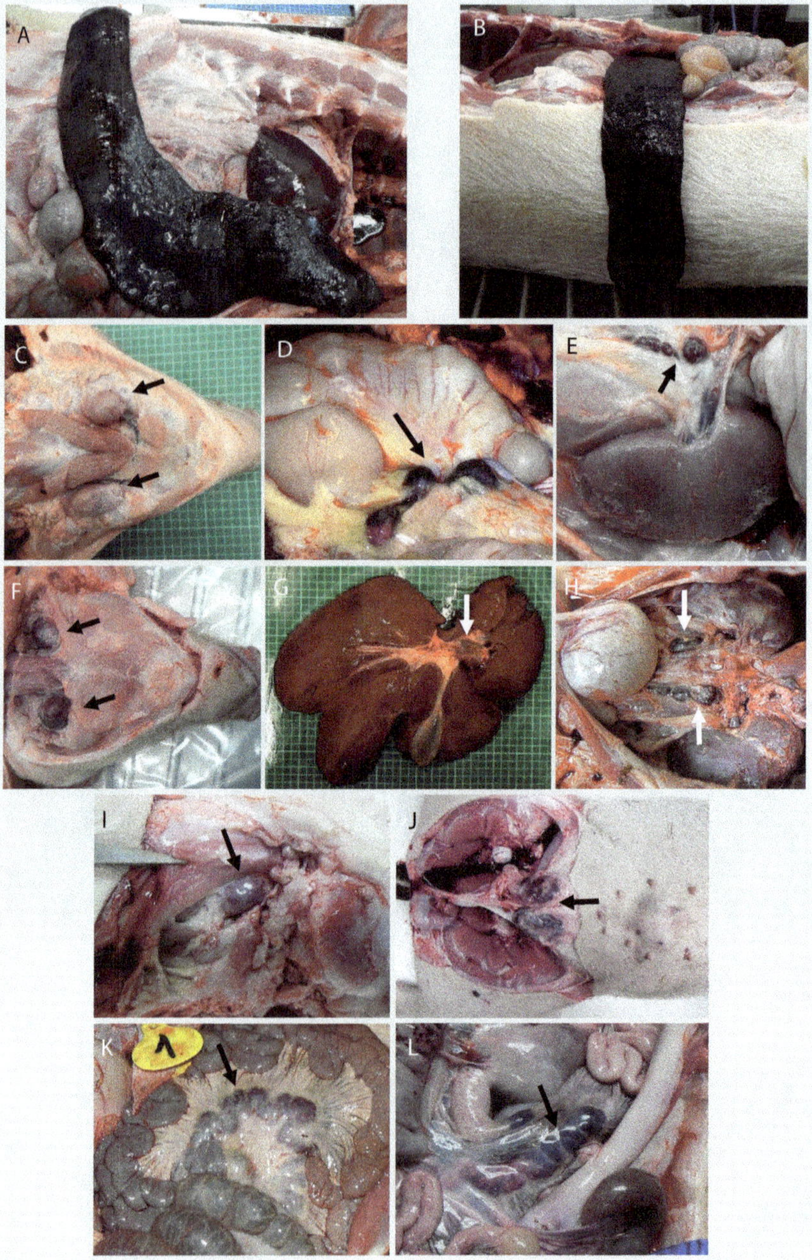

Figure 4.5. Lesions in the spleen and lymph nodes in acute and subacute forms of African swine fever (ASF). (A-B) Spleen with an increase in size, rounded edges and dark red to black colour (hyperaemic splenomegaly) from animals with acute ASF; (C) Submandibular lymph node showing severe lymphadenopathy in acute ASF; gastrohepatic (D) and renal (E) lymph nodes showing severe haemorrhagic lymphadenitis in acute ASF; Submandibular (F), gastrohepatic (G), renal (H), retropharyngeal (I), superficial inguinal (J), mesenteric (K) and ilecaecal (L) lymph nodes showing haemorrhagic lymphadenitis in subacute ASF.

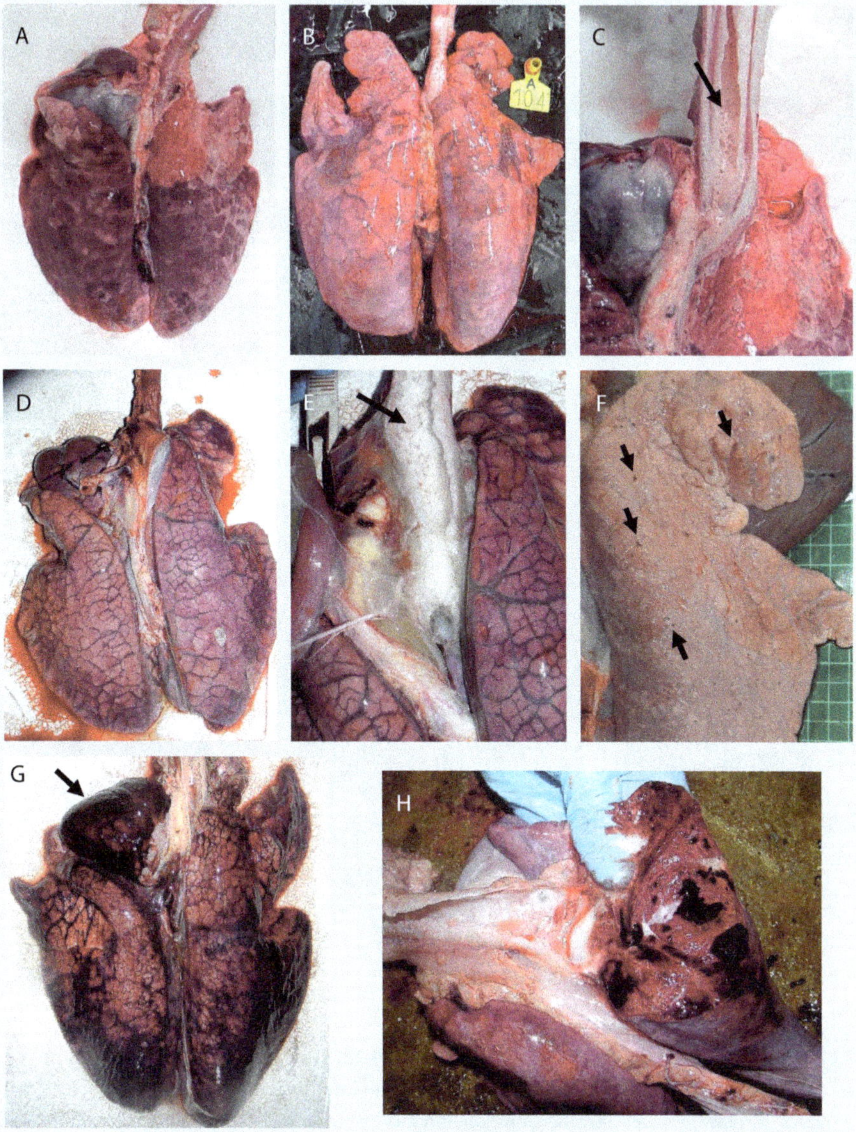

Figure 4.6. Lesions in the lungs and heart of pigs with acute, subacute and chronic forms of African swine fever (ASF). (A-C) Lungs from pigs with acute ASF displaying lack of collapse, diffuse congestion, mild distension of interlobular septa (interstitial oedema) and mild to moderate alveolar oedema with the presence of foamy material in trachea (arrow); (D-E) Lung from a pig with subacute ASF showing severe lack of collapse, congestion, severe distension of the interlobular septa with accumulation of fluid (interstitial oedema) and severe alveolar oedema with presence of abundant foamy material in the trachea lumen (arrow); (F) Lung from a pig with acute ASF showing subpleural petechial haemorrhages (arrows); (G-H) Lung from a pig with subacute ASF showing severe congestion, alveolar and interstitial oedema along with severe interstitial and alveolar haemorrhages. Observe also the severe haemorrhages almost completely covering the epicardium (arrow) in G. Cut surface in lung shows severe alveolar oedema with the presence of blood and abundant blood-stained foamy material. ***Continued on next page***.

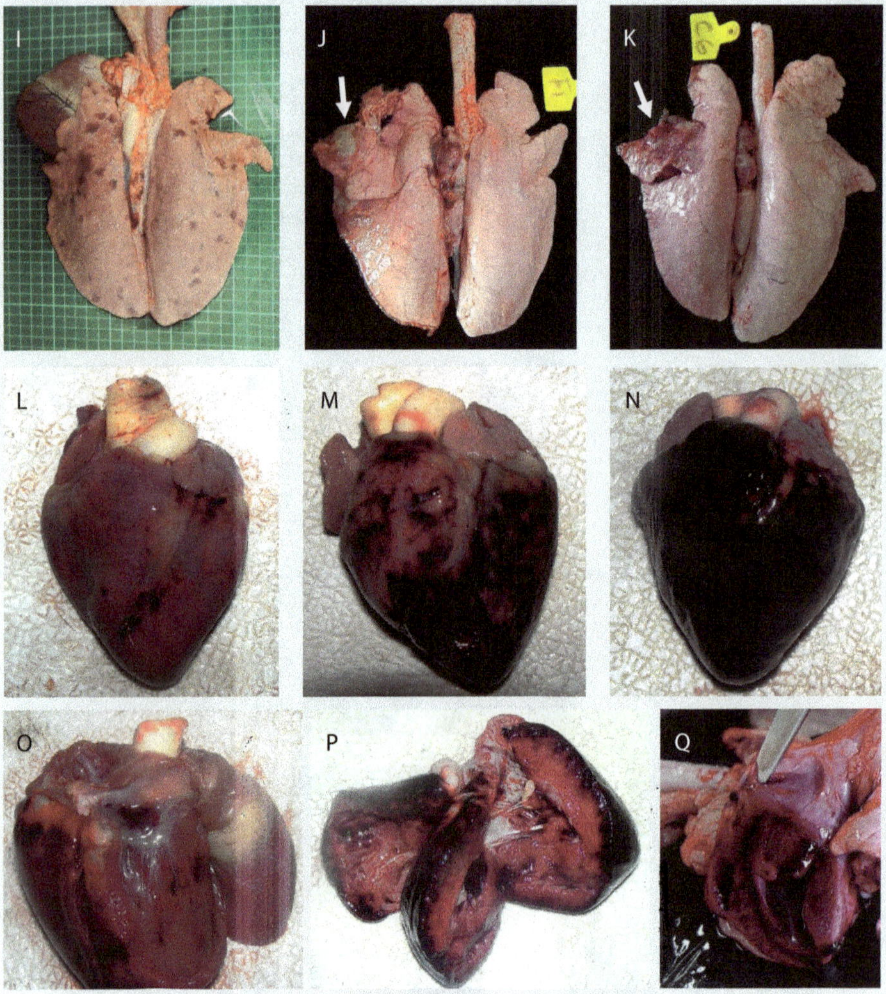

Figure 4.6. *Continued*. Lesions in the lungs and heart of pigs with acute, subacute and chronic forms of African swine fever (ASF). (I) Lung from a pig with subacute ASF with multifocal areas of consolidation; (J-K) Lungs from pigs with chronic ASF displaying fibrinonecrotic pleuropneumonia in cranial and middle pulmonary lobes (arrows); (L) Presence of petechial haemorrhages and ecchymoses in the epicardium of a pig with acute ASF; (M-N) Severe and coalescent interstitial haemorrhages (ecchymoses and suggillations) almost completely affecting the epicardium in pigs with subacute ASF; (O-P) Severe interstitial haemorrhages affecting the myocardium and endocardium in pigs with subacute ASF; (Q) Fibrous pericarditis with adherences to the epicardium in a pig with chronic ASF.

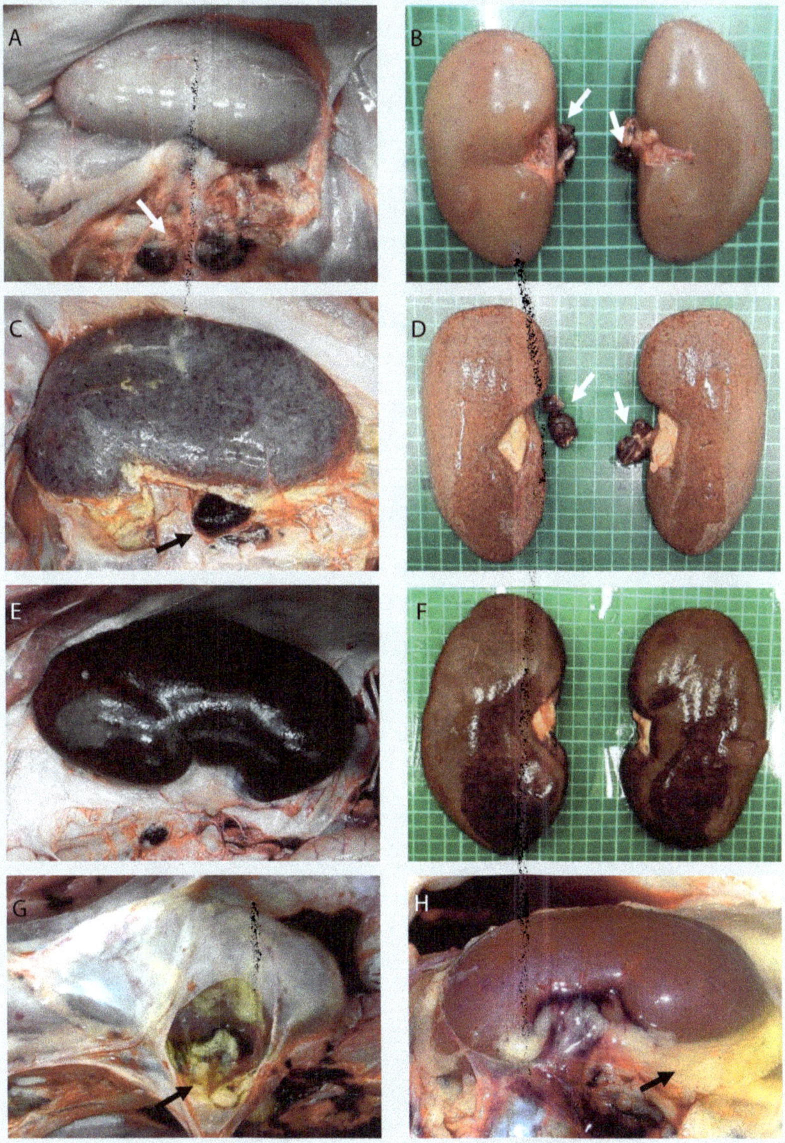

Figure 4.7. Lesions in the urinary tract of pigs with acute and subacute forms of ASF. (A-B) Mild presence of subcapsular petechiae in the renal cortex in a pig with acute ASF. Observe the severe haemorrhagic lymphadenitis affecting the renal lymph nodes (arrows); (C-D) Moderate interstitial haemorrhagic lesions (petechiae and ecchymoses) in the renal cortex in pigs with subacute ASF, renal lymph nodes showing haemorrhagic lymphadenitis (arrows); (E-F) Diffuse severe congestion and severe interstitial haemorrhages (ecchymoses and purpura) in pigs with subacute ASF; (G-H) Moderate to severe perirenal oedema in pigs with subacute ASF (arrows). ***Continued on next page***.

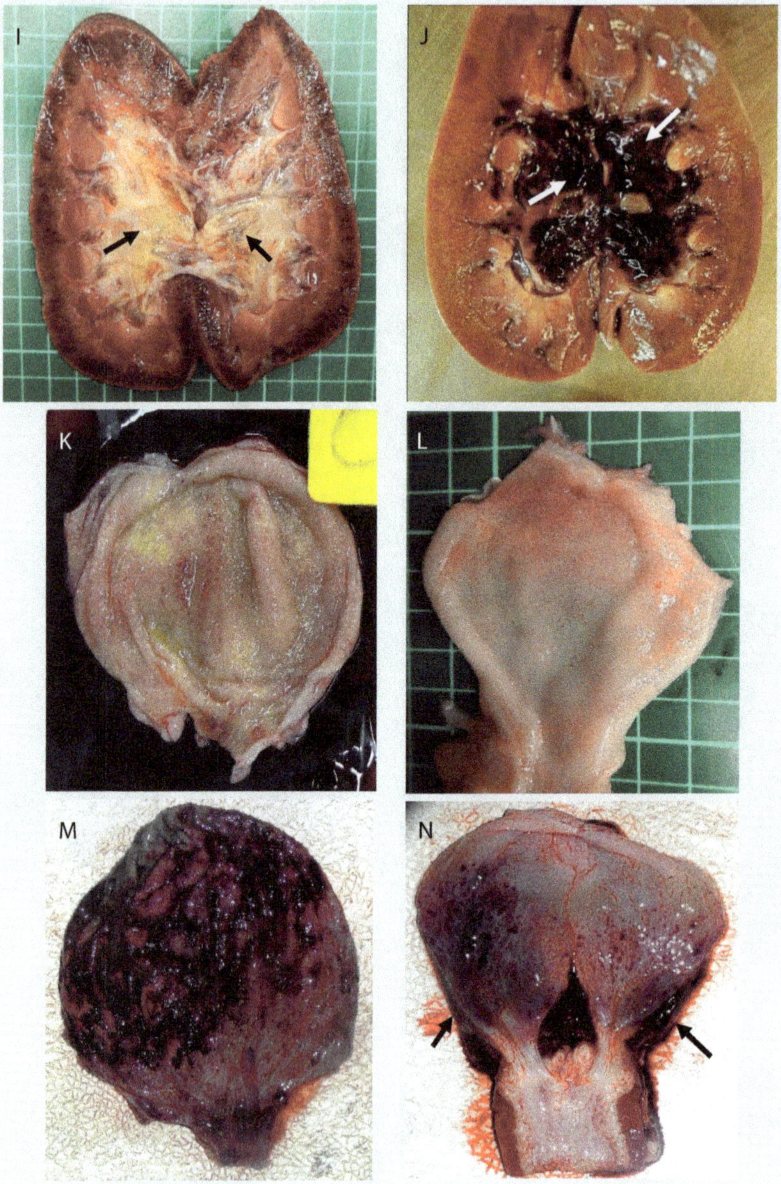

Figure 4.7. *Continued.* Lesions in the urinary tract of pigs with acute and subacute forms of ASF. (I) Multiple petechial haemorrhages in the renal cortex, medulla and renal pelvis in a pig with subacute ASF. Observe also the pelvic dilation and the oedema affecting the renal pelvis (arrows); (J) Petechiae in the renal cortex and pelvis along with pelvic dilation and severe haemorrhages in the pelvis of a pig with subacute ASF (arrow); (K-L) Occasional presence of multifocal petechial haemorrhages in the submucosa of the urinary bladder from animals with acute ASF; (M-N) Blood clots covering large areas of the mucosal surface along with multifocal severe haemorrhages (petechiae, ecchymoses and suggillations) in the submucosa of the urinary bladder from animals with subacute ASF. Observe also the urinary bladder wall thickening due to oedema (arrows).

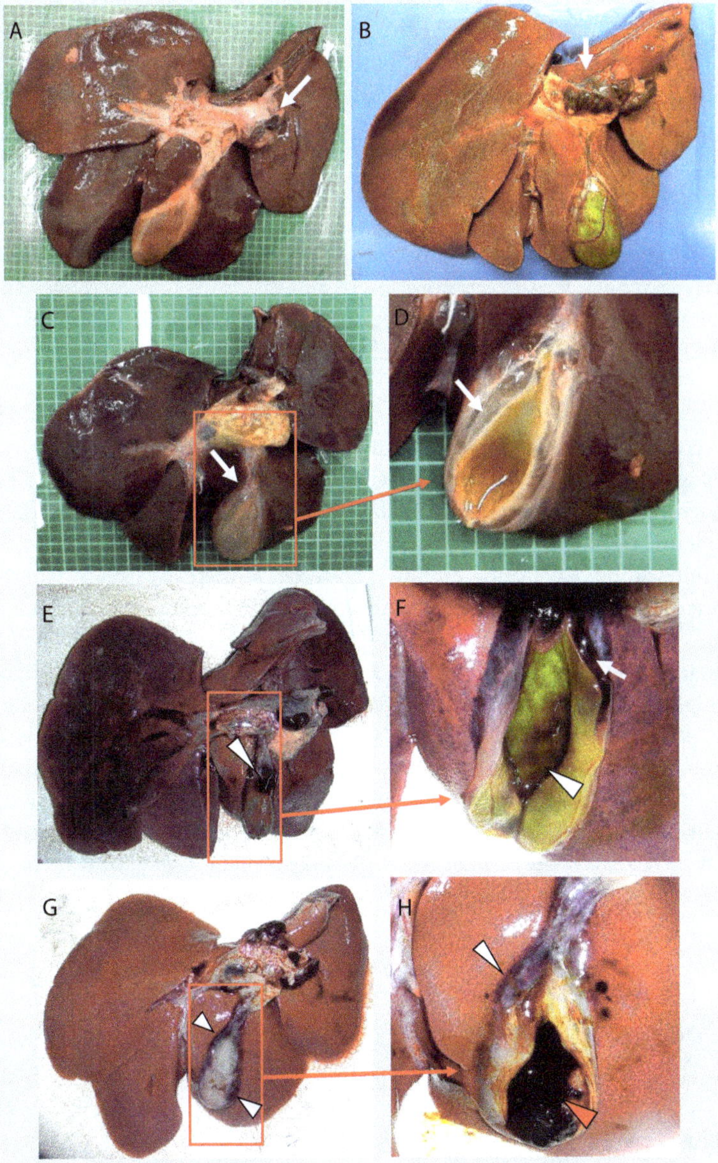

Figure 4.8. Lesions in the liver and gall bladder of pigs with acute and subacute forms of African swine fever (ASF). (A) Liver from a pig with acute ASF with mild hepatomegaly and congestion; (B) Liver from a pig with subacute ASF showing moderate to severe hepatomegaly with rounded edges and diffuse congestion. Observe in both animals (A and B) the severe haemorrhagic lymphadenitis affecting the gastrohepatic lymph nodes (arrows); (C-H) Livers from pigs with subacute ASF showing hepatomegaly and diffuse congestion along with haemorrhagic lymphadenitis of gastrohepatic lymph nodes. The gall bladder wall and bile ducts display severe oedema with wall distension which appears yellow and gelatinous (white arrows). Observe the presence of multifocal haemorrhages (petechiae and ecchymoses) on the serosa and submucosal surfaces of the gall bladders (white arrowheads), along with the presence of blood clots on the mucosa (red arrowhead).

Apart from moderate congestion observed in the meninges (Figure 4.9A) and choroid plexus, no other gross lesions are usually observed in the central nervous system (Galindo-Cardiel *et al.*, 2013; Gallardo *et al.*, 2015b; Kleiboeker, 2002; Moulton and Coggins, 1968; Sánchez-Vizcaíno *et al.*, 2015; Sehl *et al.*, 2020).

4.3.2 Subacute forms of African swine fever

Macroscopic lesions associated with subacute ASF forms are similar to those described in acute forms but are characterised by more severe and extensive haemorrhages and oedema as a consequence of longer clinical courses induced by less virulent isolates, severe and prolonged thrombocytopenia and consumption of coagulation factors.

Animals that develop subacute forms of ASF usually display abundant petechial haemorrhages along with severe ecchymoses, haematomas and oedema in the subcutis and skeletal muscles of ventral areas (Figure 4.2E-H). Haemorrhages may also appear in the diaphragm. Animals usually display severe hydrothorax (Figure 4.4G) and hydropericardium with abundant yellowish or red-tinged fluid (Figure 4.4B, 4.4D), while the pleural cavity may contain clear fluid. Severe ascites with blood-stained fluid is a common finding (Figure 4.4F). There are usually blood clots floating or adhered to serosal surfaces and fibrin strands in both the thoracic and abdominal cavities.

Most of the lymph nodes are markedly enlarged and oedematous. Besides gastrohepatic, renal and mesenteric lymph nodes, which are mainly affected in acute forms, other lymph nodes may display haemorrhagic lesions resembling blood clots (severe haemorrhagic lymphadenitis) (Figure 4.5F-L). The spleen may appear moderately enlarged but not friable, displaying minimal bleeding, or may be of almost normal size. Other animals may display partial hyperaemic splenomegaly where only some areas of the spleen show an increase in size with friable consistency and dark colour, changes that may resolve if the animal recovers and that may cause focal damage. Multifocal

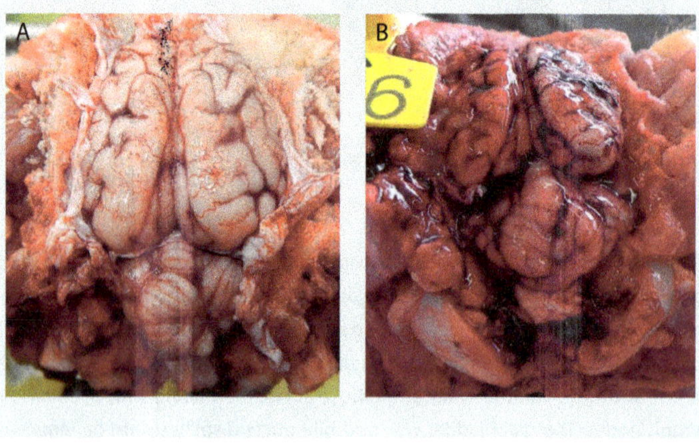

Figure 4.9. Lesions in the meninges of pigs with acute and subacute forms of African swine fever (ASF). (A) Mild congestion in the meninges in a pig with acute ASF; (B) Moderate to severe congestion and haemorrhages in the meninges of a pig with subacute ASF.

petechial haemorrhages in the thymus and diffuse erythema in the palatine tonsillars surface are also frequent lesions.

In the lungs, lesions are more severe and include failure to collapse, marked rib impressions, diffuse severe congestion, severe alveolar oedema with abundant blood oozing from cut surfaces and presence of blood-stained foamy material, which is also present in the trachea lumen. In addition, the lungs display moderate to severe distension of the interlobular septa with accumulation of fluid (interstitial oedema) along with multifocal to coalescent petechial haemorrhages, ecchymoses and purpura in the pulmonary interstitium and pleura (Figure 4.6D-E, 4.6G-H). Areas of consolidation characteristic of pneumonia with focal or lobar extension, red or grey in colour and rubbery in consistency may be present (Figure 4.6I). In the heart, lesions may range from multifocal petechial haemorrhages to severe and coalescent interstitial haemorrhages (ecchymoses and suggillations) almost completely affecting the epicardium, myocardium and endocardium (Figure 4.6M-P). Areas of myocardial necrosis may be observed.

Moderate to severe perirenal oedema is a common feature of subacute forms of ASF (Figure 4.7G-H). Haemorrhagic lesions may affect extensive areas of cortex, medulla, the renal pelvis and even the capsule (Figure 4.7C-F). The kidneys usually display pelvic dilation, diffuse congestion and more severe interstitial haemorrhagic lesions (petechiae, ecchymoses and purpura) than those described in acute forms (Figure 4.7I-J). Infarcts may be observed. The urinary bladder wall may display moderate to severe oedema characterised by the presence of a gelatinous infiltrate resulting in wall expansion. Multifocal severe haemorrhages (petechiae, ecchymoses and suggillations) in the submucosa and serosa along with blood clots covering large areas of the mucosal surface are common lesions in subacute forms (Figure 4.7M-N).

Congestive changes and haemorrhagic lesions in the gastrointestinal tract are usually more severe than in acute forms, affecting the stomach (Figure 4.10A-B) and larger segments of the small and large intestine (Figure 4.10C, 4.10E-H). Particularly noteworthy is the presence of abundant petechiae, ecchymoses and suggillations in serosal and mucosal surfaces along with clotted blood mixed with intestinal content in the stomach. Hyperaemia, oedematous wall and sometimes erosions of the stomach mucosa (cardiac gland region, region of proper gastric glands and pyloric gland region) may be present. Mild thickening of the intestinal wall is a common finding.

The liver usually displays moderate to severe hepatomegaly with rounded edges, firm consistency, diffuse congestion with marked bleeding after sectioning, red mottling and subcapsular haemorrhages (petechiae and suggillations) (Figure 4.8B-C, 4.8E, 4.8G). The gall bladder wall and bile ducts usually display severe oedema with wall distension that appears yellow and gelatinous (Figure 4.8C-H). Multifocal haemorrhages (petechiae and ecchymoses) on the serosal and submucosal surfaces of the gall bladder, along with occasional blood clots on the mucosa can be observed (Figure 4.8E-H).

In the central nervous system, the meninges (Figure 4.9B) and choroid plexus may be severely congested, oedematous and haemorrhagic. Within the endocrine system, the presence of haemorrhages is occasionally described in the inner cortical and medullary region of the adrenal gland. Extensive haemorrhages and necrotic areas in pancreas (Figure 4.10D) can be observed

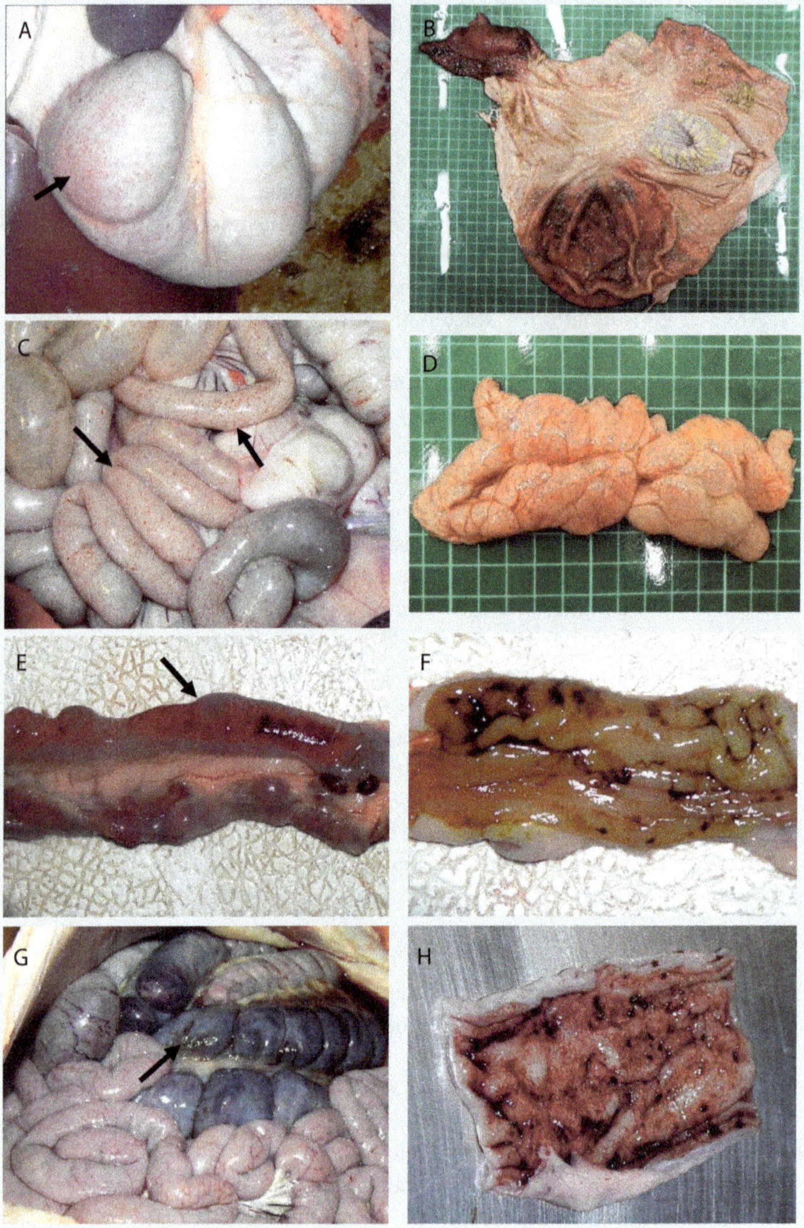

Figure 4.10. Lesions in the gastrointestinal tract of pigs with subacute forms of African swine fever (ASF). (A) Abundant petechiae in the serosa of the stomach in a pig with subacute ASF (arrow); (B) Hyperaemia and inflamed mucosal surfaces in the stomach of a pig with subacute ASF along with the presence of petechiae and ecchymoses in the mucosal surface mixed with stomach content; (C) Petechial haemorrhages in the serosa of small intestine in a pig with subacute ASF (arrows); (D) Petechial haemorrhages in the pancreas of a pig with subacute ASF. Petechiae, ecchymoses and purpura in serosa and mucosal surfaces along with clotted blood mixed with intestinal content in sections of small (arrow) (E-F) and large intestine (arrow) (G-H) in pigs with subacute ASF.

(Galindo-Cardiel *et al.*, 2013; Gallardo *et al.*, 2015b; Kleiboeker, 2002; Moulton and Coggins, 1968; Sánchez-Vizcaíno *et al.*, 2015; Sehl *et al.*, 2020).

4.3.3 Chronic forms of African swine fever

Chronic forms of ASF, traditionally described as 'atypical', are characterised by the absence of fluid exudation into body cavities and macroscopic haemorrhagic lesions, both considered hallmarks of acute and subacute forms. Most of the descriptions are based on lesions observed in natural cases. Traditionally, these animals have been considered as individuals with a compromised immune system susceptible to infection of opportunistic pathogens, mainly bacteria, which are ultimately responsible for the appearance of macroscopic lesions. Among those lesions, the following stand out: fibrinous and fibrous pericarditis (Figure 4.6Q), fibrous pleuritis and pleural adhesions to the thoracic wall, consolidated lobules in the lungs, necrotic and caseous pneumonia with mineralisation or necrotic areas on tonsils and tongue. The spleen and lymph nodes display hyperplasia without haemorrhagic changes. Lymph nodes appear enlarged, diffusely white or tan, and rubbery in consistency. Due to the presence of opportunistic pathogens that may affect the cardiorespiratory system, pulmonary lymph nodes (mainly the tracheobronchial and mediastinal) might be swollen and partially haemorrhagic. In the spleen, cicatrisation after partial hyperaemic splenomegaly may appear. Neither the liver nor the bile ducts show significant visible lesions. Foetal maceration in pregnant sows has been associated with chronic courses of infection (Moulton and Coggins, 1968; Pérez *et al.*, 1994; Sánchez-Vizcaíno *et al.*, 2015).

Experimental infections with low virulent ASFV isolates focused on studying or inducing chronic forms of ASF are scarce, normally conducted within experiments focused on assessing vaccine candidates (Boinas *et al.*, 2004; Gallardo *et al.*, 2019; Sánchez-Cordón *et al.*, 2017, 2020b). In the course of these experiments, domestic pigs inoculated by the intranasal or intramuscular route with low virulent ASFV isolates of genotype I and II, or infected by direct contact, developed chronic forms with lesions that mainly affected the cardiorespiratory system including fibrinonecrotic pleuropneumonia (Figure 4.6J-K), fibrinous pleuritis, pulmonary abscesses with areas of calcification, fibrinous pericarditis and hydropericardium with fibrin clots. The joints were swollen, showing serofibrinous periarthritis and serohaemorrhagic arthritis with abundant, red-tinged fluid. Also, some pigs showed loss of body condition, necrotic skin areas and ulcers along with necrotic areas in skeletal muscle (Figure 4.3J). Although viraemia levels in these animals were low, virus genome was detected by PCR and virus was recovered from most of the evaluated tissues. Curiously, high viral loads were detected in skin lesions and synovial fluid obtained from the joints (Gallardo *et al.*, 2015a, 2019; Sánchez-Cordón *et al.*, 2017).

4.4 Virus replication, spread and associated changes

Over the last five decades, mechanisms involved in ASFV replication and spread in the body, as well as pathogenic mechanisms of the most characteristic lesions, have been widely studied in experimentally infected domestic pigs. These animals developed acute and subacute forms of ASF induced by highly and moderately virulent ASFV isolates.

Cells belonging to the monocyte/macrophage lineages, which display a high heterogeneity in phenotype and function, have been identified as main target cells of ASFV. Some studies have suggested different susceptibilities within this cell population, so that more mature and well differentiated macrophages that express specific markers (e.g. CD163, CD107a, SLA-II) could be more susceptible to virus infection (Sánchez-Torres *et al.*, 2003), although specific receptors for ASFV have not yet been identified. If this is the case, the structural proteins p12, p72 and p54 are involved in virus attachment while early expressed protein p30 is necessary for virus internalisation (Galindo and Alonso, 2017). The first studies carried out on tissue samples taken from experimentally infected domestic pigs indicated that monocytes and macrophages are the main target cells of ASFV (e.g. Sierra *et al.*, 1987). Ultrastructural studies of infected tissue samples using transmission electron microscopy (TEM) described ASFV replication centres in infected macrophages as rounded areas free of cytoplasmic organelles. In these areas elongated membranous structures and icosahedral structures (175-190 nm diameter) with electron-dense nucleoids, corresponding to mature viral particles, were identified. Immature viral particles did not show an electron-dense nucleoid. Viral particles are released by budding. Semi-thin sections (1 μm) stained with toluidine blue dye revealed the presence of macrophages with intracytoplasmic inclusion bodies described as pale-staining juxtanuclear cytoplasmic areas in which virus replication takes place (Gómez-Villamandos *et al.*, 2013). From attachment and entry to budding of mature virus particles, the ASFV infection cycle may be completed within 24 hours post infection (Muñoz-Moreno *et al.*, 2015).

Because of ASFV replication, macrophages show a cytopathic effect. This effect is characterised ultra-structurally by the presence of enlarged macrophages with rounded morphology along with rounded cores, margination of nuclear chromatin and presence of cytoplasmic vacuoles, changes characteristic of necrosis causing massive release of viral particles (Sierra *et al.*, 1989). Apoptosis, characterised by condensation and margination of chromatin, fragmentation of cell nuclei and cytoplasm, and the appearance of apoptotic bodies (membrane-bounded fragments of cytoplasm and condensed nuclear chromatin) has also been suggested as a possible cause of macrophage destruction (Carrasco *et al.*, 1996d; Dixon *et al.*, 2017; Gómez-Villamandos *et al.*, 1995a).

Haemadsorption is another ASFV characteristic phenomenon initially described *in vitro* and used as a diagnostic technique. It is characterised by the presence of infected cells surrounded by a crown of erythrocytes. This binding is mediated by the CD2v/EP402R virus protein, which is expressed on the surface of infected cells and extracellular enveloped virions (Borca *et al.*, 1994; Rodríguez *et al.*, 1993). Haemadsorption has been observed by TEM in tissues from infected animals where infected cells, mainly macrophage populations with virus replication sites and showing cytopathic effect, displayed adhered pear-shape erythrocytes (Sierra *et al.*, 1991).

In blood from pigs infected with haemadsorbing ASFV isolates, virions can be detected free in plasma. However, a high percentage of viral particles (>90%) are found attached to erythrocytes within small invaginations of the cytoplasmic membrane (Quintero *et al.*, 1986). This mechanism is suggested to be responsible for virus spread at initial stages of the disease. In addition, haemadsorption could also be responsible for the extended persistence of the virus genome in animals vaccinated with live attenuated isolates rather than the presence of continuously replicating virus (Reis *et al.*, 2017). Virus replication sites have not been demonstrated in erythrocytes, lymphocytes (Carrasco *et al.*, 1996b) or platelets (Gómez-Villamandos *et al.*, 1996), even though intracytoplasmic viral particles have been observed in these cells.

The oronasal route is considered as the most common route of infection, with the conjunctiva, genital tract, skin abrasions and infected tick bites described as alternative routes of exposure. It is generally accepted that independently from the route of infection, the virus replicates primarily in mononuclear phagocytic cells of tonsils and submandibular, retropharyngeal and other regional lymph nodes where it is detected as early as 16-24 hours after infection. In experimental oronasal infections, virus replication in nasal mucosa is not described at this early stage of infection. Most studies point out tonsillar lympho-reticular cells, as the primary target site from where the virus spreads subsequently to regional lymph nodes (Greig, 1972; Heuschele, 1967; Howey *et al.*, 2013; Plowright *et al.*, 1968). Other authors, however, suggest that ASFV might replicate simultaneously in both tonsils and lymph nodes, without the need to go through the tonsils before replicating in lymph nodes (Colgrove *et al.*, 1969). After initial replication, the virus spreads through lymph and blood (free in plasma, adhered to erythrocytes or carried by infected monocytes). It is detectable in almost all tissues between 48-72 hours after infection, with high titres in tissues containing a large component of the mononuclear phagocyte system (MPS), such as the spleen, lymph nodes and bone marrow, as well as in liver, lung or kidney. The high viral replication rates in these organs are usually associated with a later peak of viraemia, which coincides with the appearance of pyrexia and a febrile syndrome from day 3-4 after infection. In other tissues containing few components of the MPS (central nervous system, stomach, large intestine, etc.), viral titres may be attributed to the presence of virus in the blood of these tissues, but not to viral replication. In acute and subacute forms, before the onset of clinical signs, virus is usually shed in oronasal and lachrymal secretions, urine and faeces from 1 to 7 dpi depending on the isolate and route of infection. The highest viral titres are usually reported in oronasal fluid while lower titres are detected in conjunctival and genital fluid. Shedding from the oral cavity occurs prior to systemic dissemination of the virus. Virus shedding through the faecal route varies from intermittent shedding periods with low viral titres to continuous shedding with high titres. Generally, the level of infectious ASFV excreted via these routes is lower than the level in blood (Howey *et al.*, 2013). In convalescent animals and animals with chronic forms, information regarding ASFV shedding is less clear, although some studies describe the presence of virus genome in oropharyngeal fluid and blood up to 70 days after infection (De Carvalho Ferreira *et al.*, 2012). Despite the high fatality rate that characterises ASF (most infected animals die), it is not a highly infectious disease. The virus usually spreads slowly within the herd and not all animals will be affected (see Chapter 9).

In the so-called secondary viral replication organs, replication has been observed in cells other than the MPS such as epithelial or mesenchymal cells which play a minor role in the pathogenesis of the disease. In this regard, viral replication in megakaryocytes has been suggested as the cause of the presence of viral particles in platelets (Gómez-Villamandos *et al.*, 1997a). Additionally, the presence of viral particles, but not virus replication, has been described in lymphocytes (Carrasco *et al.*, 1996b). This suggests that along with erythrocytes and monocytes, these cell populations might also contribute to virus spread in later stages of disease, but only playing a minor role. In addition, ASFV has the ability to replicate in mature and immature neutrophils (Carrasco *et al.*, 1996d) as well as in dendritic cells, which likely impairs the induction of an effective immune response (Franzoni *et al.*, 2018; Gregg *et al.*, 1995). As mentioned, many other cell types may also become infected, especially in late stages of the disease, including endothelial cells, pericytes, glomerular mesangial cells, renal collecting duct epithelial cells, tonsillar epithelium, hepatocytes, fibroblasts, reticular cells, smooth muscle cells and Ito cells (see review in Gómez-Villamandos *et al.*, 2013). This demonstrates the pantropic nature of ASFV. However, knowledge of the ASFV-entry mechanisms into cells is still limited. It has been proposed that the virus enters cells through receptor-mediated endocytosis (Alcamí *et al.*, 1989) and other studies have suggested macropinocytosis as another entrance mechanism (Sánchez *et al.*, 2012). It is likely that more than one mechanism is involved (Galindo and Alonso, 2017). To explain the early presence of replicating ASFV in cells other than MPS with little phagocytic activity, a possible role for ASF virus-specific membrane receptors, expressed in these cells (e.g. fibroblasts, smooth muscle cells, megakaryocytes, fat-storing cells or reticular cells) and induced by activated macrophage populations (infected or not) through chemical mediators, has been suggested (Gomez-Villamandos *et al.*, 1999). Such mechanisms remain unclear.

4.5 Microscopic lesions and pathogenic mechanisms

4.5.1 Integumentary and skeletal system

In animals infected with ASF, histopathological lesions in the skin are characterised by the presence of vascular changes that range from mild in acute forms to severe in subacute forms. Thus, lesions in the dermis characteristically consist of mild hyperaemia and occasional small focal haemorrhages, vasculitis and minimal perivascular and intravascular accumulation of macrophages and monocytes that frequently show karyorrhexis (Figure 4.11A). These changes are more severe in subacute forms, where additional lesions in the dermis include perivascular oedema and inflammatory infiltrates with macrophages, lymphocytes, mast cells and plasma cells. Fibrinoid vasculitis with necrosis of blood vessel walls along with thrombi and fibrin deposits may be present. Ultrastructural studies describe platelets located among the fibrin mesh, many of them showing partial or complete degranulation. Additionally, endothelial cells appear enlarged, showing cytoplasmic projections into the vascular lumen, an increase in micropinocytic vesicles, impaired intercellular cohesion, vascular degeneration and necrosis. In subacute and chronic forms, sub-corneal and intraepithelial pustules are frequent and there is occasional folliculitis along with multifocal necrotising ulcerative dermatitis.

Ultrastructural and immunohistochemical techniques (Figure 4.11B) have demonstrated virus replication and the presence of viral antigen in the dermis of skin samples from animals with acute and subacute forms of ASF. Virus material was mainly located in endothelial cells, intravascular monocytes, macrophages within perivascular infiltrates and reticular cells. The presence of macrophage infiltrates containing immunolabelled viral antigen has also been described in the epidermis. Macrophages, and occasionally some lymphocytes, appear to be the main cells with positive immunolabelling for viral antigen in the cell infiltrates described in the ulcerative lesions that are characteristic of chronic forms (Mozos *et al.*, 2003).

In the eye, oedema, hyperaemia, perivascular mononuclear cell infiltrates and vascular thrombosis affecting the limbus of the cornea, the ciliary body and subretinal tissues have been described as characteristic histopathological changes of ASF (Moulton and Coggins, 1968).

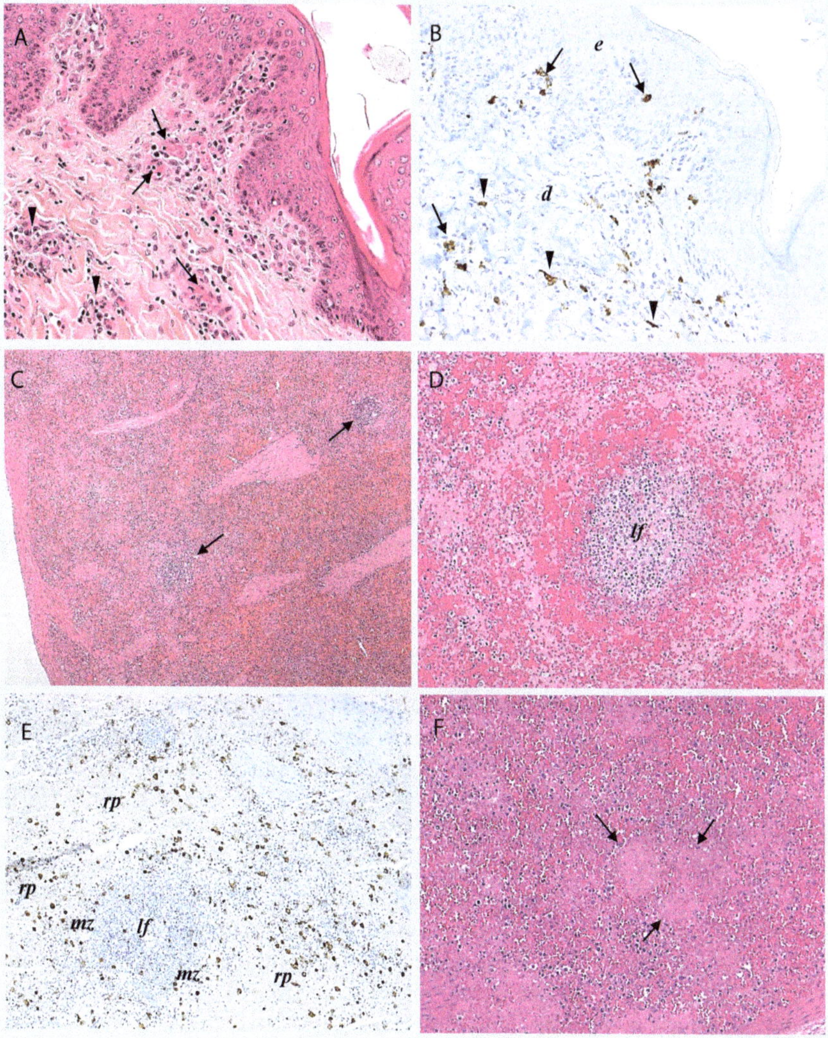

Figure 4.11. Skin and spleen. (A-B) Serial sections of skin from a pig with acute African swine fever (ASF); (A) HE stain, 20×. Capillary endothelial cells displaying hypertrophy with rounded nuclei, along with the degeneration of vascular endothelium (arrows); Mild perivascular mononuclear infiltrates (arrowheads) with presence of pyknotic cells and cell debris; (B) IHC, 20×. Macrophages (arrows) immunolabelled against viral antigen in epidermis (*e*) and within perivascular infiltrates in dermis (*d*) along with endothelial cells positively immunolabelled (arrowheads); (C) Spleen (acute ASF). HE stain, 4×. Severe, diffuse engorgement of the splenic red pulp which appears filled with erythrocytes along with the presence of small lymphoid follicles (arrows); (D) Spleen (acute ASF). HE stain, 20×. Lymphoid follicle (*lf*) infiltrated with erythrocytes and macrophages showing severe lymphocytic depletion along with abundant pyknotic cells and cell debris; (E) Spleen (acute ASF). IHC, 10×. Massive presence of cells, mainly hypertrophic macrophages, immunolabelled against viral antigen in the splenic red pulp (*rp*), marginal zone (*mz*) and within lymphoid follicles (*lf*); (F) Spleen (acute ASF). HE stain, 20×. Splenic ellipsoids (arrows) showing an increase in size, hyalinisation and massive destruction of macrophages. Haematoxylin-eosin staining (HE); immunohistochemistry against P30 protein (IHC); original magnification (number ×).

Histologically, skeletal muscles from animals with acute and subacute forms of ASF show mononuclear cell infiltrates mainly composed of macrophages and lymphoplasmacytic cells that may show karyorrhexis. Other lesions include hyperaemia, haemorrhages and vasculitis with thrombi and fibrin deposits. Zenker's degeneration (coagulative necrosis) in the muscle fibres, sometimes with calcification, may be present. Mononuclear infiltrates may likewise be observed in the subcutis and fat. Macrophages within cell infiltrates and endothelial cells are the main target cells for ASFV. Fibrous swellings of joints described macroscopically in animals with chronic forms of ASF histologically correspond to granulation tissue infiltrated by mononuclear cells (Moulton and Coggins, 1968).

4.5.2 Lymphoid system

In the course of an infection with ASFV, the most significant histopathological lesions affect the lymphoid system. Lymphoid organs usually display higher levels of tissue destruction, lymphocytic depletion and number of ASFV infected cells. These histopathological changes are more prominent in infections with highly virulent ASFV isolates than in acute stages of infections with moderately virulent isolates. The degree of lymphocyte destruction has been directly related to the number of ASFV infected macrophages (Oura *et al.*, 1998b).

4.5.2.1 Spleen

In the spleen, considered one of the main target organs for ASFV, the severity of histopathological lesions depends on the virulence of the isolate involved and on the effect of the virus on the resident splenic macrophage populations (Gómez-Villamandos *et al.*, 2013). In acute forms of ASF, the splenic red pulp shows severe, diffuse engorgement, appearing filled with erythrocytes, fibrin deposits and cell debris (Figure 4.11C). A massive infiltration of mononuclear cells, often hypertrophic and necrotic, and the presence of megakaryocytes are observed. Schweigger-Seidel sheaths (or splenic ellipsoids) typically show an increase in size. Splenic ellipsoids are often hyalinised, necrotic and obliterated, showing massive destruction of macrophages (Figure 4.11F). The marginal zone, the lymphoid follicles and periarteriolar lymphoid sheaths (PALS) are usually small and difficult to discern, appearing infiltrated by erythrocytes and macrophages, many of the latter with abundant cytoplasm containing phagocytised cell debris (tingible bodies). The germinal centres typically show severe lymphocytic depletion along with abundant pyknotic cells, karyorrhexis and cell debris, changes that are usually less severe in PALS (Figure 4.11C-D). Small and medium vessels commonly show endothelial activation denoted by hypertrophic endothelial cells with rounded nuclei, along with the degeneration of vascular endothelium, fibrinoid vasculitis and thrombosis of blood vessels. Histopathological lesions are usually less severe and frequent in the spleen of animals that develop subacute forms of ASF (Moulton and Coggins, 1968; Salguero *et al.*, 2002; Sehl *et al.*, 2020).

ASFV primarily targets areas where macrophage populations are abundant, such as the splenic red pulp, splenic ellipsoids and peripheral areas of splenic lymphoid follicles (marginal zones). Macrophages with viral antigen indicated by positive immunolabelling in the splenic red pulp usually display increase in size (hypertrophy) (Figure 4.11E). The presence of lymphocytes positively immunolabelled for viral antigen is also frequent in this area. Cells containing immunolabelling are further described in peripheral areas of lymphoid follicles, and less frequently within lymphoid follicles. Additionally, endothelial cells and neutrophils may show immunolabelling indicative of viral antigen. Occasionally, star-shaped cells revealing labelled viral antigen, likely reticular cells or follicular dendritic cells, have been observed within splenic lymphoid follicles (Carrasco *et al.*, 1997a; Pérez *et al.*, 1994; Salguero *et al.*, 2002, 2005; Sehl *et al.*, 2020).

4.5.2.2 Pathogenesis of hyperaemic splenomegaly

The structure of porcine splenic cords is constituted by a network of smooth muscle cells surrounded by a population of fixed macrophages, which constitute one of the main targets for ASFV (Carrasco *et al.*, 1995). ASFV replication in the population of resident macrophages induces cytopathic effect, progressive detachment of resident macrophages from muscle cells due to the loss of intercellular junctions and, finally, necrosis of macrophages. These changes result from the exposure of the collagen of the basal lamina of smooth muscle to blood components. This exposure promotes activation of the coagulation cascade, activation and aggregation of platelets and the appearance of fibrin deposits, giving rise to accumulation of erythrocytes within splenic cords. Consequently, blood clearance, one of the main functions of the spleen, is impaired and hypoxia induces massive necrosis of lymphocytes within the lymphoid structures, thus enhancing lymphoid depletion, especially in late stages of the disease (Carrasco *et al.*, 1997a).

4.5.2.3 Lymph nodes

In late stages of the disease, especially in subacute forms, most of the lymph nodes in infected animals may display lesions. However, the submandibular, retropharyngeal, gastrohepatic, renal, tracheobronchial, mediastinal and mesenteric lymph nodes are usually the focus of histopathological evaluations carried out in the course of acute and subacute forms of ASF. Lymph nodes usually display diffuse, moderate to severe hyperaemia. Sinuses are typically infiltrated with mononuclear cells, pyknotic cells with karyorrhexis and cell debris. Diffuse, moderate to severe haemorrhages that mainly affect medullary cords and diffuse lymphoid tissue in the cortex are usually observed. Marginal zones of lymphoid follicles in the cortex and germinal centres are typically infiltrated with blood and macrophages (Figure 4.12A-C). In addition, moderate to severe lymphocytic depletion affecting both lymphoid follicles and diffuse lymphoid tissue, usually accompanied by a massive presence of pyknotic cells, cell fragmentation and tingible body macrophages (TBM), is frequently observed (Figure 4.12B, 4.12E). Megakaryocytes are usually observed among haemorrhages and cell debris. Capillary and small-vessel endothelial cells display cellular hypertrophy and rounded, prominent nuclei characteristic of endothelial activation. Vasculitis with intramural inflammatory infiltrates, necrosis of endothelial cells, perivascular oedema, and fibrin deposits in the vascular lumen can be observed. Damage of blood vessels and thrombosis may contribute to destruction of lymphocytes (Moulton and Coggins, 1968; Salguero *et al.*, 2002; Sehl *et al.*, 2020).

The macroscopic hyperplasia that is characteristic for lymph nodes in chronic forms of ASF is the result of reticular and MPS cell proliferation in the sinuses and germinal centres. Histologically, the sinuses appear as a solid sheet of reticular and MPS cells, often with obliteration of sinus lumens. The germinal centres appear enlarged and occupied by reticular, hyperplastic MPS and numerous

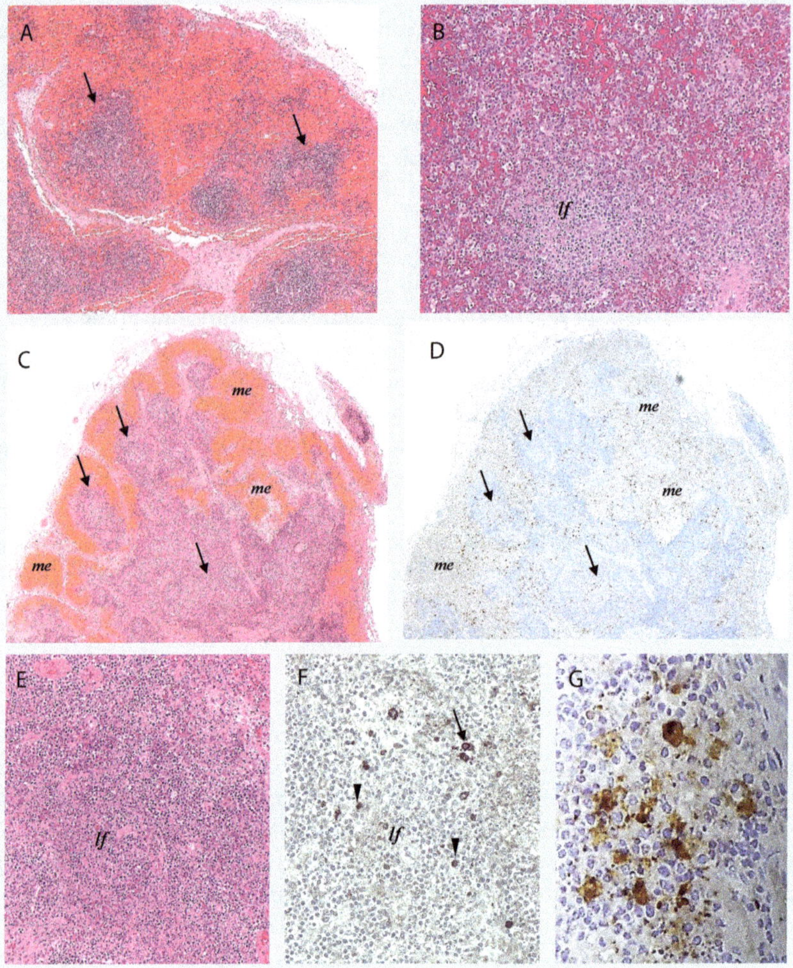

Figure 4.12. Lymph nodes. (A) Gastrohepatic lymph node from a pig with acute African swine fever (ASF). HE stain, 10×. Moderate to severe lymphocytic depletion affecting both lymphoid follicles and diffuse lymphoid tissue (arrows) accompanied by severe haemorrhages in medullary cords and presence of blood infiltrating the marginal zones and germinal centres of lymphoid follicles in the cortex; (B) Gastrohepatic lymph node from a pig with acute ASF. HE stain, 20×. Severe lymphocytic depletion affecting both lymphoid follicle (*lf*) and diffuse lymphoid tissue accompanied by the presence of pyknotic cells, cell fragmentation and tingible body macrophages; (C-D) Serial sections of the renal lymph node from a pig with acute ASF. HE stain and IHC, 4×. Moderate lymphocytic depletion affecting both lymphoid follicles and diffuse lymphoid tissue accompanied by severe haemorrhages. Observe the distribution of cells immunolabelled positively for viral antigen mainly in the medulla (*me*) and interfollicular areas of the cortex, and the presence of lower number of cells immunolabelled within lymphoid structures (arrows); (E-F) Serial sections of the renal lymph node from a pig with subacute ASF. HE stain and IHC, 20×. Lymphoid follicle showing severe lymphoid depletion along with the presence of macrophages (arrow) and lymphocytes (arrowheads) immunolabelled against viral antigen; (G) Renal lymph node from a pig with acute ASF. IHC, 40×. Detail of macrophages immunolabelled for viral antigen in the marginal zone. Haematoxylin-eosin staining (HE); immunohistochemistry against P30 protein (IHC); original magnification (number ×).

immature lymphoid cells so that, except for a peripheral thin layer of lymphocytes, reticular and MPS cells occupy almost the entire follicle (Moulton and Coggins, 1968; Pérez *et al.*, 1994).

Traditionally, ultrastructural and immunohistochemical studies have identified lymph nodes as target organs with a high rate of ASFV replication. Virus replication is usually higher in lymph nodes from animals infected with highly virulent compared to moderately virulent ASFV isolates (Oura *et al.*, 1998b). Immunolabelled cells, mainly monocytes and macrophages along with occasional lymphocytes and neutrophils, are usually observed in the medulla and interfollicular areas of the cortex, while the presence of positive cells within the lymphoid follicles is usually low (Figure 4.12D, 4.12F-G). Star-shaped, immunolabelled cells have occasionally been observed in interfollicular areas and within lymphoid follicles. In areas with severe haemorrhages and lymphoid depletion, the presence of immunolabelled cells is usually very low (Figure 4.12F). Virus replication has been described in sinus and capillary endothelial cells, as well as in other cell populations including pericytes, fibroblasts, smooth muscle fibres and reticular cells (Carrasco *et al.*, 1997b; Salguero *et al.*, 2002; Sehl *et al.*, 2020).

4.5.2.4 Tonsils

Histopathological changes in palatine tonsils are characterised by the presence of diffuse, moderate to severe lymphocytic depletion affecting both lymphoid follicles and interfollicular areas. Lymphocytic depletion is accompanied by the presence of pyknotic cells, karyorrhexis, cell fragmentation and TBM infiltrates, the latter especially abundant within the lymphoid follicles. Mononuclear cell infiltrates, often with marked karyorrhexis, and connective tissue proliferation are common findings observed in interfollicular areas. The epithelium of the crypts usually shows moderate to severe infiltrates of mononuclear cells displaying pyknosis and cell fragmentation (Figure 4.13A-C). Tonsillar epithelial cells that overlie the tonsil and line the crypts may display necrosis as well as a cytopathic effect characterised by pyknosis and karyorrhexis. Microthrombi may be observed in interstitial blood vessels along with mild congestion and haemorrhages (Fernández de Marco *et al.*, 2007; Howey *et al.*, 2013; Moulton and Coggins, 1968; Rodríguez *et al.*, 1996a; Sehl *et al.*, 2020).

The palatine tonsils constitute one of the first sites for ASFV replication and are one of the locations where the viral titres remain high for longer periods of time (Oura *et al.*, 1998a). In acute forms of ASF, in addition to macrophages, virus replication is described by ultrastructural studies in epithelial cells, fibroblasts and reticular cells from early stages after infection (Gómez-Villamandos *et al.*, 1997b). Using immunohistochemistry, the highest number of immunolabelled cells positive for viral antigen, mainly macrophages and some lymphocytes, are observed infiltrating the crypt epithelium and the diffuse lymphoid tissue surrounding the crypts. Viral antigen could also be visualised in the tonsillar epithelium. A high number of immunolabelled cells, mainly macrophages and some lymphocytes, are observed in interfollicular areas, while the presence of antigen positive cells inside lymphoid follicles is usually lower (Figure 4.13D-F). Immunoreactive star-shaped, likely reticular or dendritic cells, are occasionally observed in interfollicular areas and within lymphoid follicles (Fernández de Marco *et al.*, 2007; Howey *et al.*, 2013; Rodríguez *et al.*, 1996a; Sehl *et al.*, 2020). As described for other lymphoid tissues, virus replication levels in tonsils are higher in animals infected with highly virulent ASFV isolates compared to moderately

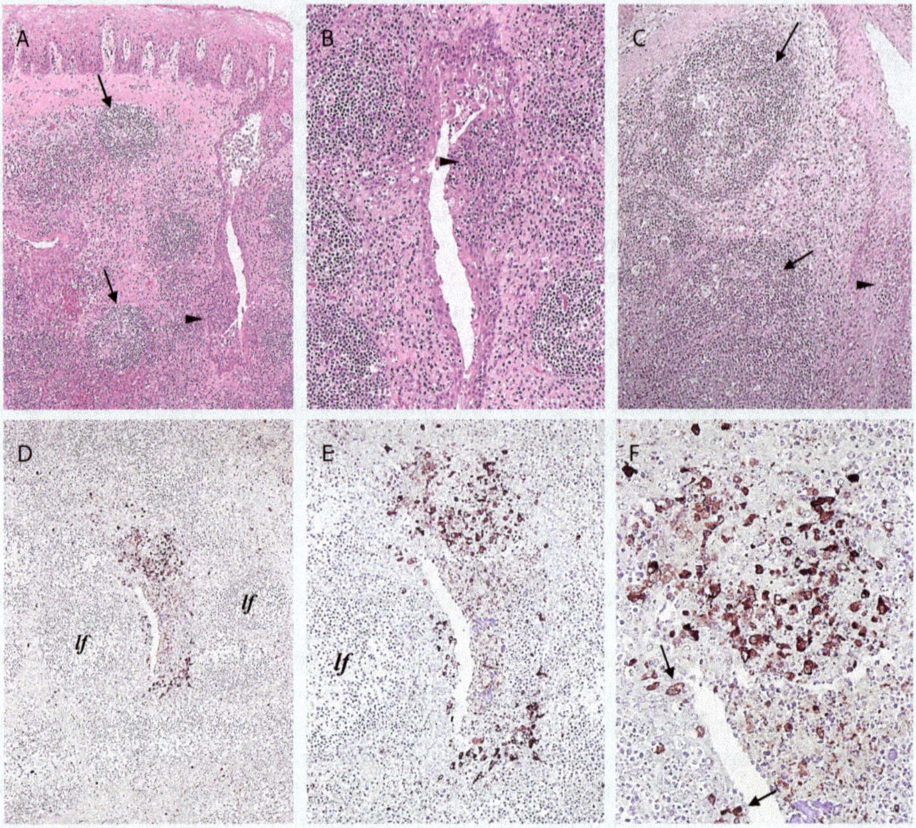

Figure 4.13. Palatine tonsil. (A-C) Palatine tonsil from a pig with acute African swine fever (ASF). HE stain, 4× (A), 10× (B, C). Moderate lymphocytic depletion affecting both lymphoid follicles (arrows) and interfollicular areas. Lymphocytic depletion is accompanied by the presence of pyknotic cells, cell fragmentation and tingible body macrophage infiltrates, the latter especially abundant within the lymphoid follicles. The epithelium of the crypts usually shows moderate to severe infiltrates of mononuclear cells displaying pyknosis and cell fragmentation (arrowheads); (D-F) Palatine tonsil from a pig with acute ASF. IHC, 4× (D), 10× (E), 20× (F). Cells immunolabelled positively for viral antigen, mainly macrophages and lymphocytes, infiltrating the crypts epithelium and the diffuse lymphoid tissue surrounding the crypts. Viral antigen could also be visualised in the tonsillar epithelium (arrows). A high number of immunolabelled cells, mainly macrophages and some lymphocytes, are observed in interfollicular areas, while the presence of antigen positive cells inside lymphoid follicles (*lf*) is scarce. Haematoxylin-eosin staining (HE); immunohistochemistry against P30 protein (IHC); original magnification (number ×).

virulent isolates. Different patterns of cell tropism, characterised by the presence of infected cells in more extensive areas in the epithelium of tonsillar crypts and interfollicular areas in animals infected with highly virulent isolates, have been described (Oura *et al.*, 1998b).

4.5.2.5 Thymus

Histopathological lesions in the thymus of pigs infected with highly virulent ASFV isolates are characterised by mild, diffuse lymphocytic depletion initially observed in the medulla and later in the cortex. The presence of pyknotic cells, karyorrhexis and cell fragmentation, along with the presence of TBM with abundant cytoplasm containing cell debris, are mainly identified in the cortex from initial stages after infection. As disease progresses, TBM in the cortex progressively increase, giving rise to the so-called 'starry sky' appearance (Figure 4.14A-B). In parallel, there is progressive, diffuse, lymphocytic depletion in the cortex accompanied by increased pyknosis and cell fragmentation that, in severe cases, may induce cortical atrophy and result in prominence of Hassall's corpuscles (structures usually observed in the medulla of the thymus constituted by epithelial reticular cells concentrically arranged) at the surface of the organ. Vascular changes in acute forms of ASF are not present in thymus, and only mild perivascular mononuclear infiltrates in vessels of the cortex and capsule are observed (Salguero *et al.*, 2005; Salguero *et al.*, 2004). In subacute forms, the lesions observed in the cortex and medulla are similar to those described in animals infected with highly virulent isolates, although these lesions usually appear later. However, mild hyperaemia and small haemorrhages in both cortex and medulla are frequently observed. Endothelial cells display cellular hypertrophy and rounded, prominent

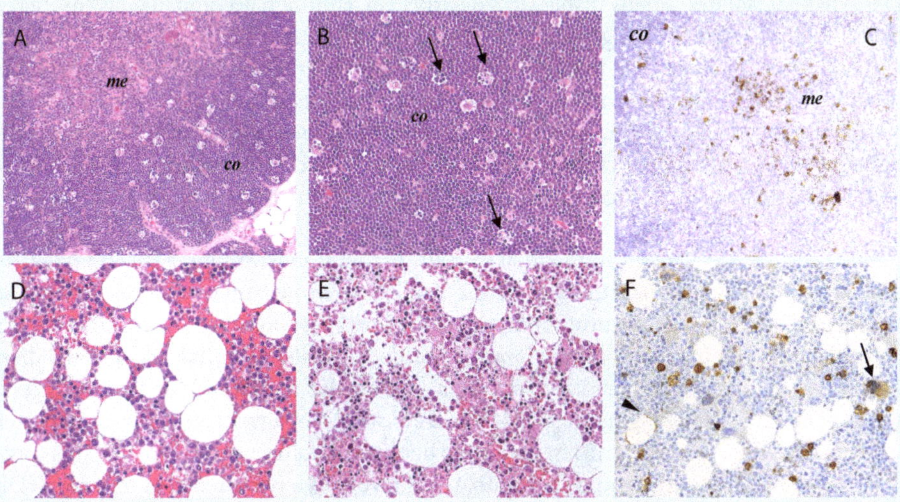

Figure 4.14. Thymus and bone marrow. (A-C) Thymus from a pig with acute African swine fever (ASF). HE stain and IHC, 10× (A), 20× (B), 10× (C). Presence of pyknotic cells and cell fragmentation, along with the presence of tingible body macrophages (arrows) with abundant cytoplasm containing cell debris in the cortex (*co*). Observe the presence of cells immunolabelled, mainly macrophages and star-shaped cells, in the medulla (*me*) and corticomedullary junction, and occasionally in the cortex (*co*); (D-F) Bone marrow from a pig with acute ASF. HE stain and IHC, 20× (D, E, F). Observe the presence of interstitial haemorrhages (D) and necrotic mononuclear cells (E). Viral antigen (F) is mainly detected in monocytes and macrophages. Also observe the occasional presence of immunolabelled megakaryocytes (arrow) and capillary endothelial cells (arrowhead); Haematoxylin-eosin staining (HE); immunohistochemistry against P30 protein (IHC); original magnification (number ×).

nuclei characteristic of endothelial activation. In addition, perivascular mononuclear infiltrates in vessels of the cortex, capsule and trabeculae are more severe, displaying cell fragmentation (P. Sánchez-Cordón, unpublished data).

In the thymus, macrophages show an increase in size and are the main target cells of ASFV in both the cortex and medulla. Using immunohistochemistry, lymphocytes and star-shaped cells occasionally contain viral antigen. The presence of immunolabelled cells is especially high in the corticomedullary junction and medulla, and to a lesser extent in the cortex and subcapsular areas (Figure 4.14C). Differences between animals that developed acute and subacute forms of ASF in the distribution and number of cells immunolabelled against viral antigen have not been observed at final stages of the disease (Salguero *et al.*, 2004, 2005).

4.5.2.6 Pathogenesis of lymphoid depletion

As discussed above, lymphoid tissue in primary and secondary lymphoid organs (tonsils, lymph nodes, spleen and thymus) constitutes the main target for ASFV replication in the initial stages after infection. Lymphoid organs usually display a range of histopathological changes which may be caused both by the direct action of the virus on specific cell populations and by the virus triggering indirect destructive mechanisms. Infections with highly virulent ASFV isolates usually result in massive destruction with widespread lymphoid tissue karyorrhexis, whereas infections with less virulent isolates usually display less destruction, although accompanied by more severe histopathological vascular changes and haemorrhages affecting a higher number of lymphoid organs. As a consequence of the massive lymphoid tissue destruction, the total leukocyte count is normally reduced in acute forms of ASF. Differential counts reveal lymphopenia that affects B and T lymphocytes and a decrease in the number of monocytes. In pigs that survive long enough to manifest subacute or chronic disease, the leukocyte count normalises, probably owing to increased haematopoiesis (Blome *et al.*, 2013; Gómez-Villamandos *et al.*, 2013).

The target cells for ASFV replication are MPS cells, predominantly monocytes and fixed-tissue macrophages, with key roles in activation of innate and adaptive immune responses. Manipulation of the function of these cells can profoundly affect the host's response to the infection. The ASFV genome encodes proteins that inhibit apoptosis in infected cells. At initial stages of the disease, inhibition of apoptosis will favour virus replication, while at the middle and late stages of infection, apoptosis of target cells belonging to the MPS will constitute an advantage for virus persistence. The presence of ASFV particles within apoptotic bodies facilitates virus spread. In this way, the virus will also avoid the immune response that will favour the persistence of infection (Dixon *et al.*, 2017). Cytopathic effect in infected cells of the MPS has been described in ultrastructural studies on lymphoid tissue samples at middle and late stages of ASF acute forms (Carrasco *et al.*, 1996c; Gómez-Villamandos *et al.*, 1995a; Sierra *et al.*, 1989).

In acute forms of ASF the virus can induce massive apoptosis of uninfected cells through indirect mechanisms, particularly affecting B and T lymphocytes, from early stages after infection. In a study by Gomez-Villamandos and colleagues (1999) it was shown that ASFV replication in macrophages induces the secretory activation of these cells and that of neighbouring non-infected macrophages, because of an autocrine effect. These findings suggest that virus replication does not inhibit secretory activity. Ultrastructural studies described the presence of these activated cells (increased in size and with lysosomal proliferation) close to areas with intense apoptosis of lymphocytes, suggesting that these cells might have an indirect effect on uninfected lymphocytes by way of secretion of chemical mediators (Carrasco *et al.*, 1996c; Gómez-Villamandos *et al.*, 1995a,b). Subsequent immunohistochemical studies corroborated the existence of a 'cytokine storm' where proinflammatory cytokines, mainly tumour necrosis factor (TNF)-α and interleukin (IL)-1, secreted by infected and non-infected macrophages might play a key role in the mechanisms of apoptosis of neighbouring lymphocytes (Fernández de Marco *et al.*, 2007; Salguero *et al.*, 2002, 2004, 2005). Another chemokine expressed in macrophages, CXCL10, also increased markedly in blood after infection with a highly virulent ASFV isolate, changes that have been associated with apoptosis of bystander lymphocytes (Fishbourne *et al.*, 2013).

4.5.2.7 Bone marrow

The evaluation of histopathological lesions in bone marrow during the course of infectious diseases is usually complex. Changes can generally be classified quantitatively (hyperplasia and hypoplasia of different cell lines) or qualitatively, including e.g. morphological aberrations of precursor cells or necrotic changes. Histopathological evaluations of bone marrow sections taken from the femur of animals experimentally infected with highly virulent ASFV isolates showed moderate hyperaemia in venous sinuses and capillaries along with increase of bone marrow cellularity due to an increase in myeloid cells before 5 dpi. Megakaryocytes showed no morphological changes. After 5 dpi, hyperaemia became more intense in venous sinuses and capillaries, with formation of extensive haemorrhages appearing in the interstitium and necrotic foci (Figure 4.14D). A high number of macrophages displayed signs of activation with phagocytosis of erythrocytes and cellular debris. In addition, macrophages also showed degenerative changes characterised by margination of the chromatin, pyknosis and karyorrhexis. A higher, but moderate, number of megakaryocytes displayed condensed chromatin, pyknosis or karyorrhexis and a thin rim of cytoplasm. The presence of cell debris and fibrin deposits within the vascular lumen was frequent (Gómez-Villamandos *et al.*, 1997a; Pérez *et al.*, 1997).

In other experimental infections with moderately virulent ASFV isolates, histopathological evaluations of bone marrow described the presence of some fixed and some circulating mononuclear cells exhibiting margination of chromatin, and some containing cell debris and phagocytised erythrocytes at 4 dpi. These changes of activation and erythrophagocytosis by monocyte/macrophages were more evident from 8 dpi. A high presence of necrotic mononuclear cells (Figure 4.14E) and numerous megakaryocytes showing margination of condensed chromatin, pyknosis and apparent loss of cytoplasm, was observed in animals euthanised between 6 and 10 dpi. In addition, there was intense cell destruction in haematopoietic foci, with numerous erythrocytes. Fibrin and cell debris further appeared within the vascular lumen. By 12 days after inoculation, megakaryocytic hyperplasia along with a substantial number of megakaryocytes undergoing necrosis were observed (Pérez *et al.*, 1997; Rodríguez *et al.*, 1996a,b).

In acute forms of ASF, immunoreactivity against viral antigen in bone marrow is detected from 3-4 dpi in monocytes and macrophages that show cytopathic effect. The number of immunolabelled monocytes and macrophages increases after 5 dpi, with presence of occasional immature cells, neutrophils and megakaryocytes (less than 10%) also positive for viral antigen (Figure 4.14F). Both infected and uninfected megakaryocytes show degenerative changes characterised by peripheral margination of condensed chromatin, pyknosis and karyorrhexis (Pérez *et al.*, 1997). Ultrastructural studies in bone marrow samples taken from animals experimentally infected with highly virulent isolates describe viral replication and cytopathic effect in monocytes, macrophages, reticular cells, immature neutrophils, myelocytes and promonocytes from 3 dpi. Only at the final stage of the disease, at 7 dpi, was viral replication described, and at that time only in a small number of megakaryocytes, sinus and capillary endothelial cells, as well as in pericytes of arterioles and venules (Gómez-Villamandos *et al.*, 1997a).

In subacute forms of ASF, immunoreactivity to viral antigen is observed in a small number of intravascular mononuclear cells at 2 dpi. From 4 dpi onwards, an increase in the number of fixed and circulating mononuclear cells, and in the number of osteoclasts showing cytoplasmic immunoreactivity is noted. From 6 dpi onwards, a diffuse or globular cytoplasmic positive reaction is observed both in numerous MPS cells and megakaryocytes (25-30%) (Rodríguez *et al.*, 1996a,b). In other studies, the percentage of immunolabelled megakaryocytes positive for viral antigen after 7 dpi was lower, ranging between less than 1% (Pérez *et al.*, 1997) and 2-10% (Edwards *et al.*, 1985). A positive diffuse or granular cytoplasmic reaction was observed in a small number of granulocytic cells (Rodríguez *et al.*, 1996b). Ultrastructural studies carried out in bone marrow from pigs experimentally infected with moderately virulent isolates confirmed the presence of viral infections in macrophages, promonocytes, neutrophilic myelocytes, megakaryocytes and megakaryoblasts from 5 dpi. Viral replication was not detected in either reticular cells or endothelial cells (Bautista *et al.*, 1998; Gómez-Villamandos *et al.*, 1998).

4.5.2.8 Pathogenesis of thrombocytopenia

In acute forms of ASF, thrombocytopenia generally occurs in the final phase, often undetected due to the sudden deterioration and death of affected animals, although lesions and virus replication are observed in bone marrow from early stages. In subacute forms, transitory thrombocytopenia occurs between the initial and middle phases of the disease. Different mechanisms have been suggested to explain the thrombocytopenia in ASF. These include: (1) decreased thrombocytopoiesis due to infection, (2) impairment and destruction of megakaryocytes by ASFV (Rodríguez *et al.*, 1996b) and (3) consumption of platelets due to platelet aggregation during disseminated intravascular coagulation (Anderson *et al.*, 1987). The destruction of megakaryocytes could be the consequence of the accelerated maturation of these cells caused by the action of cytokines, the peripheral platelet consumption, or both. Soluble chemical mediators are powerful modulators of megakaryocyte maturation and may play a key role in megakaryocyte destruction during ASF. In this regard, degenerative changes in megakaryocytes could be due to the effects of cytokines released by infected macrophages in the bone marrow, and to the high serum concentration of these chemical mediators released by infected MPS cells from other locations. Altogether, this may indicate that indirect damage of cytokines to megakaryocytes could be more important than direct damage caused by viral infection (Bautista *et al.*, 1998; Gómez-Villamandos *et al.*, 1997a). On the other hand, peripheral platelet consumption coincides with the infection, activation of monocytes/macrophages in different locations, and secretion of platelet activation factors. Attempts of the bone marrow to compensate for this consumption may induce accelerated maturation of megakaryocytes, producing large quantities of platelets. This fact would prevent thrombocytopenia at initial stages of the disease, but result in subsequent megakaryocyte exhaustion giving rise to numerous apoptotic megakaryocytes and thus to the severe and transient thrombocytopenia observed in pigs inoculated with moderately virulent isolates (Bautista *et al.*, 1998; Gómez-Villamandos *et al.*, 1997a).

4.5.3 Gastrointestinal tract

Histopathological lesions affecting the gastrointestinal tract during ASF have scarcely been evaluated. Lesions are usually more severe in subacute forms of ASF with more intense involvement of the large intestine, especially the caecum and colon, than the small intestine, where the ileum appears most commonly affected. Occasionally, sub-epithelial oedema, hyperaemia, and mononuclear cell infiltrates have been described in the tongue during histopathological evaluations. The salivary glands may display hyperaemia, thrombosis and mononuclear cell infiltrates in acute forms of ASF. Microscopically, the stomach shows intravascular and perivascular accumulation of mononuclear cells along with hyperaemia and oedema in the lamina propria, submucosa and serosa. These lesions are usually more severe in subacute forms of ASF and are also accompanied by haemorrhages in the mucosa, submucosa and serosa, and the presence of mucosal necrotic ulcers. Animals usually display mild shortening and thickening of the mucosal villi along with mononuclear cell infiltrates in the mucosa and lamina propria in the small intestine (duodenum and jejunum). Serosal and mucosal hyperaemia and small haemorrhages may be present. In the distal ileum there is mononuclear cell infiltration in the mucosa and lamina propria, submucosal diffuse hyperaemia, endothelial activation and vasculitis characterised by intramural inflammatory infiltrates in small and medium vessels. Often, there is mild diffuse lymphocytic depletion of Peyer's patches and interfollicular areas with lymphocyte pyknosis and cell fragmentation (Figure 4.15A). The same lesions are usually described in the ileocaecal valve, although hyperaemia, lymphocytic depletion and cellular fragmentation are typically more extensive and severe (Figure 4.15C, 4.15E). Histopathological lesions in the large intestine (colon and cecum) are characterised by the presence of mild, diffuse hyperaemia with occasional haemorrhages in the epithelium, presence of mild, multifocal mononuclear infiltrates in the lamina propria and severe hyperaemia and oedema in the submucosa (Figure 4.15G). Intravascular and perivascular mononuclear cell infiltrates in the submucosa showing pyknosis and cell fragmentation have been described. In the colon, isolated lymphoid follicles show lymphocytic depletion with pyknosis and cell fragmentation (Moulton and Coggins, 1968; P. Sánchez-Cordón, unpublished data).

In the small intestine, viral antigen is mainly observed in macrophages and, occasionally, in lymphocytes. Immunolabelled cells usually occur in infiltrates in the lamina propria, interfollicular areas and peripheral areas of lymphoid follicles of the distal ileum and ileocaecal valve. In both intestinal regions, the presence of immunolabelled cells within lymphoid follicles is scarce (Figure 4.15B, 4.15D, 4.15F). Viral antigen is also detected in capillary endothelial cells of the lamina propria and submucosa. In the large intestine, immunolabelled mononuclear infiltrates, mainly constituted by macrophages, are frequent in lamina propria and submucosa of the colon (Figure 4.15H), while immunolabelled cells within the lymphoid follicles are scarce. Capillary endothelial cells of the lamina propria and submucosa also appear positive for viral antigen by immunohistochemistry (P. Sánchez-Cordón, unpublished data).

Ultrastructural studies on intestinal samples from pigs experimentally inoculated with highly and moderately virulent ASFV isolates, as well as from naturally infected pigs, describe the presence of

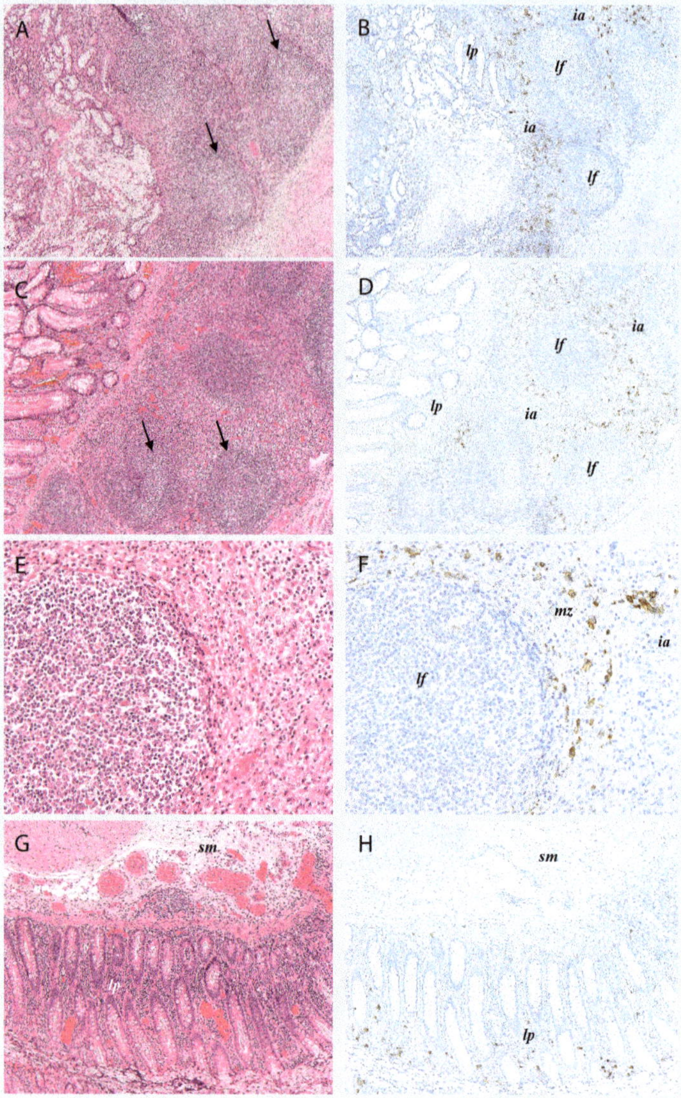

Figure 4.15. Gastrointestinal tract. Serial sections of distal ileum (A-B, 10×), ileocaecal valve (C-D, 10×; E-F, 20×) and colon (G-H, 10×) from pigs with subacute ASF. HE stain and IHC. Observe the mild lymphocytic depletion affecting the lymphoid follicles of the Peyer's patches (arrows) and interfollicular areas of the distal ileum and ileocaecal valve (A and C) along with diffuse hyperaemia in lamina propria and submucosa in the ileocaecal valve (C); Immunolabelled cells against viral antigen usually infiltrate the lamina propria (*lp*), interfollicular areas (*ia*) and peripheral areas of lymphoid follicles (*lf*) of the distal ileum (B) and ileocaecal valve (D); (E-F) Detail of a lymphoid follicle (*lf*) in the ileocaecal valve showing mild lymphocytic depletion and presence of cells, mainly macrophages, immunolabelled for viral antigen in the marginal zone (*mz*) and interfollicular area (*ia*); (G-H) Moderate mononuclear infiltrates in the lamina propria (*lp*) and severe hyperaemia and oedema in the submucosa of colon (*sm*). Observe the presence of cells immunolabelled positively for viral antigen mainly in lamina propria. Haematoxylin-eosin staining (HE); immunohistochemistry against P30 protein (IHC); original magnification (number ×).

amniotic fluid, and foetal heart blood. Although viral antigen was mainly detected in maternal tissues, specific immunoreactions against viral antigen were present in placental tissues in only a few sows. In the same study, histopathological lesions in foetal tissues were inconsistent and mild, including focal placentitis, interstitial pneumonia as well as hepatic degeneration and necrosis. In marked contrast to these changes in the foetal tissues, maternal lymphoid tissues had high titres of virus with marked necrosis of lymphoid tissues and many cells with ASF viral antigen (Schlafer and Mebus, 1987).

4.5.11 Endocrine system

Haemorrhages, occasional mononuclear cell accumulations and a few thrombosed vessels have sometimes been observed in the inner cortical and medullary regions of the adrenal glands. The presence of interstitial hyperaemia and thrombosis of capillaries has been described in the pituitary gland. Histopathological lesions in the pancreas are characterised by interstitial hyperaemia, haemorrhages and parenchymal necrosis. Haemorrhagic lesions are usually more frequent and severe in animals that develop subacute forms of ASF (Moulton and Coggins, 1968).

4.5.12 Central nervous system

The central nervous system (CNS) has been poorly studied in pigs infected with ASFV. Therefore, the nature and distribution of lesions in different areas of the CNS, possible differences in lesion patterns induced by isolates of different virulence, the pathogenesis of lesions and ASFV target cells remain unclear.

Inflammatory lesions have been described in almost all areas of the CNS in animals that developed acute, subacute and chronic forms of ASF. In animals that developed subacute forms of ASF induced by moderately virulent isolates, histopathological lesions in CNS are more severe and generalised than those described in pigs infected with highly virulent isolates. Animals show non-purulent meningoencephalomyelitis characterised by the presence of inflammatory cell infiltrates in the meninges of most brain areas and spinal cord constituted by lymphoplasmacytic cells, macrophages and occasional eosinophils. Pyknotic cells and cell debris are frequently observed within the infiltrates. The presence of small- and medium-sized blood vessels in the meninges displaying endothelial activation and vasculitis with intramural inflammatory infiltrates have been described (Figure 4.19A, 4.19C). Haemorrhages, oedema and fibrin deposits may be present in the meninges. Perivascular cuffs of different size have been described in almost all areas of the brain and spinal cord of infected animals. Perivascular cuffs are usually composed of lymphoplasmacytic cells, macrophages, occasional eosinophils and cell debris with pyknotic cells. Vessels in the centre of the cuffs may show endothelial cell activation and vasculitis (Figure 4.19E). Haemorrhages in the Virchow-Robins space may be observed. In neuroparenchymal areas, the presence of oedema, neuronal necrosis and astrocytic infiltrates have been described. Inflammatory infiltrates in the choroid plexus along with the presence of haemorrhages, endothelial activation and necrosis of epithelium are common findings in acute and subacute forms of ASF. Viral antigen is detected by immunohistochemistry in circulating monocytes, endothelial cells and occasional macrophage-like cells within the meningeal infiltrates, within the perivascular cuffs and also in the inflammatory infiltrates of the neuroparenchyma (Figure 4.19B, 4.19D, 4.19F) (Moulton and Coggins, 1968; Sánchez-Cordón *et al.*, 2020b; Sehl *et al.*, 2020).

oedema and vacuolisation in endothelial cells with extravasation along with fibrin deposits within vascular lumens. In general, changes were more pronounced in animals that developed subacute forms of ASF. The presence of single virions or virus aggregates free in plasma or attached to erythrocytes has also been observed (Krauthausen *et al.*, 1992).

4.5.4 Liver and gall bladder

In the liver, histopathological changes described consist of diffuse congestion, oedema, multifocal to coalescing haemorrhages and diffuse interstitial infiltrates formed by viable and degenerate macrophages, some lymphoplasmacytic cells, and occasional neutrophils and eosinophils in portal spaces and the interlobular septa (Figure 4.16A-C). Scattered, small foci of mononuclear infiltrates are present in hepatic sinusoids, which may appear congested and expanded. Changes are usually more severe in subacute forms. Sinusoidal leucocytosis (Figure 4.16D), defined as an increased number of circulating cells, particularly monocytes, neutrophils and fewer lymphocytes, along with enlarged Kupffer cells (KC) containing engulfed cell debris are also present. KC and monocytes typically display rounded nuclei and chromatin margination, changes characteristic of cytopathic effect. Activated, hypertrophic KC with large cytoplasmic vacuoles and protrusions into the sinusoids are present (Figure 4.16D). Isolated necrotic hepatocytes or necrotic foci of hepatocytes with mononuclear cell infiltration are common findings (Figure 4.16D). Areas with hepatocyte degeneration are observed adjacent to the central veins, being more common in acute than subacute forms. Endothelial activation is usually accompanied by vasculitis with intramural inflammatory mononuclear infiltrates and mononuclear perivasculitis that affects mainly arteries and veins in portal spaces. Intraluminal fibrin thrombi are occasionally observed. Sinusoidal endothelial cells may also appear swollen. In chronic forms, interlobular foci of lymphoid cells or myeloid cells and interlobular fibrosis have been described (Konno *et al.*, 1971; Moulton and Coggins, 1968; Sánchez-Cordón *et al.*, 2008; Sehl *et al.*, 2020).

In the liver, viral antigen against ASF has been detected from early stages of infection (3-4 dpi) with highly and moderately virulent ASFV isolates. Circulating monocytes and KC within hepatic sinusoids, macrophages and neutrophils within interstitial infiltrates observed in portal spaces and interlobular areas, along with a high number of hepatocytes, are the main cells immunolabelled (Figure 4.16E-F). Immunolabelled endothelial cells are observed mainly in capillaries and small-sized vessels in portal spaces (Figure 4.16E), hepatic sinusoids and, occasionally, in central veins. Ultrastructural studies corroborate viral infection of these cell populations. In addition, viral replication has been observed in fibroblasts and in smooth muscle cells of hepatic arterioles and venules, and fat-storing cells (Gómez-Villamandos *et al.*, 1995a,b; Sierra *et al.*, 1987).

In the gall bladder wall, connective tissue fibres and muscle fibres are usually separated by oedema (Figure 4.16G). Congested and dilated blood vessels are also present. These histopathological lesions are usually more severe in subacute forms where haemorrhages are present and mucosal necrosis observed. Bile duct hyperplasia is a characteristic feature of chronic forms (Konno S, 1971; Moulton and Coggins, 1968; Sánchez-Cordón *et al.*, 2008; Sehl *et al.*, 2020).

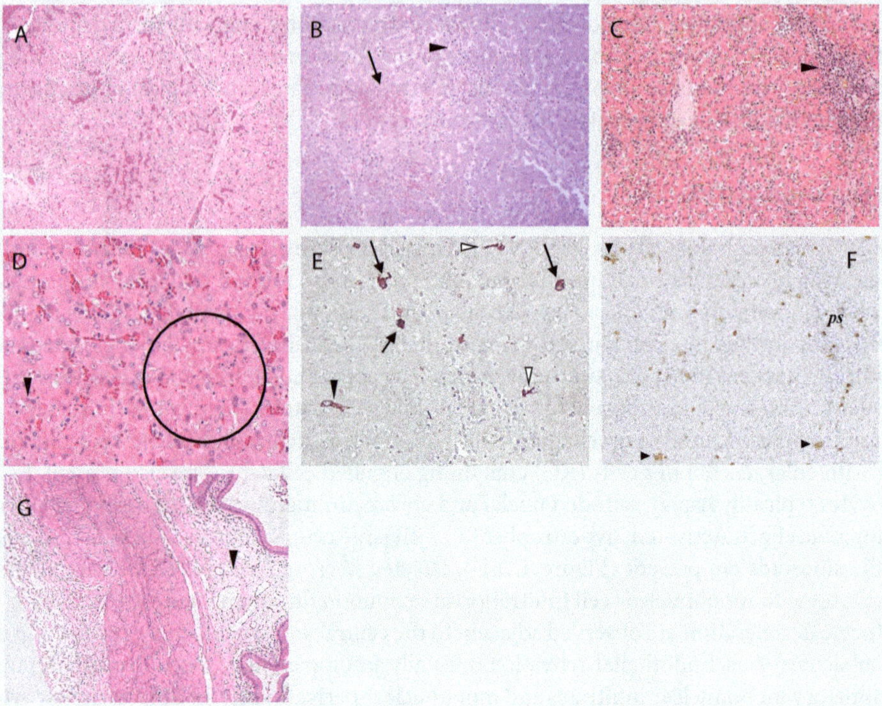

Figure 4.16. Liver and gall bladder. (A-D) Liver from pigs with subacute ASF. HE stain, 10× (A, B, C), 20× (D). (A) Diffuse moderate congestion; (B) Focal haemorrhages (arrow) and interstitial infiltrates (arrowhead); (C) Sinusoidal leucocytosis and interstitial infiltrates (arrowhead) composed of macrophages, some lymphoplasmacytic cells and occasional neutrophils in portal spaces and the interlobular septa; (D) Sinusoidal leucocytosis along with enlarged Kupffer cells (arrowhead) and presence of necrotic hepatocytes (circle); (E) Liver. IHC, 40×. Presence of hepatocytes (arrows), Kupffer cells (black arrowhead) and endothelial cells (white arrowheads) immunolabelled positively for viral antigen; (F) Liver. IHC (serial section from Figure 4.16C), 10×. Observe the presence of circulating monocytes within hepatic sinusoids, macrophages within interstitial infiltrates in portal spaces (*ps*) and the presence of immunolabelled hepatocytes (arrowheads); (G) Gall bladder wall (subacute ASF). HE stain, 10×. Separation of connective tissue fibres (oedema; arrowhead). Haematoxylin-eosin staining (HE); immunohistochemistry against P30 protein (IHC); original magnification (number ×).

4.5.5 Acute phase response in acute African swine fever

Proinflammatory cytokines induce acute phase protein synthesis by hepatocytes, a mechanism in which KC play an intermediary role. Serum amyloid A (SAA) protein and C-reactive protein (CRP) (classified as type 1 acute phase proteins) are predominantly induced by IL-1 and TNF-α, while type 2 acute phase proteins such as haptoglobin (Hp), are mainly induced by IL-6 (Petersen *et al.*, 2004). The characterisation of the acute phase response was carried out in the course of experimental infections with highly virulent ASFV isolates. In this study, high serum concentrations of SAA and CRP coincided with high serum concentrations of IL-1 and expression of IL-1 and TNF-α by KC and other hepatic macrophage populations. The highest

Hp concentrations coincided with the largest number of KC and other hepatic macrophage populations secreting IL-6 (Sánchez-Cordón *et al.*, 2007). Hepatic malfunction along with the anorexia and reduced food/protein intake may contribute to the development of oedema (ascites, hydrothorax and hydropericardium), which may become especially severe in subacute forms of ASF.

4.5.6 Cardiorespiratory system

Histopathological evaluations of the pericardial sac indicate that thickening is due to the presence of hyperaemia, mild to moderate oedema and mononuclear cell infiltrates with few eosinophils and neutrophils. In chronic forms, pericardial sac thickening may become more severe, composed of diffuse, severe mononuclear cell infiltrates and granulation tissue along with an overlying layer of fibrin. The same lesions can be observed in the epicardium. In the heart, hyperaemia and haemorrhages are common histopathological lesions observed among the cardiac muscle fibres in the epicardium, myocardium and endocardium, very extensive and severe in subacute forms of ASF (Figure 4.17A-B). Experimental infections with highly virulent isolates show the presence of small, focal haemorrhages among cardiac muscle fibres, attributed to an increased permeability of vessels at 4 dpi. From 6 dpi, diffuse, severe haemorrhages have been observed among cardiac muscle fibres along with medium- and large-sized vessel damage. At final stages of the disease, myocardial ischaemia may induce the appearance of necrotic cardiac muscle fibres with cytoplasmic hypereosinophilia and loss of striations (Figure 4.17B) (Moulton and Coggins, 1968; Semerjyan *et al.*, 2018).

The lungs of ASF infected pigs display a high number of histopathological lesions, more severe and extensive in animals that develop subacute forms of ASF. In chronic forms, especially in natural cases, pneumonic lesions are frequently associated with secondary opportunistic pathogens. The lungs usually show hyperaemia, congestion and haemorrhages at the perivascular, peribronchial and alveolar levels (Figure 4.17C). Vessels of different sizes display endothelial activation characterised by endothelial cell hypertrophy and rounded nuclei. Vessels may show oedema in the tunica media as well as vasculitis with intramural and perivascular inflammatory infiltrates, necrosis and/or thrombosis (Figure 4.17D). Alveolar and interstitial proteinaceous oedema is a characteristic pulmonary lesion (Figure 4.17C-F). Haemorrhages, cell debris, fibrin deposits, macrophage infiltrates and occasional megakaryocytes are common histopathological findings in the alveolar lumen. Alveolar septal thickening is a consequence of capillary dilation and the presence of inflammatory cells (mainly mononuclear cells) inside pulmonary capillaries (Figure 4.17C-D). Septal thickening and alveolar infiltrates may contribute to the appearance of areas with mild, focal consolidation, with necrotic cells frequently seen in both alveolar septa and alveolar lumen. In subacute forms of ASF where consolidation areas are more extensive and severe, mononuclear cells are increased in number in the alveolar septa and lumens. Many of these cells are undergoing necrosis, particularly within alveolar lumens, but also in the alveolar septa. Eosinophils and occasionally a few neutrophils are found within necrotic areas. Peribronchial and peribronchiolar mononuclear infiltrates are usually described. The bronchi and bronchioles may contain cell debris, fibrin and detached epithelium. In cases in which bronchus-associated lymphoid tissue (BALT) is present, the presence of pyknotic cells, cell debris and macrophage infiltrates is observed (Figure 4.17G) (Carrasco *et al.*, 1996a; Moulton and Coggins, 1968; Sehl *et al.*, 2020). In pneumonic lesions described in animals experimentally infected with

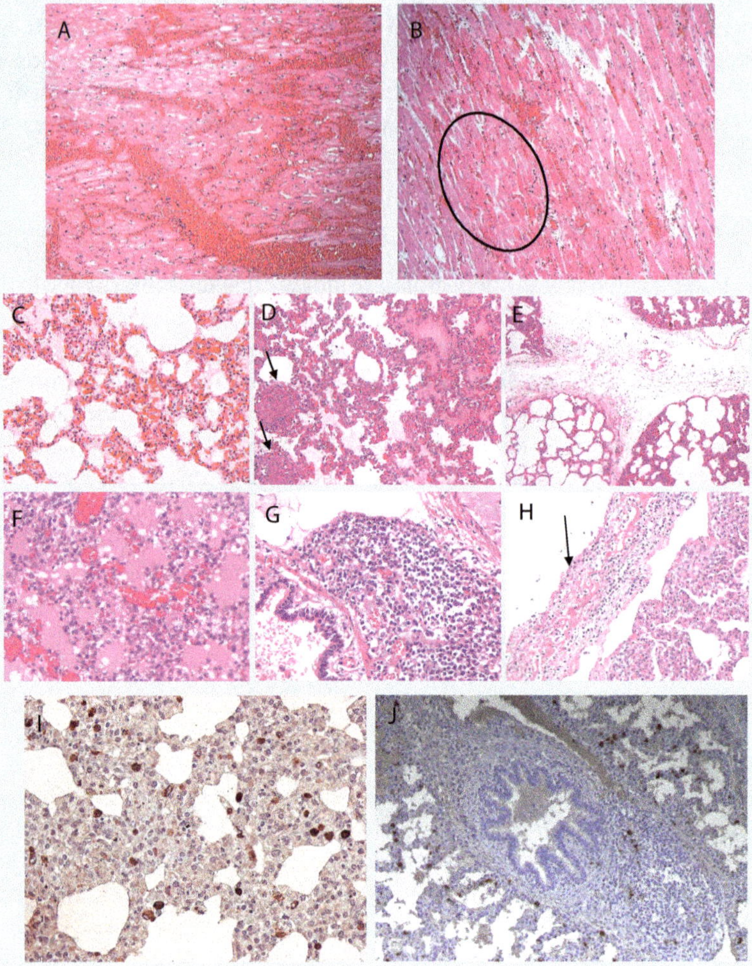

Figure 4.17. Cardiorespiratory system. (A-B) Heart from pigs with subacute African swine fever (ASF). HE stain, 20×. Severe hyperaemia and haemorrhages among the cardiac muscle fibres in myocardium. Observe the presence of areas with necrotic cardiac muscle fibres showing cytoplasmic hypereosinophilia and loss of striations (circle); (C-D) Lung from pigs with acute ASF. HE stain, 10×. Mild to moderate hyperaemia, septal thickening due to capillary dilation and the presence of inflammatory cells and alveolar proteinaceous oedema. Observe the presence of thrombi along with vasculitis and necrosis of vessels (arrows); (E) Lung from a pig with subacute ASF. HE stain, 4×. Severe interstitial oedema with distension of the interlobular septa; (F) Lung from a pig with subacute ASF. HE stain, 10×. Severe alveolar proteinaceous oedema and congestion. Observe the septal thickening and the necrosis of cells in the alveolar septa; (G) Lung from a pig with acute ASF. HE stain, 20×. Bronchus-associated lymphoid tissue (BALT) showing mild lymphocytic depletion, pyknotic cells and cell debris; (H) Pleural thickening (arrow) due to connective tissue and inflammatory cells proliferation in a pig with subacute ASF. HE stain, 20×; (I) Presence of pulmonary intravascular macrophages, interstitial macrophages and endothelial cells immunolabelled for viral antigen in lung samples from a pig with subacute ASF. IHC, 20×. (J) Presence of macrophages immunoreactive against viral antigen infiltrating the BALT and in the alveolar septa. IHC, 10×. Haematoxylin-eosin staining (HE); immunohistochemistry against P30 protein (IHC); original magnification (number ×).

low virulent ASF isolates, severe necrosis is first observed in alveolar lumen monocytes and later in neighbouring structures. Necrotic areas might be surrounded by mononuclear cells, including plasma cells, and may become mineralised and walled off by fibrosis, giving rise to sequestra similar to those observed in other respiratory diseases (Moulton and Coggins, 1968; Moulton *et al.*, 1975).

In the pleura, histopathological findings range from occasional subpleural petechial haemorrhages in acute forms, to pleural or subpleural diffuse petechial haemorrhages, ecchymoses and suggillations along with pleural thickening due to connective tissue proliferation in subacute forms (Figure 4.17H). Fibrin deposits on the pleural surface are also observed (Moulton and Coggins, 1968; Moulton *et al.*, 1975).

Ultrastructural and immunohistochemical studies show pulmonary intravascular macrophages (PIMs) as the main target cells for ASFV replication in the lungs from initial stages after infection (Figure 4.17I), both in acute and subacute forms of ASF (Carrasco *et al.*, 1996a; Carrasco *et al.*, 2002; Sierra *et al.*, 1990). However, viral replication and viral antigen are observed only in a small number of pulmonary alveolar macrophages (PAMs) respectively. While the number of immunoreactive PIMs increases as the disease progresses, the number of PAMs remains constant (Carrasco *et al.*, 1996a; Carrasco *et al.*, 2002). Viral replication and viral antigen immunostaining are observed in interstitial macrophages (Figure 4.17I) as well as in occasional fibroblasts and neutrophils (Carrasco *et al.*, 1996a). The presence of positively immunolabelled pneumocytes indicating viral antigen has been described. Capillary endothelial cells and, to a lesser extent, endothelial cells of larger-sized vessels typically show immunolabelling for viral antigen in late stages of acute and subacute forms. The presence of macrophages immunoreactive against viral antigen infiltrating the BALT is also observed (Figure 4.17J) (Carrasco *et al.*, 2002; Pérez *et al.*, 1994; Rodríguez *et al.*, 1996a; Sehl *et al.*, 2020; Sierra *et al.*, 1990).

4.5.7 Pathogenesis of pulmonary oedema

The image of pigs with foam in the mouth and nose during the agonal period due to severe pulmonary oedema, especially in acute but also in subacute forms of ASF, constitutes one of the most characteristic hallmarks of the disease. From a functional point of view, PAMs are involved in removing particles that have been inhaled and are present on the alveolar surfaces, while PIMs would be involved in removing blood-circulating debris as well as in triggering pulmonary inflammatory mechanisms. Ultrastructural studies suggest that circulating cell debris, likely originated in primary ASFV replication organs that suffer severe tissue damage, induce phagocytic activation of PIMs and morphological changes characterised by the presence of enlarged and rounded PIMs rather than the original spindle-shaped PIMs. In addition, PIMs increase in numbers, likely induced by circulating chemical mediators. These changes, which were described before the detection of ASFV in any pulmonary macrophage population, make blood circulation difficult and increase the vascular pressure, increasing vascular permeability and contributing to the appearance of initial interstitial oedema. Both the increase in vascular permeability and the interstitial oedema further contribute to separating the alveolar epithelium from the capillary endothelium, giving rise to the initial alveolar oedema. As disease progresses, viral replication is observed in pulmonary macrophages, especially in PIMs. As a result, infected but also non-infected PIMs show increase in size and biosynthetic activation, changes that influence the physiology of pulmonary capillaries through the release of pro-inflammatory mediators such as arachidonic acid metabolites and cytokines. These chemical mediators display marked chemotactic activity, being able to attract and activate inflammatory cells, and thus playing an important role in the mechanisms leading to increased endothelial permeability. In fact, the most severe pulmonary lesions including the presence of cell infiltrates, microthrombi in septal capillaries as well as severe interstitial and alveolar oedema, coincide with the presence of a high number of infected and non-infected PIMs secreting IL-1α and TNF-α. Even though viral replication is also observed in PAMs, the number of PAMs showing biosynthetic activation does not change significantly after infection when more severe vascular changes are described, so that their role in the pathogenesis of pulmonary oedema is considered to be minor (Carrasco *et al.*, 1996a, 2002).

4.5.8 Urinary system

The presence of lesions in the kidneys, especially severe and extensive in subacute forms induced by moderately virulent ASFV isolates, is one of the most representative features of animals with ASF. In acute forms, animals usually display interstitial congestion and oedema from the initial stages after infection in the renal cortex. As disease progresses, lesions become more severe and extensive, and multifocal haemorrhages and interstitial inflammatory mononuclear infiltrates, mainly constituted by macrophages, some lymphocytes and occasional neutrophils and eosinophils appear (Figure 4.18A-B). Pyknotic cells and cell fragmentation are described within inflammatory infiltrates. Endothelial activation and necrotising vasculitis of interstitial renal capillaries and arterioles along with the presence of microthrombi are common histopathological lesions described in late stages of acute forms (Figure 4.18C-D). In subacute forms, there is a delay in the appearance of interstitial lesions. As disease progresses, haemorrhages become more severe than described in acute forms and the interstitial infiltrates may be constituted by lymphohistiocytic and plasma cells (Figure 4.18I-J).

In acute forms, glomerular hyperaemia and haemorrhages as well as enlarged hypercellular glomeruli (glomerulonephritis), endothelial activation, necrosis of glomerular elements, intraglomerular fibrin deposits and hyaline casts are observed (Figure 4.18E-G). Bowman's capsules and tubules often contain protein-rich fluid due to leakage from damaged glomeruli. Segmental tubular necrosis with presence of tubular hyaline casts is a common finding (Figure 4.18H). In subacute forms, glomerulonephritis characterised by hyalinised, thickened, PAS-positive glomerular basement membranes is observed. The medulla may display interstitial diffuse congestion, oedema, multifocal haemorrhages and mononuclear infiltrates that are usually less severe than those observed in the interstitium (Gómez-Villamandos *et al.*, 1995d; Hervás *et al.*, 1996; Moulton and Coggins, 1968).

Replication of virulent ASFV isolates has been described by ultrastructural methods in circulating monocytes and interstitial macrophages at initial stages of disease, with later involvement of endothelial cells of interstitial capillaries, glomerular mesangial cells and epithelium of collector renal ducts. Virus replication has further been observed in fibroblasts and in the smooth muscle cells of arterioles and venules. No evidence of viral replication has been reported in any interstitial cell infiltrates during subacute forms (Gómez-Villamandos *et al.*, 1995c,d; Hervás *et al.*, 1996; Sierra *et al.*, 1989). From initial stages of the disease, the detection of viral antigen

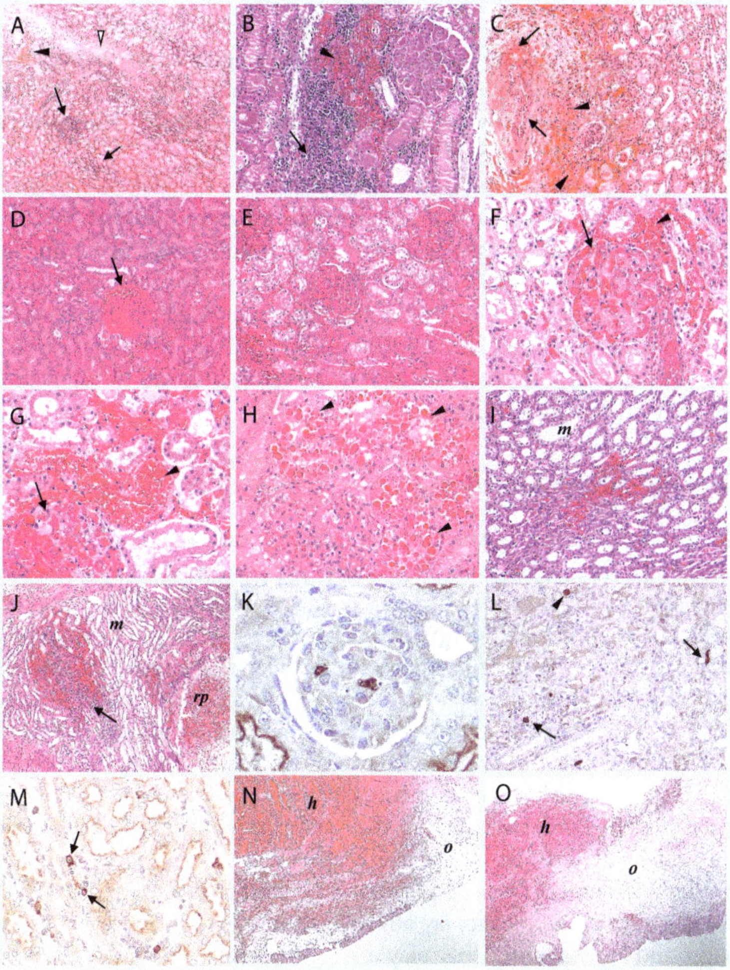

Figure 4.18. Urinary system. (A-B) Kidney from pigs with acute ASF. HE stain, 10× (A), 20× (B). Diffuse interstitial congestion, oedema (white arrowhead), haemorrhages (black arrowheads) and interstitial inflammatory mononuclear infiltrates (arrows), mainly constituted by macrophages and lymphocytes, in renal cortex; (C) Kidney from a pig with acute ASF. HE stain, 10×. Severe interstitial congestion, haemorrhages (arrowheads) and vasculitis (arrows) in renal cortex; (D) Kidney, acute ASF. HE stain, 10×. Interstitial congestion and presence of microthrombus (arrow) in renal cortex; (E) Kidney, acute ASF. HE, stain 10×. Glomerular hyperaemia and enlarged hypercellular glomeruli; (F-G) Kidney, acute ASF. HE stain, 20×. Interstitial haemorrhages (arrowhead), intraglomerular fibrin deposits and hyaline casts (arrow); (H) Kidney, acute ASF. HE stain, 20×. Tubular necrosis along with tubules containing protein-rich fluid and hyaline casts (arrowheads); (I-J) Kidney from pigs with subacute ASF. HE stain, 10×. Interstitial haemorrhages in medulla (*m*) and renal pelvis (*rp*) along with mononuclear interstitial infiltrates (arrow); (K-M) Kidney. IHC, 20× (K), 10× (L, M). Detection of viral antigen by immunohistochemistry in capillary endothelial cells within the glomeruli (K), interstitial macrophages (arrows) and circulating monocytes (arrowhead) in the renal cortex (L) and epithelial cells (arrows) of renal ducts (M); (N-O) Urinary bladder from pigs with subacute ASF. HE, 4×. Diffuse haemorrhages (*h*) and oedema (*o*) among connective tissue fibres in submucosa. Haematoxylin-eosin staining (HE); immunohistochemistry against P30 protein (IHC); original magnification (number ×).

by immunohistochemistry has been described mainly in circulating monocytes and interstitial macrophages in the renal cortex and medulla. As disease progressed, viral antigen has also been described in endothelial cells of interstitial vessels and, occasionally, capillary endothelial cells within the glomeruli, mesangial cells and epithelial cells of renal ducts (Figure 4.18K-M) (Pérez *et al.*, 1994; Rodríguez *et al.*, 1996a; Sehl *et al.*, 2020).

Histopathological descriptions of lesions in the urinary bladder from animals infected with ASFV are scarce. In acute and subacute forms, the urinary bladder shows hyperaemia and haemorrhages in the mucosa, submucosa and serosa, lesions that are much more severe in animals infected with moderately virulent isolates. Oedema in submucosal connective tissue is also a frequent change observed in subacute forms (Figure 4.18N-O). Small-sized vessels may display endothelial activation, vasculitis and thrombosis. Viral antigen is mainly detected in circulating monocytes and endothelial cells (Moulton and Coggins, 1968).

4.5.9 Pathogenesis of haemorrhages

Regarding the pathogenic mechanisms of haemorrhagic lesions in the kidneys, some studies suggest that highly virulent ASFV isolates produces renal haemorrhages as a result of intense endothelial damage. This is influenced by the phagocytic activation that occurs in capillary endothelial cells to ensure that cell debris from elsewhere is removed from blood circulation. This mechanism has been suggested for explaining haemorrhages in other organs without a fixed vascular macrophage population such as the renal and gastrohepatic lymph nodes. Endothelial activation and hypertrophy coincided with virus replication and activation in nearby macrophages, which suggests that some cytokines released may induce their activation. Consequently, a loss of endothelial cells and disorganisation of basement membranes occurs, contributing to erythrocytes and cell debris moving into the interstitium and causing haemorrhages. In addition, the exposure of the capillary basement membrane induces platelet aggregation and triggers disseminated intravascular coagulation. Thus, virus replication in endothelial cells, only detected in the last stage of the disease once haemorrhages are already present, do not seem to be the primary cause of haemorrhages and consequently only play a secondary role for the occurrence of this sign. However, in the case of moderately virulent ASFV isolates, haemorrhages are not the consequence of endothelial cell disruption or intravascular coagulation phenomena, but of vasodilation and increased vascular permeability. These changes lead to the appearance of haemorrhages, interstitial oedema and perivascular infiltrates formed by macrophages, lymphocytes and plasma cells due to diapedesis. The transmigration of cells from capillaries may be mediated by activated macrophages, immune mechanisms including the immune-complex deposits, or both (Carrasco *et al.*, 1997b; Gómez-Villamandos *et al.*, 1995c, 2013).

4.5.10 Reproductive system

Information regarding lesions and ASFV distribution in the reproductive system is scarce. In males, interstitial hyperaemia, haemorrhage and thrombosis of the epididymis is described in pigs with acute forms of disease (Moulton and Coggins, 1968). On the other hand, pregnancy failure seems to result from the direct effects of the infection on the sow rather than from viral damage to the placenta or foetus. In pregnant sows experimentally infected with moderately virulent ASFV isolates, the foetal tissues most frequently yielding virus were the foetal placenta,

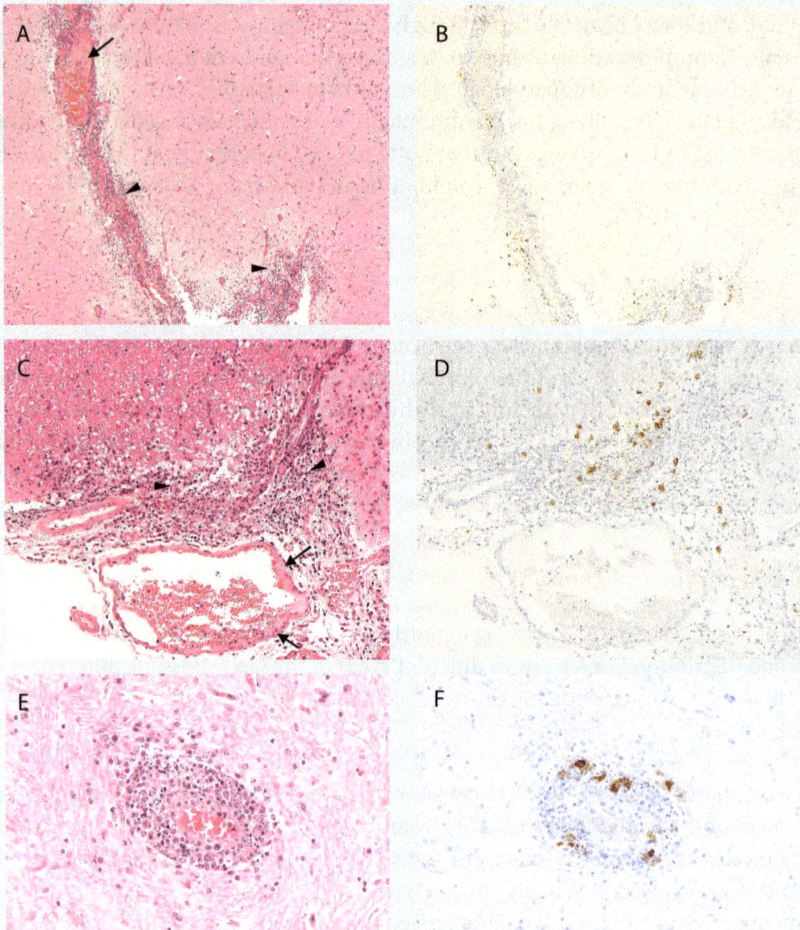

Figure 4.19. Central nervous system. (A-B) Brain. Serial sections of cortex from a pig with subacute ASF. HE and IHC, 4×. Congested vessels (arrow) and severe mononuclear cell infiltrates (arrowheads) in the meninges (non-purulent meningitis). Observe the presence of cells, mainly macrophages, immunolabelled positively for viral antigen within the meningeal infiltrates; (C-D) Brain. Serial sections of cortex from a pig with subacute ASF. HE and IHC, 10×. Haemorrhages, pyknotic cells and cell debris within the meningeal infiltrates (arrowheads) and vasculitis (arrows). The presence of immunolabelled macrophages in meningeal infiltrates can also be observed; (E-F) Brain. Serial sections of caudal medulla from a pig with subacute ASF. HE and IHC, 20×. Perivascular cuff composed of lymphoplasmacytic cells, macrophages and pyknotic cells. The vessel in the centre of the cuff shows endothelial cell activation. Viral antigen is observed mainly in macrophages within the perivascular cuff. Haematoxylin-eosin staining (HE); immunohistochemistry against P30 protein (IHC); original magnification (number ×).

Acknowledgements

This publication is based upon work from COST Action ASF-STOP CA15116, supported by COST (European Cooperation in Science and Technology), https://www.cost.eu. PJSC would like to dedicate this book chapter to the four generations of veterinary pathologists from the Veterinary Faculty at University of Córdoba (Spain) that from the early '80s worked on ASF. Especially, he would like to thank Prof. Sierra, his scientific mentors (Prof. Gómez-Villamandos and Prof. Carrasco) and his colleagues Dr Ruiz-Villamor, Dr Salguero and Dr Núñez. '*The triumph of one is the triumph of all*'. Macroscopic and microscopic pictures have been obtained from different experiments carried out at The Pirbright Institute (UK), APHA (UK), CISA-INIA (Spain) and CRESA (Spain).

References

Achenbach, J.E., Gallardo, C., Nieto-Pelegrín, E., Rivera-Arroyo, B., Degefa-Negi, T., Arias, M., Jenberie, S., Mulisa, D.D., Gizaw, D., Gelaye, E., Chibssa, T.R., Belaye, A., Loitsch, A., Forsa, M., Yami, M., Diallo, A., Soler, A., Lamien, C.E. and Sánchez-Vizcaíno, J.M., 2017. Identification of a new genotype of African swine fever virus in domestic pigs from Ethiopia. Transboundary and Emerging Diseases 64: 1393-1404. https://doi.org/10.1111/tbed.12511

Alcamí, A., Carrascosa, A.L. and Viñuela, E., 1989. The entry of African swine fever virus into Vero cells. Virology 171: 68-75. https://doi.org/10.1016/0042-6822(89)90511-4

Anderson, E.C., Williams, S.M., Fisher-Hoch, S.P. and Wilkinson, P.J., 1987. Arachidonic acid metabolites in the pathophysiology of thrombocytopenia and haemorrhage in acute African swine fever. Research in Veterinary Science 42: 387-394.

Bautista, M.J., Gómez-Villamandos, J.C., Carrasco, L., Ruíz-Villamor, E., Salguero, F.J. and Sierra, M.A., 1998. Ultrastructural pathology of the bone marrow in pigs inoculated with a moderately virulent strain (DR'78) of African swine fever virus. Histology and Histopathology 13: 713-720. https://doi.org/10.14670/hh-13.713

Blome, S., Gabriel, C. and Beer, M., 2013. Pathogenesis of African swine fever in domestic pigs and European wild boar. Virus Research 173: 122-130. https://doi.org/10.1016/j.virusres.2012.10.026

Boinas, F.S., Hutchings, G.H., Dixon, L.K. and Wilkinson, P.J., 2004. Characterization of pathogenic and non-pathogenic African swine fever virus isolates from *Ornithodoros erraticus* inhabiting pig premises in Portugal. Journal of General Virology 85: 2177-2187. https://doi.org/10.1099/vir.0.80058-0

Borca, M.V., Kutish, G.F., Afonso, C.L., Irusta, P., Carrillo, C., Brun, A., Sussman, M. and Rock, D.L., 1994. An African swine fever virus gene with similarity to the T-lymphocyte surface antigen CD2 mediates hemadsorption. Virology 199: 463-468. https://doi.org/10.1006/viro.1994.1146

Carrasco, L., Bautista, M.J., Gomez-Villamandos, J.C., Martin de las Mulas, J., Chacon, M.d.L.F., Wilkinson, P.J. and Sierra, M.A., 1997a. Development of microscopic lesions in splenic cords of pigs infected with African swine fever virus. Veterinary Research 28: 93-99.

Carrasco, L., Bautista, M.J., Martin de las Mulas, J., Gomez-Villamandos, J.C., Espinosa de los Monteros, A. and Sierra, M.A., 1995. Description of a new population of fixed macrophages in the splenic cords of pigs. Journal of Anatomy 187 (Part 2): 395-402.

Carrasco, L., Chàcón, M.d.L.F., Martín de Las Mulas, J., Gómez-Villamandos, J.C., Sierra, M.A., Villeda, C.J. and Wilkinson, P.J., 1997b. Ultrastructural changes related to the lymph node haemorrhages in acute African swine fever. Research in Veterinary Science 62: 199-204. https://doi.org/10.1016/s0034-5288(97)90190-9

Carrasco, L., de Lara, F.C., Gómez-Villamandos, J.C., Bautista, M.J., Villeda, C.J., Wilkinson, P.J. and Sierra, M.A., 1996a. The pathogenic role of pulmonary intravascular macrophages in acute African swine fever. Research in Veterinary Science 61: 193-198. https://doi.org/10.1016/s0034-5288(96)90062-4

Carrasco, L., de Lara, F.C., Martín de las Mulas, J., Gómez-Villamandos, J.C., Hervás, J., Wilkinson, P.J. and Sierra, M.A., 1996b. Virus association with lymphocytes in acute African swine fever. Veterinary Research 27: 305-312.

Carrasco, L., de Lara, F.C., Martín de las Mulas, J., Gómez-Villamandos, J.C., Pérez, J., Wilkinson, P.J. and Sierra, M.A., 1996c. Apoptosis in lymph nodes in acute African swine fever. Journal of Comparative Pathology 115: 415-428.

Carrasco, L., Gómez-Villamandos, J.C., Bautista, M.J., Martín de las Mulas, J., Villeda, C.J., Wilkinson, P.J. and Sierra, M.A., 1996d. *In vivo* replication of African swine fever virus (Malawi '83) in neutrophils. Veterinary Research 27: 55-62.

Carrasco, L., Núñez, A., Salguero, F.J., Díaz San Segundo, F., Sánchez-Cordón, P., Gómez-Villamandos, J.C. and Sierra, M.A., 2002. African swine fever: expression of interleukin-1 alpha and tumour necrosis factor-alpha by pulmonary intravascular macrophages. Journal of Comparative Pathology 126: 194-201. https://doi.org/10.1053/jcpa.2001.0543

Colgrove, G.S., Haelterman, E.O. and Coggins, L., 1969. Pathogenesis of African swine fever in young pigs. American Journal of Veterinary Research 30: 1343-1359.

De Carvalho Ferreira, H.C., Weesendorp, E., Elbers, A.R., Bouma, A., Quak, S., Stegeman, J.A. and Loeffen, W.L., 2012. African swine fever virus excretion patterns in persistently infected animals: a quantitative approach. Veterinary Microbiology 160: 327-340. https://doi.org/10.1016/j.vetmic.2012.06.025

Dixon, L.K., Sánchez-Cordón, P.J., Galindo, I. and Alonso, C., 2017. Investigations of pro- and anti-apoptotic factors affecting African swine fever virus replication and pathogenesis. Viruses 9: 241. https://doi.org/10.3390/v9090241

Edwards, J.F., Dodds, W.J. and Slauson, D.O., 1985. Megakaryocytic infection and thrombocytopenia in African swine fever. Veterinary Pathology 22: 171-176. https://doi.org/10.1177/030098588502200212

Fernández de Marco, M., Salguero, F.J., Bautista, M.J., Núñez, A., Sánchez-Cordón, P.J. and Gómez-Villamandos, J.C., 2007. An immunohistochemical study of the tonsils in pigs with acute African swine fever virus infection. Research in Veterinary Science 83: 198-203. https://doi.org/10.1016/j.rvsc.2006.11.011

Fishbourne, E., Hutet, E., Abrams, C., Cariolet, R., Le Potier, M.F., Takamatsu, H.H. and Dixon, L.K., 2013. Increase in chemokines CXCL10 and CCL2 in blood from pigs infected with high compared to low virulence African swine fever virus isolates. Veterinary Research 44: 87. https://doi.org/10.1186/1297-9716-44-87

Franzoni, G., Graham, S.P., Sanna, G., Angioi, P., Fiori, M.S., Anfossi, A., Amadori, M., Dei Giudici, S. and Oggiano, A., 2018. Interaction of porcine monocyte-derived dendritic cells with African swine fever viruses of diverse virulence. Veterinary Microbiology 216: 190-197. https://doi.org/10.1016/j.vetmic.2018.02.021

Galindo-Cardiel, I., Ballester, M., Solanes, D., Nofrarías, M., López-Soria, S., Argilaguet, J.M., Lacasta, A., Accensi, F., Rodríguez, F. and Segalés, J., 2013. Standardization of pathological investigations in the framework of experimental ASFV infections. Virus Research 173: 180-190. https://doi.org/10.1016/j.virusres.2012.12.018

Galindo, I. and Alonso, C., 2017. African swine fever virus: a review. Viruses 9: 103. https://doi.org/10.3390/v9050103

Gallardo, C., Nieto, R., Mur, L., Soler, A., Pelayo, V., Bishop, R., Sánchez-Cordón, P.J., Martins, C., Sánchez-Vizcaíno, J.M. and Arias, M., 2012. African swine fever (ASF) in Africa. The role of the African indigenous pigs in the transmission of the disease. In: Proceedings of the 6th Annual Meeting Epizone Brighton, United Kingdom, p. 15.

Gallardo, C., Nurmoja, I., Soler, A., Delicado, V., Simón, A., Martin, E., Perez, C., Nieto, R. and Arias, M., 2018. Evolution in Europe of African swine fever genotype II viruses from highly to moderately virulent. Veterinary Microbiology 219: 70-79. https://doi.org/10.1016/j.vetmic.2018.04.001

Gallardo, C., Soler, A., Nieto, R., Sánchez, M.A., Martins, C., Pelayo, V., Carrascosa, A., Revilla, Y., Simón, A., Briones, V., Sánchez-Vizcaíno, J.M. and Arias, M., 2015a. Experimental transmission of African swine fever (ASF) low virulent isolate NH/P68 by surviving pigs. Transboundary and Emerging Diseases 62: 612-622. https://doi.org/10.1111/tbed.12431

Gallardo, C., Soler, A., Rodze, I., Nieto, R., Cano-Gómez, C., Fernandez-Pinero, J. and Arias, M., 2019. Attenuated and non-haemadsorbing (non-HAD) genotype II African swine fever virus (ASFV) isolated in Europe, Latvia 2017. Transboundary and Emerging Diseases 66: 1399-1404. https://doi.org/10.1111/tbed.13132

Gallardo, M.C., Reoyo, A.T., Fernández-Pinero, J., Iglesias, I., Muñoz, M.J. and Arias, M.L., 2015b. African swine fever: a global view of the current challenge. Porcine Health Management 1: 21. https://doi.org/10.1186/s40813-015-0013-y

Gómez-Villamandos, J.C., Bautista, M.J., Carrasco, L., Caballero, M.J., Hervás, J., Villeda, C.J., Wilkinson, P.J. and Sierra, M.A., 1997a. African swine fever virus infection of bone marrow: lesions and pathogenesis. Veterinary Pathology 34: 97-107. https://doi.org/10.1177/030098589703400202

Gómez-Villamandos, J.C., Bautista, M.J., Carrasco, L., Chacón-Manrique de Lara, F., Hervás, J., Wilkinson, P.J. and Sierra, M.A., 1998. Thrombocytopenia associated with apoptotic megakaryocytes in a viral haemorrhagic syndrome induced by a moderately virulent strain of African swine fever virus. Journal of Comparative Pathology 118: 1-13. https://doi.org/10.1016/s0021-9975(98)80023-6

Gómez-Villamandos, J.C., Bautista, M.J., Hervás, J., Carrasco, L., de Lara, F.C., Pérez, J., Wilkinson, P.J. and Sierra, M.A., 1996. Subcellular changes in platelets in acute and subacute African swine fever. Journal of Comparative Pathology 115: 327-341. https://doi.org/10.1016/s0021-9975(96)80069-7

Gómez-Villamandos, J.C., Bautista, M.J., Sánchez-Cordón, P.J. and Carrasco, L., 2013. Pathology of African swine fever: the role of monocyte-macrophage. Virus Research 173: 140-149. https://doi.org/10.1016/j.virusres.2013.01.017

Gomez-Villamandos, J.C., Carrasco, L.B., Bautista, M.J. and Sierra, M.A., 1999. Pathogenesis of African swine fever. The role of monokines. Recent Research Developments in Virology 1: 7-17.

Gómez-Villamandos, J.C., Hervás, J., Méndez, A., Carrasco, L., Martín de las Mulas, J., Villeda, C.J., Wilkinson, P.J. and Sierra, M.A., 1995a. Experimental African swine fever: apoptosis of lymphocytes and virus replication in other cells. Journal of General Virology 76 (Part 9): 2399-2405. https://doi.org/10.1099/0022-1317-76-9-2399

Gómez-Villamandos, J.C., Hervás, J., Méndez, A., Carrasco, L., Villeda, C.J., Sierra, M.A. and Wilkonson, P.J., 1995b. A pathological study of the perisinusoidal unit of the liver in acute African swine fever. Research in Veterinary Science 59: 146-151. https://doi.org/10.1016/0034-5288(95)90049-7

Gómez-Villamandos, J.C., Hervás, J., Méndez, A., Carrasco, L., Villeda, C.J., Wilkinson, P.J. and Sierra, M.A., 1995c. Pathological changes in the renal interstitial capillaries of pigs inoculated with two different strains of African swine fever virus. Journal of Comparative Pathology 112: 283-298. https://doi.org/10.1016/s0021-9975(05)80081-7

Gómez-Villamandos, J.C., Hervás, J., Méndez, A., Carrasco, L., Villeda, C.J., Wilkinson, P.J. and Sierra, M.A., 1995d. Ultrastructural study of the renal tubular system in acute experimental African swine fever: virus replication in glomerular mesangial cells and in the collecting ducts. Archives in Virology 140: 581-589. https://doi.org/10.1007/bf01718433

Gómez-Villamandos, J.C., J. Hervás, C., Moreno, L., Carrasco, M.J., Bautista, J.M., Caballero, P.J., Wilkinson, P.J. and Sierra, M.A., 1997b. Subcellular changes in the tonsils of pigs infected with acute African swine fever virus. Veterinary Research 28: 179-189.

Gregg, D.A., Mebus, C.A. and Schlafer, D.H., 1995. Early infection of interdigitating dendritic cells in the pig lymph node with African swine fever viruses of high and low virulence: immunohistochemical and ultrastructural studies. Journal of Veterinary Diagnostic Investigation 7: 23-30. https://doi.org/10.1177/104063879500700104

Greig, A., 1972. Pathogenesis of African swine fever in pigs naturally exposed to the disease. Journal of Comparative Pathology 82: 73-79. https://doi.org/10.1016/0021-9975(72)90028-x

Hervás, J., Gómez-Villamandos, J.C., Méndez, A., Carrasco, L., Pérez, J., Wilkinson, P.J. and Sierra, M.A., 1996. Structural and ultrastructural study of glomerular changes in African swine fever. Journal of Comparative Pathology 115: 61-75. https://doi.org/10.1016/s0021-9975(96)80028-4

Heuschele, W.P., 1967. Studies on the pathogenesis of African swine fever. I. Quantitative studies on the sequential development of virus in pig tissues. Archiv für die Gesamte Virusforschung 21: 349-356. https://doi.org/10.1007/bf01241735

Howey, E.B., O'Donnell, V., de Carvalho Ferreira, H.C., Borca, M.V. and Arzt, J., 2013. Pathogenesis of highly virulent African swine fever virus in domestic pigs exposed via intraoropharyngeal, intranasopharyngeal, and intramuscular inoculation, and by direct contact with infected pigs. Virus Research 178: 328-339. https://doi.org/10.1016/j.virusres.2013.09.024

Kleiboeker, S.B., 2002. Swine fever: classical swine fever and African swine fever. Veterinary Clinics of North America: Food Animal Practice 18: 431-451. https://doi.org/10.1016/s0749-0720(02)00028-2

Konno S, T.W., Hess WR, Heuschele WP, 1971. Liver pathology in African swine fever. Cornell Veterinarian 61: 125-150.

Krauthausen, E., Drommer, W., Sierra, M.A. and Jover, A., 1992. [Light and electron microscopic findings in the intestine of spontaneous and experimentally-produced African swine fever]. Deutsche Tierarztliche Wochenschrift 99: 54-59.

Leitão, A., Cartaxeiro, C., Coelho, R., Cruz, B., Parkhouse, R.M.E., Portugal, F.C., Vigário, J.D. and Martins, C.L.V., 2001. The non-haemadsorbing African swine fever virus isolate ASFV/NH/P68 provides a model for defining the protective anti-virus immune response. Journal of General Virology 82: 513-523. https://doi.org/10.1099/0022-1317-82-3-513

Moulton, J. and Coggins, L., 1968. Comparison of lesions in acute and chronic African swine fever. Cornell Veterinarian 58: 364-388.

Moulton, J.E., Pan, I.C., Hess, W.R., DeBoer, C.J. and Tessler, J., 1975. Pathologic features of chronic pneumonia in pigs with experimentally induced African swine fever. American Journal of Veterinary Research 36: 27-32.

Mozos, E., Herraez, P., Perez, J., Fernandez, A., Blanco, A., Martin, M.P. and Jover, A., 2003. Cutaneous lesions in experimental acute and subacute African swine fever: an immunohistopathological and ultrastructural study. Deutsche Tierarztliche Wochenschrift 110: 150-154.

Muñoz-Moreno, R., Galindo, I., Cuesta-Geijo, M., Barrado-Gil, L. and Alonso, C., 2015. Host cell targets for African swine fever virus. Virus Research 209: 118-127. https://doi.org/10.1016/j.virusres.2015.05.026

Oura, C.A., Powell, P.P., Anderson, E. and Parkhouse, R.M., 1998a. The pathogenesis of African swine fever in the resistant bushpig. Journal of General Virology 79: 1439-1443. https://doi.org/10.1099/0022-1317-79-6-1439

Oura, C.A., Powell, P.P. and Parkhouse, R.M., 1998b. African swine fever: a disease characterized by apoptosis. Journal of General Virology 79: 1427-1438. https://doi.org/10.1099/0022-1317-79-6-1427

Pan, I.C. and Hess, W.R., 1984. Virulence in African swine fever: its measurement and implications. American Journal of Veterinary Research 45: 361-366.

Penrith, M.L. and Vosloo, W., 2009. Review of African swine fever: transmission, spread and control. Journal of the South African Veterinary Association 80: 58-62. https://doi.org/10.4102/jsava.v80i2.172

Pérez, J., Bautista, M.J., Rodríguez, F., Wilkinson, P.J., Sierra, M.A. and Martín de las Mulas, J., 1997. Double-labelling immunohistochemical study of megakaryocytes in African swine fever. Veterinary Record 141: 386-390. https://doi.org/10.1136/vr.141.15.386

Pérez, J., Rodríguez, F., Fernández, A., Martín de las Mulas, J., Gómez-Villamandos, J.C. and Sierra, M.A., 1994. Detection of African swine fever virus protein VP73 in tissues of experimentally and naturally infected pigs. Journal of Veterinary Diagnostic Investigation 6: 360-365. https://doi.org/10.1177/104063879400600314

Petersen, H.H., Nielsen, J.P. and Heegaard, P.M., 2004. Application of acute phase protein measurements in veterinary clinical chemistry. Veterinary Research 35: 163-187. https://doi.org/10.1051/vetres:2004002

Plowright, W., Parker, J. and Staple, R.F., 1968. The growth of a virulent strain of African swine fever virus in domestic pigs. Journal of Hygiene 66: 117-134. https://doi.org/10.1017/s0022172400040997

Post, J., Weesendorp, E., Montoya, M. and Loeffen, W.L., 2017. Influence of age and dose of African swine fever virus infections on clinical outcome and blood parameters in pigs. Viral Immunology 30: 58-69. https://doi.org/10.1089/vim.2016.0121

Quintero, J.C., Wesley, R.D., Whyard, T.C., Gregg, D. and Mebus, C.A., 1986. *In vitro and in vivo* association of African swine fever virus with swine erythrocytes. American Journal of Veterinary Research 47: 1125-1131.

Reis, A.L., Netherton, C. and Dixon, L.K., 2017. Unraveling the armor of a killer: evasion of host defenses by African swine fever virus. Journal of Virology 91: e02338-16. https://doi.org/10.1128/jvi.02338-16

Rodríguez, F., Fernández, A., Martín de las Mulas, J.P., Sierra, M.A. and Jover, A., 1996a. African swine fever: morphopathology of a viral haemorrhagic disease. Veterinary Record 139: 249-254. https://doi.org/10.1136/vr.139.11.249

Rodríguez, F., Martín de las Mulas, J., Herráez, P., Sánchez Vizcaíno, J.M. and Fernández, A., 1996b. Immunohistopathological study of African swine fever (strain E-75)-infected bone marrow. Journal of Comparative Pathology 114: 399-406. https://doi.org/10.1016/s0021-9975(96)80015-6

Rodríguez, J.M., Yáñez, R.J., Almazán, F., Viñuela, E. and Rodriguez, J.F., 1993. African swine fever virus encodes a CD2 homolog responsible for the adhesion of erythrocytes to infected cells. Journal of Virology 67: 5312-5320. https://doi.org/10.1128/jvi.67.9.5312-5320.1993

Salguero, F.J., Ruiz-Villamor, E., Bautista, M.J., Sánchez-Cordón, P.J., Carrasco, L. and Gómez-Villamandos, J.C., 2002. Changes in macrophages in spleen and lymph nodes during acute African swine fever: expression of cytokines. Veterinary Immunology and Immunopathology 90: 11-22. https://doi.org/10.1016/s0165-2427(02)00225-8

Salguero, F.J., Sánchez-Cordón, P.J., Núñez, A., Fernández de Marco, M. and Gómez-Villamandos, J.C., 2005. Proinflammatory cytokines induce lymphocyte apoptosis in acute African swine fever infection. Journal of Comparative Pathology 132: 289-302. https://doi.org/10.1016/j.jcpa.2004.11.004

Salguero, F.J., Sánchez-Cordón, P.J., Sierra, M.A., Jover, A., Núñez, A. and Gómez-Villamandos, J.C., 2004. Apoptosis of thymocytes in experimental African swine fever virus infection. Histology and Histopathology 19: 77-84. https://doi.org/10.14670/hh-19.77

Sánchez, E.G. Quintas, A., Pérez-Núñez, D., Nogal, M., Barroso, S., Carrascosa Á, L. and Revilla, Y., 2012. African swine fever virus uses macropinocytosis to enter host cells. PLoS Pathogens 8: e1002754. https://doi.org/10.1371/journal.ppat.1002754

Sánchez-Cordón, P.J., Chapman, D., Jabbar, T., Reis, A.L., Goatley, L., Netherton, C.L., Taylor, G., Montoya, M. and Dixon, L., 2017. Different routes and doses influence protection in pigs immunised with the naturally attenuated African swine fever virus isolate OURT88/3. Antiviral Research 138: 1-8. https://doi.org/10.1016/j.antiviral.2016.11.021

Sánchez-Cordón, P.J., Jabbar, T., Chapman, D., Dixon, L.K. and Montoya, M., 2020a. Absence of long-term protection in domestic pigs immunized with attenuated African swine fever virus isolate OURT88/3 or BeninΔMFG correlates with increased levels of regulatory T cells and IL-10. Journal of Virology 94: e00350-20. https://doi.org/10.1128/jvi.00350-20

Sánchez-Cordón, P.J., Montoya, M., Reis, A.L. and Dixon, L.K., 2018. African swine fever: A re-emerging viral disease threatening the global pig industry. Veterinary Journal 233: 41-48. https://doi.org/10.1016/j.tvjl.2017.12.025

Sanchez-Cordon, P.J., Nunez, A., Neimanis, A., Wikstrom-Lassa, E., Montoya, M., Crooke, H. and Gavier-Widen, D., 2019. African swine fever: disease dynamics in wild boar experimentally infected with ASFV isolates belonging to genotype I and II. Viruses 11: 852. https://doi.org/10.3390/v11090852

Sánchez-Cordón, P.J., Romero-Trevejo, J.L., Pedrera, M., Sánchez-Vizcaíno, J.M., Bautista, M.J. and Gómez-Villamandos, J.C., 2008. Role of hepatic macrophages during the viral haemorrhagic fever induced by African Swine Fever Virus. Histology and Histopathology 23: 683-691. https://doi.org/10.14670/hh-23.683

Sánchez-Cordón, P.J., Vidaña, B., Dixon, L., Crooke, H. and Núñez, A., 2020b. Neuropathology and viral antigen distribution in the central nervous system of domestic pigs experimentally infected with African swine fever virus. In: Final international conference of the COST Action ASF-STOP – Understanding and combating African swine fever in Europe Brescia, Italy, 29-30 January 2020, pp. 17.

Sánchez-Torres, C., Gómez-Puertas, P., Gómez-del-Moral, M., Alonso, F., Escribano, J.M., Ezquerra, A. and Domínguez, J., 2003. Expression of porcine CD163 on monocytes/macrophages correlates with permissiveness to African swine fever infection. Archives of Virology 148: 2307-2323. https://doi.org/10.1007/s00705-003-0188-4

Sánchez-Vizcaíno, J.M., Mur, L., Gomez-Villamandos, J.C. and Carrasco, L., 2015. An update on the epidemiology and pathology of African swine fever. Journal of Comparative Pathology 152: 9-21. https://doi.org/10.1016/j.jcpa.2014.09.003

Schlafer, D.H. and Mebus, C.A., 1987. Abortion in sows experimentally infected with African swine fever virus: pathogenesis studies. American Journal of Veterinary Research 48: 246-254.

Schulz, K., Staubach, C. and Blome, S., 2017. African and classical swine fever: similarities, differences and epidemiological consequences. Veterinary Research 48: 84. https://doi.org/10.1186/s13567-017-0490-x

Sehl, J., Pikalo, J., Schäfer, A., Franzke, K., Pannhorst, K., Elnagar, A., Blohm, U., Blome, S. and Breithaupt, A., 2020. Comparative pathology of domestic pigs and wild boar infected with the moderately virulent African swine fever virus strain "Estonia 2014". Pathogens 9: 662. https://doi.org/10.3390/pathogens9080662

Semerjyan, A.B., Tatoyan, M.R., Karalyan, N.Y., Nersisyan, N.H., Hakobyan, L.H., Arzumanyan, H.H. and Karalyan, Z.A., 2018. Cardiopathology in acute African swine fever. Annals of Parasitology 64: 253-258. https://doi.org/10.17420/ap6403.161

Sierra, M.A., Bernabe, A., Mozos, E., Mendez, A. and Jover, A., 1987. Ultrastructure of the liver in pigs with experimental African swine fever. Veterinary Pathology 24: 460-462. https://doi.org/10.1177/030098588702400516

Sierra, M.A., Carrasco, L., Gómez-Villamandos, J.C., Martin de las Mulas, J., Méndez, A. and Jover, A., 1990. Pulmonary intravascular macrophages in lungs of pigs inoculated with African swine fever virus of differing virulence. Journal of Comparative Pathology 102: 323-334. https://doi.org/10.1016/s0021-9975(08)80021-7

Sierra, M.A., Gomez-Villamandos, J.C., Carrasco, L., Fernandez, A., Mozos, E. and Jover, A., 1991. *In vivo* study of hemadsorption in African swine fever virus infected cells. Veterinary Pathology 28: 178-181. https://doi.org/10.1177/030098589102800213

Sierra, M.A., Quezada, M., Fernandez, A., Carrasco, L., Gomez-Villamandos, J.C., Martin de las Mulas, J. and Sanchez-Vizcaino, J.M., 1989. Experimental African swine fever: evidence of the virus in interstitial tissues of the kidney. Veterinary Pathology 26: 173-176. https://doi.org/10.1177/030098588902600211

Zani, L., Forth, J.H., Forth, L., Nurmoja, I., Leidenberger, S., Henke, J., Carlson, J., Breidenstein, C., Viltrop, A., Höper, D., Sauter-Louis, C., Beer, M. and Blome, S., 2018. Deletion at the 5'-end of Estonian ASFV strains associated with an attenuated phenotype. Scientific Reports 8: 6510. https://doi.org/10.1038/s41598-018-24740-1

5. Methods for African swine fever diagnosis in clinical and environmental samples

C. Gallardo[1], P. Sastre[2], P. Rueda[2], A.P. Gerilovych[3], M. Kit[3], I. Nurmoja[4] and M.F. Le Potier[5*]

[1]European Union Reference Laboratory for African Swine Fever (EURL), Centro de Investigación en Sanidad Animal, (INIA-CISA), Valdeolmos, 28130 Madrid, Spain; [2]Eurofins INGENASA, 28037 Madrid, Spain; [3]National Scientific Center 'Institute of Experimental and Clinical Veterinary Medicine' (NSC IECVM), 61023 Kharkiv, Ukraine; [4]Estonian Veterinary and Food Laboratory (VFL), Kreutzwaldi 30, 51006 Tartu, Estonia; [5]French Agency for Food, Environmental and Occupational Health & Safety (ANSES), 22440 Ploufragan, France; marie-frederique.lepotier@anses.fr

Abstract

Several infectious diseases may manifest similar clinical signs to African swine fever (ASF). Differential diagnosis can only be confirmed by laboratory testing. The combination of at least two different methods of laboratory testing to confirm an ASF suspicion is recommended, e.g. detection of viral genome and detection of antibodies in suspected animals. Fully validated methods for detection of the ASF virus (ASFV) genome by PCR are available and some of them are commercialised. They are easy to use in diagnostic laboratories, and suitable for both active and passive surveillance. However, the virus isolation (VI) required for further characterisation of a viral strain requires specific skills and a highly-equipped laboratory. Both methods (PCR, VI) can be applied to assess the contamination of the environment or pork products, pig feed, etc. Fully validated methods for the detection of antibodies to ASFV by ELISA are also commercially available and easy to run. However, for serological test confirmation, the most recommended method (IPT) needs the same equipment as for VI and is mainly run only in national reference laboratories. Pen-side tests have been developed for rapid virus or antibody detection at the herd level. They would be sufficient for viral screening among susceptible animals, although their performances are currently under evaluation.

Keywords: differential diagnosis, diagnostic methods, diagnosis confirmation, shipment, biosafety

This publication is based upon work from COST Action CA15116, ASF-STOP, supported by COST (European Cooperation in Science and Technology).
www.cost.eu

EUROPEAN COOPERATION IN SCIENCE & TECHNOLOGY

Funded by the Horizon 2020 Framework Programme of the European Union

DOI 10.3920/978-90-8686-910-7_5, © C. Gallardo *et al.*, 2021

5.1 Introduction

A definitive diagnosis of African swine fever (ASF) is based on the results of laboratory tests, data from the farm or field and information regarding the epidemiological situation. Laboratory testing is important for confirmation of the diagnosis, because clinical signs of the disease as well as pathological lesions might be similar to other haemorrhagic diseases of pigs, such as classical swine fever (CSF), erysipelas, and septicaemic salmonellosis (MacLachlan and Dubovi 2017; Sánchez-Vizcaíno *et al.*, 2019). Since the clinical manifestation of ASF may vary substantially, especially in case of the subacute or chronic form of the disease, ASF cannot be diagnosed only based on clinical signs or pathological findings. A clinical course of the disease without characteristic clinical signs has been reported from the field and from experimental infections.

Currently, control and eradication of ASF is highly challenging, since neither a vaccine nor treatment is available (Dixon *et al.*, 2020). To prevent spread of the virus in domestic pig herds or wild boar populations, early detection and rapid declaration of the disease are crucial. However, they have to be implemented in combination with strict sanitary measures and in the European Union a stamping out policy in pig herds (Council Directive 2002/60/EC).

A variety of accurate and internationally recognised laboratory tests, including fully validated commercial kits, are available for the diagnosis of ASF (Arias *et al.*, 2018; Gallardo *et al.*, 2015).

Availability of reliable and accurate diagnostic assays is a prerequisite for efficient disease control, as any clinical suspicion of ASF in domestic pigs and wild boar has to be verified by laboratory diagnostic methods. At the international level, laboratory methods as well as sampling and shipping guidelines can be found in the World Organisation for Animal Health (OIE) Manual of Diagnostic Tests and Vaccines for Terrestrial Animals (Chapter 3.8.1, 2019 edition) and the respective EU Diagnostic Manual (European Commission Decision EC 2003/422/EC). The selection of which test to use depends on available matrices, the purpose of the testing (surveillance, eradication, diagnosis, confirmation), as well as the ASF epidemiological status of the country (region) or stage of the epidemic in the region.

Due to the high persistence of the ASF virus (ASFV) in the environment and in pork meat, the need for methods to detect the virus in alternative samples has been identified. Indeed, transmission via raw meat and processed pork products is considered to be one of the most important and probable ways for ASFV spread as proven by documented transboundary and transcontinental introductions. Introduction of African genotype II of ASFV to Georgia in 2007 is a vivid example of the virus' ability to jump over a long distance, even across the sea. Food waste containing contaminated pork from the Black Sea port of Poti is suspected to be the source of the virus in this case (Rowlands *et al.*, 2009). Later, after ASFV introduction to Asian countries, the virus has been detected in pork products confiscated from travellers in airports of ASF-free countries, such as Japan, Taiwan, South Korea, Australia, United Kingdom, etc. Usually, such products are tested by polymerase chain reaction (PCR) assay, which gives no information about virus infectivity (Kim *et al.*, 2019; Wang *et al.*, 2019).

5.2 Sample collection

The starting point for any laboratory investigation on ASF is sample collection. An important consideration is the purpose of the investigation, for example disease diagnosis, disease surveillance, or health certification. Which animals to sample will depend on the objective of the sampling. For example, when investigating an outbreak (passive surveillance), sick and dead animals should be targeted, while the oldest animals should be sampled when checking if animals have been exposed to the disease (active surveillance) (Beltrán-Alcrudo *et al.*, 2017). To be effective, appropriate samples combined with the selection of diagnostic methods are of fundamental importance in order to make a rapid and reliable diagnosis. Samples collected from live pigs should include anti-coagulated whole blood for the detection of virus or viral nucleic acid and serum for the detection of antibodies, whereas samples collected from dead pigs or wild boar should comprise tissues for both virus and antibody detection. Several recommendations for usual or alternative matrices are described below.

Whatever the diagnostic purpose, because of the ASFV persistence, samples must be handled in a safe way to prevent any further environmental contamination, whatever their origins.

5.2.1 Blood

If it is not possible to collect blood from live animals, post-mortem sampling of whole blood should be performed immediately after slaughter. A minimum of 1 ml of blood should be taken from the heart or veins, preferably with ethylene-diamine-tetra-acetic acid disodium salt (EDTA) as anticoagulant. Sterile vacutainers with purple stoppers indicating EDTA can be used for that purpose. Heparin (green stoppers) should be avoided as it affects the performance of both PCR (false-negative results) and haemadsorbing (HAD) test (false-positive results) (Beltrán-Alcrudo *et al.*, 2017; OIE, 2019).

When sampling in remote locations in poor technical conditions, a dried blood spot sampling (DBS) technique can be used. A few drops of blood are applied on a special absorbing Whatman 903 filter paper (for both DNA and antibody detection) or FTA card (for nucleic acids detection only), using a lancet, capillary pipette or syringe needle. The card must not be touched by the instrument when applying the sample. Once the blood drop has soaked into the paper, the card should dry for about 2 hours (time depends on card type) avoiding extreme heat, humidity or direct sunlight. The card should be stored in a sealable air-tight plastic bag, if possible a desiccant package should be added to reduce humidity. Specimens sampled by DBS technique can be stored at room temperature for up to 2 months (Beltrán-Alcrudo *et al.*, 2017; Majumdar *et al.*, 2011; Randriamparany *et al.*, 2016).

Dry blood swabs can also be used as a rapid and simple sampling technique. Routine commercially available swabs immersed in blood, once dried up, can be stored at room temperature for up to 2 months. Obtained samples can be used for both genome and antibodies detection. Although the sensitivity of genome and antibodies detection tests using specimens sampled in such a way is slightly lower than that of conventional whole blood samples, this technique is very convenient and useful, especially if sampling is performed by a non-professional (hunter, forester, etc.) (Carlson *et al.*, 2017).

5.2.2 Serum

Whole blood collected from the jugular vein, the inferior vena cava, the auricular vein on live animals or immediately after euthanasia at necropsy, is used to obtain serum. The minimum amount needed is 1 ml. Blood collected without anticoagulants (e.g. in a vacutainer with red stopper) should be left overnight at 4±3 °C. After removing the clot, serum should be centrifuged for 10-15 minutes to separate cell components. Obtained serum should be clear; red colour indicates that the sample was haemolysed, which may result in a false-positive enzyme-linked immunosorbent assay (ELISA) test (Beltrán-Alcrudo *et al.*, 2017; OIE, 2019).

5.2.3 Organs and tissues

Target organs are spleen, lymph nodes, tonsil, heart, lung, liver and kidney. Of these, spleen, tonsil and lymph nodes are the most important as they usually contain the highest amounts of virus, whereas liver and kidney can cause problems for virus detection because of the presence of PCR inhibitors. Bone marrow is also useful in incidents involving dead wild animals, as it might be the only available tissue if an animal has been dead for some time. Intra-articular fluid or tissues of joints can be tested to check for the presence of low virulent isolates. Tissue fluids are also useful for serological investigations (Gallardo *et al.*, 2019b).

The minimum recommended amount of tissue collected should be about 5 g. Samples should be kept at 4 °C during transport. If a cold chain is unavailable, samples can be stored in 10% buffered formalin, but this limits the choice of tests that can be performed afterwards. Dry swabs of organs and bone marrow can also be taken in case of absence of trained staff (Beltrán-Alcrudo *et al.*, 2017; Carlson *et al.*, 2017; OIE, 2019).

5.2.4 Faeces

The optimal amount of faeces is about 50 g, depending on the assay used for further testing. Faecal material should be placed into a sterile plastic tube using a clean and dry collection spoon (Gaur *et al.*, 2017). Faeces can be collected directly from the rectum using swabs. Samples collected on swabs should be placed in preservative sterile saline solution (i.e. standard sterile PBS) or transport media. Double packaging of tubes in sealable plastic bags prevents cross-contamination of both samples and packaging materials. Faeces samples should be transported in a refrigerator at +4 °C or on ice and tested as soon as possible, especially if they were collected in the environment (OIE, 2019). However, faeces can be a less than optimal medium for ASF surveillance due to the short-term survival of viable ASFV, the temperature-dependence, and the differences found among ASFV strains regarding their virulence (Davies *et al.*, 2017; De Carvalho Ferreira *et al.*, 2014). Analysis of faecal samples is not considered to be sensitive enough for ASF surveillance.

5.2.5 Oral fluids

Mouchantat *et al.* (2014) showed the possibility of using a rope-in-a-bait sampling method ('pSWAB': pathogen sampling wild animal with baits) for oral fluid collection from wild boars. Cotton ropes 8 mm in diameter embedded in a cereal-based bait matrix placed on the floor of a pig pen were used for that purpose. The bait matrix consists of cornmeal, milk powder, almond aroma, paraffin wax, and coconut oil. After chewing, the wet part of the rope is placed in a syringe with further squeezing of oral fluid into a sterile tube.

Oral fluid sampling using pSWAB from wild boars under field conditions is carried out by placing the ropes embedded in baits in a hole dug in the ground, or leaving them on the ground just covered with 1-2 cm of soil or tying them to trees (Figure 5.1). If ropes are left for a few days and dry out, chewed parts should be placed into plastic tubes or sealable bags. At reception in the laboratory, they must be incubated in phosphate buffered saline solution (PBS) for 1 hour at 4 °C before extracting the oral fluids (Kit, unpublished data; Männistö, 2018). Other types of baits for oral fluid collection, such as swabs incorporated in maize cobs, can also be used (Khomenko *et al.*, 2013).

5.2.6 Soil

Different types of sampling equipment can be used for soil sampling; however, spoons and spatulas are the most appropriate as they can be easily sterilised. About 50 g (wet weight) of soil should be collected in a clean and dry plastic tube. Samples should be preserved in wet ice (H_2O) or in a refrigerator, especially if specimens will be used for virus isolation. Freezing or using dry ice (CO_2) results in a considerable loss in viral titre in soil (Hurst and Reynolds, 2007).

Figure 5.1. Rope in a bait used for oral fluid sampling from wild boar under field conditions.

5.2.7 Feed

The choice of sampling technique for feed depends on the material type and amount, as a variety of special tools can be used for this purpose. Special slotted grain probes are usually used for sampling of a bulk feed consisting of grain and meal. The grain probe should be long enough to reach at least 3/4 of the total feed depth. It usually consists of two tubes one inside the other about 35 mm in diameter. The inner tube may be divided into compartments to enable separate sampling at different levels of the total feed amount. At least 10 samples from evenly located places should be collected from each bulk carrier. Bag triers of different shapes, often sharp-pointed, are used to collect powdered and granular feeds packaged into bags. A trier is inserted diagonally from one corner to the other. Samples size should be at least 500 g per bag. A bomb sampler can be used for collecting liquids from bulk carriers. At least 500 ml should be sampled from a container. In cases when feed`s consistency and structure make the use of special samplers unfeasible, manual sampling can be used (applicable for silage, dry and green forage etc.). Since the total amount of sampled material is usually much bigger than is needed for laboratory tests, sample reduction is performed. It can be done either using special equipment (a riffler, the Boerner divider) or by the quartering method (Herrman, 2001; Menegat *et al.*, 2019).

Solid feeds should be packed into clean and dry sealable plastic bags or plastic containers with lids. Liquid components of feed can be placed in tubes or other sterile containers made of plastic, glass or stainless steel (for liquid fats). Labels should be placed outside the bag, without any contact with the material. High-moisture feedstuff must be frozen immediately after collection, whereas dry materials should be kept cool and dry (Herrman, 2001; Menegat *et al.*, 2019).

5.2.8 Water

Water must be collected in sterile sample containers. For sampling water from ponds, lakes or tanks, the container should be put under the water surface at a depth of 30 cm holding its neck down, then turn it up to let the container fill. When sampling well water, the collection should be started only after two or more minutes of pumping to avoid sampling water that has been stagnating in pipes (Herrman, 2001).

5.2.9 Surface sampling

Cotton- or polyester-tipped swabs are commonly used for sampling from contaminated surfaces and fomites. Water, saline solutions or different transport media may be used as eluent in the sampling process. With swabs pre-wetted with the eluent, the swab is rubbed across the surface horizontally, vertically and diagonally with moderate pressure. Parts of the surface that are obviously stained with blood, faeces, etc. should be given preference for sampling. Then swabs should be placed in capped tubes and after further packaging (see Section 5.7) should be transported on ice (Julian *et al.*, 2011).

5.2.10 Blood-stained materials

Blood-stained material such as soil and parts of plants can be sampled for laboratory testing. About 5-10 g of material have to be sampled. Blood-stained soil can be placed in a clean plastic tube, whereas for the sampling of stained plant parts sealable plastic bags may be used (Gaur *et al.*, 2017).

5.3 Internationally prescribed African swine fever diagnostic tests

5.3.1 African swine fever virus detection tests

For agent identification, nucleic acid detection tests by real-time polymerase chain reaction (rPCR), or gel-based PCR, virus isolation and HAD assays or antigen detection tests, such as direct fluorescent antibody test (DIF) on fixed cryosections of organ material and ELISAs detecting p72 antigen, are available (OIE, 2019) (Table 5.1). Additionally, point of care tests for antigen detection are valuable tools to be used at field level, especially in situations where laboratory infrastructure and skilled personnel are limited and where in many cases first evidence of the disease is based only on clinical symptoms. Lateral flow assay (LFA) for antigen detection provides a rapid and easy way to identify ASFV infection at individual level (Sastre *et al.*, 2016a).

For screening a large number of animals Antigen ELISA is a rapid method that can be fully automated; however, its sensitivity is rather low especially because sample quality can have a

Table 5.1. Overview of validated African swine fever virus and antibody detection tests.

Detection	Available tests	Type: in house/commercial	Recommended use
Virus	genome detection	PCR (OIE TaqMan probe[1], OIE UPL probe[1] or OIE conventional PCR[1], and commercial kits[2])	suspicion; surveillance; individual and herd testing
	virus isolation	VI/haemadsorption (HAD) test[1] (i.h.)	confirmation of primary outbreak
	antigen detection	Direct Immuno fluorescence (DIF)[1] (i.h.)	individual testing (acute forms)
		Antigen ELISA commercial kit INgezim PPA DAS, Double Ab Sandwich	surveillance; herd testing (acute forms)
	pen-side test	Lateral flow assay (LFA) commercial kit (INgezim ASF CROM Ag)	herd testing (acute forms)
Antibody	ELISA	ELISA (OIE, commercial kits[3])	surveillance; herd testing
	confirmatory test	Immunoblot (IB) test[1] (i.h.)	confirmatory; herd testing
		Immunofluorescence Antibody (IFAT) test[1] (i.h.)	confirmatory; herd testing
		Indirect Immunoperoxidase test[1] (IPT) (i.h.)	confirmatory; herd testing
	pen-side test	LFA commercial kit INgezim PPA CROM	herd testing

[1] Included in the OIE Terrestrial Manual for Diagnostic Test and Vaccines, 2019; i.h. = in house methods.

[2] PCR Commercial Kits currently validated: INgene q PPA, INGENASA. 11.PPA.K.5TX/Q; Tetracore TC-9017-064; Virotype ASFV PCR Kit, INDICAL BIOSCIENCE; LSI VetMAXTM Thermo Fisher Scientific; IDEXX RealPCR ASFV Mix, IDEXX; ID Gene® African Swine Fever Duplex – IDVet; ADIAVET ASFV REAL TIME 100R, BIO-X DIAGNOSTICS.

[3] Antibody ELISA Commercial Kits currently validated: INgezim PPA COMPAC competition-ELISA, INGENASA; IDScreen® ASF Indirect ELISA, IDVET; ID Screen® ASF Competition-ELISA, IDVET; SVANOVIR® ASFV Indirect-ELISA, SVANOVA.

significant impact on final assay results (Gallardo *et al.*, 2019b; Oura *et al.*, 2013), therefore it is not used for testing individual animals. ASF antigen detection by DIF yields quick results, and is a highly sensitive test for cases of peracute and acute ASF. It is a robust test, but it has been largely replaced by PCR and reagents are no longer widely available. It is important to note that in subacute and chronic disease, where antibodies are present, the sensitivity of this method is also limited (40%) and interpretation of test results is difficult and requires well-trained and experienced laboratory staff (Gallardo *et al.*, 2019b).

A characteristic feature of infected cells with ASFV field isolates is the ability to adsorb swine erythrocytes (haemadsorption) on its surface, forming erythrocyte rosettes, which enables virus isolation (VI) and identification by HAD tests (Figure 5.2). This property is used to differentiate the ASFV from other agents causing diseases with symptoms similar to ASF and is recommended as a reference confirmatory test in the event of a primary outbreak (EC, 2003). In theory, all of the ASFVs collected from natural outbreaks can be isolated in susceptible primary leukocyte cultures of swine origin, either from blood or lung (alveolar) monocytes or from macrophages. However, growing ASFV isolates is a critical step for diagnosis at the national reference laboratories, indeed it is more expensive than other techniques as it requires both specialised facilities and training, is time consuming and cannot be adapted to high throughput. In addition, attempts to isolate infectious virus from field-derived samples provide irregular results. The reason lies in the poor state of samples received, which affects the virus viability, especially on samples obtained from dead or hunted animals, such as wild boar (Gallardo *et al.*, 2019b). Additionally, some field strains do not produce HAD, but only a cytopathic effect (Boinas *et al.*, 2004; Gallardo *et al.*, 2019a; Leitão *et al.*, 2001); these non-HAD viruses are not easily isolated and require further confirmation using PCR or DIF test on the sediments of the cell cultures (Oura *et al.*, 2013). Despite these constraints, virus isolation is essential to obtain virus stocks for further molecular and biological characterisation studies. The use of established cell lines, such as COS-1 (African green monkey kidney fibroblast-like cell line), IPAM (immortalised porcine alveolar macrophage cells) or WSL (wild boar lung cells), can overcome the difficulty in obtaining the primary cells. However, these cell lines are not always suitable for the ASFV isolation from field samples without a previous initial isolation on porcine macrophages (Carrascosa *et al.*, 2011). Therefore, further evaluation studies are required for the potential use of established cell lines in routine diagnosis.

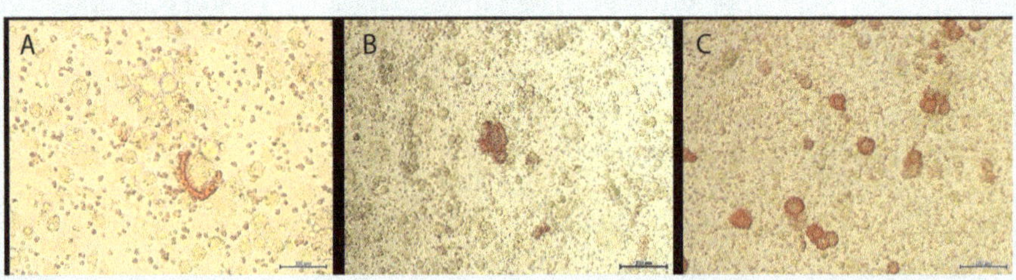

Figure 5.2. Haemadsorption with rosette formation in ASFV infected PBM cells with a virulent genotype I ASF strain. Details of the haemadsorption on the surface of swine peripheral blood monocytes infected with the virulent genotype I ASFV Spanish strain E70 at: (A) 12 hours after the infection, (B) 24 hours after the infection, and (C) 48 hours after the infection where complete rosette formation on the periphery of the cells can be observed.

Among the above listed tests, the PCR is by far the most sensitive method for the detection of the agent and should be regarded as the method of choice for first-line laboratory diagnosis. A variety of PCR tests, including both conventional and real time (rPCR), have been developed and validated to detect a wide range of ASF isolates belonging to different known virus genotypes, non-HAD strains, and isolates of diverse virulence (Gallardo *et al.*, 2019b; OIE, 2019; Oura *et al.*, 2013). All of them have been designed to target the P72-coding region, a highly conserved gene coding the major viral protein, assuring the (potential) detection of any ASFV isolate (Gallardo *et al.*, 2019b; Oura *et al.*, 2013). The OIE rPCR developed by King *et al.* (2003; OIE, 2019) and the OIE Universal probe library (UPL) developed by Fernández-Pinero *et al.* (2013; OIE, 2019) are the most widely used for routine diagnosis at international level. Both methods are able to provide a reliable ASF diagnosis, although the UPL-PCR has greater diagnostic sensitivity for detecting animals with low viral load (Fernández-Pinero *et al.*, 2013; Gallardo *et al.*, 2019b). The rPCR developed by Tignon *et al.* (2011) including an internal endogen control targeting swine beta-actin is also well fitted for diagnostic purposes. Finally, the number of commercial kits for ASFV genome detection based on published rPCRs has greatly increased over recent years. These represent an alternative that can guarantee a certain homogeneity in results, which is paramount for standardisation and replicability of results among laboratories. For the same reasons, each new ASF-commercial assay must be evaluated and validated following international criteria to ensure it is specific, sensitive, reproducible, precise, robust and accurate. In summary, the PCR is now established as the gold standard method for virus detection and used as a basic diagnostic tool for both passive and active surveillance, considering the long-term viraemia and the high viral load present in the infected animals suffering acute or subacute clinical courses. It is quick and can be used for individual as well as pooled samples, although for the latter with some limitations (Gallardo *et al.*, 2019b).

5.3.2 African swine fever antibody detection tests

Whenever the suspicion is raised that ASFV is circulating in a pig population, serological assays must also be used for the diagnosis of the disease. Moreover, serology should be applied for surveillance purposes as it is a valuable tool for further epidemiological investigations, for example, for determining the time point of agent introduction into a pig herd or into a wild boar population. As shown on Figure 5.3, anti-ASFV antibodies appear soon after infection and persist for up to several months or even years (Arias *et al.*, 2018; Penrith *et al.*, 2004, Schulz *et al.*, 2020). Additionally, as long as no vaccine is available against ASFV, the presence of anti-ASFV antibodies always indicates infection. Antibody-based surveillance is therefore essential for the detection of surviving animals, to elucidate the epidemiological characteristics of the epidemics, i.e. time since the virus introduction into a farm, and for detecting incursions involving low virulence ASFV isolates (Beltrán-Alcrudo *et al.*, 2017, Sánchez-Cordón *et al.*, 2018). Sero-surveillance can be used to demonstrate the disease-free status in a country, provided that it is combined with active or passive surveillance based on virus detection. Recently, a multiplex test for antibody detection, using the Luminex platform and including different target antigens of both ASFV and CSF virus, showed the increased sensitivity of the assay in comparison with the individual ELISAs used as reference in the study (Aira *et al.*, 2019). This could especially be of interest when analysing field samples, where animals can react differently to viral exposure and present diverse levels of antibodies to each of the virus proteins.

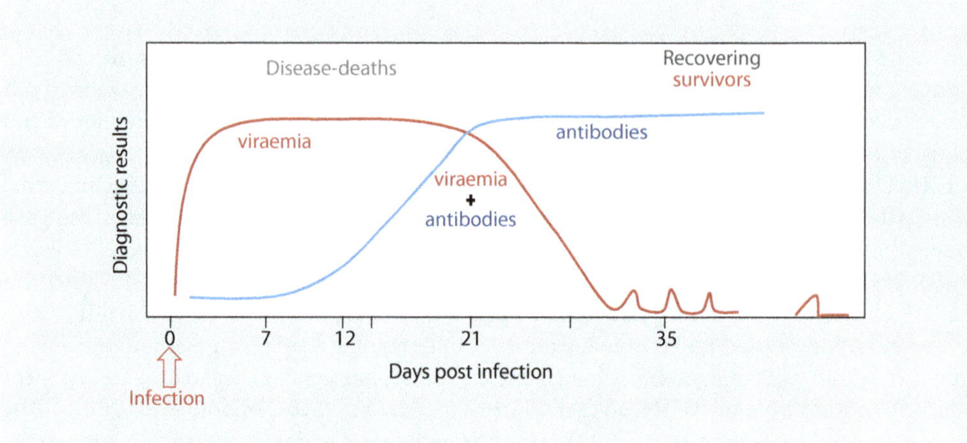

Figure 5.3. Dynamics of African swine fever virus infection.

The current EU (2003/422/EC) and OIE (2019) recommendations for ASFV antibody detection involve the use of an ELISA for antibody screening, backed up by immunoblotting (IB), indirect immunofluorescence test (IFAT) or the indirect immunoperoxidase tests (IPT) as confirmatory tests (OIE, 2019) (Table 5.1). There are a number of validated ELISAs for the detection of antibodies against ASFV, including several (OIE) 'in house' versions of the test, based on the use of live virus as antigen, recombinant ELISAs (Cubillos *et al.*, 2013), and commercial ELISA kits. The ELISA remains the most useful method for large-scale serological studies: it is fast, easy to perform and economical. For these reasons, the OIE recommends ELISA for international trade. However, only serum can be analysed, which restricts its application range, especially in case of passive surveillance of wild boar when animals are usually found dead. In addition, haemolysed serum samples could provide either false positive or negative results depending on the ELISA format employed. Therefore, positive ELISA results should always be confirmed by additional methods, such as IPT, IFAT or IB tests, as recommended by the OIE (2019). The IB is a rapid and sensitive assay but, similarly to what is described above, only serum samples can be tested. On the contrary, IPT or IFAT can easily be used for analysing all types of porcine samples, including tissue exudates, whole blood and even bone marrow (Beltrán-Alcrudo *et al.*, 2017; Gallardo *et al.*, 2019b). These tests have high specificity and sensitivity but the interpretation of the results can be subjective and well-trained staff is required. It would thus be important to develop and validate standardised ELISAs for the detection of specific ASFV antibodies in tissue extracts for an easy and more reliable evaluation of the epidemiological situation in affected areas.

5.4 Front-line African swine fever diagnostic tests

When investigating disease outbreaks in animals, the earlier the clinical signs of disease are recognised by the farmer and the earlier the clinical diagnosis is confirmed by laboratory tests, along with rapid reporting to the relevant veterinary authorities, the better the disease will be controlled. Success partly relies on sending samples to a reference laboratory to test for the presence of ASFV or antibodies against it. The availability of 'front-line' diagnostic tests would have the advantage of providing additional support to the clinical assessment in the first instance and could reduce the time needed to confirm the test results in secondary cases of the disease. The use of rapid diagnostic tests that can be conducted in the field, where the infected animal populations are, would therefore facilitate earlier and more effective disease control.

Molecular technologies such as the rPCR assays might not be available in resource-poor regions because they require specialised laboratories, trained personnel, and expensive reagents. Several challenges are associated with performing PCR in the field beyond the need for a portable, battery-operated thermocycler, which include performing nucleic acid extractions without a centrifuge/electricity, protecting samples against cross-contamination, and maintaining a cold chain for materials that may require refrigeration. Loop-mediated isothermal amplification (LAMP) could be a cheaper diagnostic alternative to PCR, and useful in field conditions. Its sensitivity is appropriate in cases where clinical signs are present and enough for detection of acute cases. However, despite being currently available, LAMP still lacks field validation data and is not recommended in cases of low ASFV genome content due to significantly lower sensitivity when compared with both conventional and real time PCR (Oura *et al.*, 2013).

An alternative is the use of the pen-side tests that offer a first-line diagnosis for rapid application in case of sanitary emergency. Currently, only lateral flow assays (LFAs) are commercially available for the detection of antibodies or the viral antigen in blood (Table 5.1). This kind of test is especially useful in field conditions, where no equipment is available, since the performance of the test and results interpretation are simple. The LFA for antigen detection (INgezim ASF CROM Ag) described by Sastre *et al.* (2016a) is based on the use of a monoclonal antibody against P72 protein of ASFV and provides similar sensitivity to that of the commercial antigen-ELISA when samples from experimental infections are tested. The LFA was demonstrated to be positive for animals with circulating virus levels exceeding 10^4 haemadsorbing units. When testing field samples, the LFA showed 60% positivity versus the 48% of the antigen-ELISA, although it exhibited lower sensitivity when compared to the rPCR.

The LFA for antibody detection (INgezim PPA CROM) under field conditions in hunted wild boar in Sardinia (Cappai *et al.*, 2017) showed a sensitivity of 82% and a specificity of 96% when compared with the commercial antibody INGENASA ELISA and the IPT (better performance under laboratory conditions). Finally, multiplex LFA tests have been developed for detection of both ASF and classical swine fever specific antibodies, facilitating surveillance for both diseases in the field (Sastre *et al.*, 2016b).

It is important to point out that antibody LFA are not useful for the detection of acute forms of the disease. Moreover, the LFAs should not be used alone due to a very limited sensitivity compared to the gold standard methods, especially in the case of the LFA for antigen detection (Gallardo *et al.*, 2019b). They need to be confirmed by virus/antigen detection techniques, since antibodies cannot be detected before 12-14 days post infection using these tests. The analysis of suspicious samples by both virus and antibody detection techniques/pen-side tests will give a better picture of the epidemiological situation. Test sensitivity needs to be high, whereas specificity is less critical, since any positive result will need to be verified by the competent National reference laboratory. However, the possible contribution of pen-side tests to the control of notifiable diseases is still controversial and their application may be limited due to national and international regulations.

5.5 Some considerations on African swine fever diagnosis

Taken together, sensitive, specific and robust laboratory diagnostic assays are available but, as for any other disease, there is no single test that is 100% reliable (sensitive and specific). For this reason, final diagnosis should be based on the interpretation of the results derived from the use of appropriate samples and validated tests in combination with the information coming from disease epidemiology, and the clinical signs. A thorough understanding of the time course of viraemia and antibody seroconversion during the ASFV infection is a prerequisite to obtain relevant information about the dynamics of the infection in the investigated areas and to support control and eradication programmes (Figure 5.3). An appropriate diagnosis therefore should always include the detection and identification of ASFV-specific antigens or DNA and antibodies (EC, 2003; Gallardo *et al.*, 2019b).

In free areas bordering infected territories, measures should be taken to enhance passive surveillance and PCR testing proportionately to the risk assessment results. In acute phases of the disease, the ASFV is easily detected in any kind of porcine sample by rPCR, VI and even using the antigen detection techniques (DIF or ELISA). However, false positive PCR results, although rare, can occur (e.g. due to lab contamination or other factors) and thus a primary outbreak (or case) of ASF should be confirmed by isolation of ASFV and HAD assay. However, this might not always be possible due to time or technical limitations and the reduced sensitivity, particularly on samples obtained from altered dead or hunted wild boar, or in weak positive PCR samples (Gallardo *et al.*, 2019b). If HAD is not possible, PCR results should be confirmed by pathology examination (see Chapter 4) and at least two distinct virus or antibody detection tests on the same-suspected pig. In wild boar, if virus isolation is not possible, a primary case of ASF can be confirmed when at least two virus or antibody detection tests have given a positive result (EC, 2003).

It is important to note that, since animals usually develop antibodies within the second week after infection, they can test positive for both ASFV and antibodies simultaneously for at least two months (Gallardo *et al.*, 2019b). Samples from animals surviving beyond this period are usually positive for ASFV-specific antibodies, but negative for ASFV and its genome. Therefore, whenever the suspicion is raised that ASFV is circulating in a swine population, serological assays must also be used for the diagnosis. Moreover, animals infected with attenuated strains seroconvert after the first week of the infection even in absence of clinical signs or viraemia (Gallardo *et al.*, 2019a; Leitão *et al.*, 2001). Positive results for both virus and antibodies indicate that the tested animal was infected at the time of sampling, whereas a positive ASFV antibody test in absence of virus indicates an ongoing or past infection, where the animal has recovered or could be chronically or sub-clinically infected with attenuated strains (Table 5.2).

5.6 African swine fever virus detection in raw and processed pork products

5.6.1 African swine fever virus survival in pork products

ASFV is highly resistant to a variety of temperatures and pH conditions especially in a protein rich environment, such as meat, blood, organs and processed pork products. Thus, in these matrices,

Table 5.2. Interpretation of the ASF diagnostic results.

Assay	Result	Probable scenarios
PCR Ab-ELISA IPT	Weak (ct>35) Negative Negative	The animal was recently infected and it has not yet seroconverted (<7 days). Clinical signs are not yet evident.
PCR Ab-ELISA IPT	Positive Negative Positive	The animal was recently infected, is developing clinical signs and is initiating the seroconversion (7-10 days).
PCR Ab-ELISA IPT	Positive Positive Positive	a. Infection in course. The animal is still viraemic with clinical signs and has already seroconverted (>10 days). b. Resurgence of the infection in an animal with preformed antibodies from a previous infection (survivor).
PCR Ab-ELISA IPT	Weak (Ct>35) or negative Positive Positive	a. Past infection. The animal has recovered from acute or subacute infection and does not present clinical signs. b. The animal has been infected with attenuated strain (with or w/o clinical signs). c. Reinfection of an animal with preformed antibodies from a previous infection (survivor).

the virus remains viable for months and even years (in refrigerated raw meat – for 110 days, in frozen meat – up to 1000 days, in lard for 300 days, in blood for 18 months) (Beltrán-Alcrudo *et al.*, 2017; EFSA, 2010; Mazur-Panasiuk *et al.*, 2019).

Survival time of ASFV in processed pig products varies greatly depending on product and processing procedure (salted, smoked, dried). Mebus *et al.* (1993, 1997) showed that ASFV remained viable for up to 140 days in traditional Spanish dry-cured pork products such as Serrano and Iberian ham, loin and shoulder. However, the commercial curing time for these types of meat products is longer than the period of viral viability, which means that the virus would be inactivated by the end of the processing. In another study on ASFV survival in traditional Italian dry-cured meat products, different results were obtained (Petrini *et al.*, 2019). Pork belly and loin samples were found to be ASFV positive respectively between 60 or 80 to 137 days, which is more than total time of curing and shelf-life of these types of products. However, salami samples were found to be ASFV negative 26 days after processing, before the end of the curing period. Thus, salami is regarded to be a very unlikely matrix for the virus spread, whereas loin and pork belly could potentially pose a threat if fed to suids (Petrini *et al.*, 2019). Sindryakova *et al.* (2016) tested contaminated corned meat and pig fat processed by salting and stored at room temperature (22-25 °C) or in a household fridge (4-6 °C). The virus was detected in corned meat after 16 days of storage at room temperature and after 60 days of storage in a fridge. Salted fat contained no infectious virus; however, viral DNA was detected in both products during the whole study period (56 days for storage at room temperature, 60 days for storage in a fridge). Canned meat, which was exposed to a high temperature during processing, contained neither virus nor viral genome from the first day of the study.

The risk of ASFV distribution with raw meat and processed pork products thus depends on the processing. Heat-treated products such as canned meat could be regarded as safe, because the virus gets inactivated by temperatures higher than 60 °C (20 minutes exposure). However, salted or dried pork products present some risk of spread. Although there are many studies regarding ASFV survival in pork products, extrapolation of this data for risk assessment is hardly possible, because it varies greatly depending on duration and type of processing and matrix used (Petrini *et al.*, 2019).

5.6.2 Methods for African swine fever virus detection in pork products

A variety of methods can be used for ASFV detection in raw and processed pig products, depending on the purpose of the test.

PCR is the most widely used diagnostic tool for the viral genome detection since it is rapid, specific, able to detect all the known genotypes and can be used for testing a variety of samples, including those in which the virus might be inactivated during transport. A few PCR-based assays are recommended by OIE, including conventional and real-time PCR (OIE, 2019). These assays are usually used for testing of imported or confiscated meat products (Kim *et al.*, 2019; Wang *et al.*, 2019). However, PCR technique cannot prove the presence of the infectious virus in the sample, which limits its use.

Loop-mediated isothermal amplification (LAMP) is an additional technique that can be used for ASFV genome detection. It is a rapid and specific diagnostic tool, which has been developed and improved at the beginning of this century (Nagamine *et al.*, 2002; Notomi *et al.*, 2000). The advantage of the method is that it uses a DNA-polymerase (usually *Bst*) with strand displacing activity in addition to the replication activity that can be performed at constant temperature with simple and cheap equipment (such as a solid-state thermostat or water bath). The reaction mixture includes three pairs of primers forming loop structures on the DNA and that act as starting points for DNA amplification. The abundance of loops allows the generation of detectable amounts of target DNA in 20-30 minutes. LAMP products can be visualised directly in the test tube by adding an intercalating dye and exposing to UV-lighting or by conventional agarose gel electrophoresis (James *et al.*, 2010; Kit, unpublished data; Wang *et al.*, 2020). Although LAMP has not been used for the detection of ASFV DNA in meat products yet, the technique is widely applied in microbiological quality assessment of different meat products. So far pathogens, such as *Yersinia enterocolitica, Listeria monocytogenes, Streptococcus suis, Salmonella* spp. and others can be detected in pork using LAMP technique (Kumar *et al.*, 2017), making the routine application of the method for ASFV detection in these samples promising.

However, viral genome detection, either by PCR or LAMP, does not give information on the virus infectiousness, and therefore on the risk of further contamination. Virus isolation based on the HAD test is the most widely used to establish the presence of viable ASFV, despite being laborious and time-consuming (see Section 5.3.1). Bioassay (or *in vivo* assay) is a far less used technique for infectious virus detection. The approach consists in administering a homogenate of the tested samples to healthy pigs via oral, oronasal route or by intradermal injection. If the pigs become sick, samples obtained from them are used for further testing via virus isolation, PCR, serological tests, etc. If the pigs show no clinical signs, they are either tested for antibodies or viral genome presence or they are slaughtered for post-mortem examination. A bioassay should only be considered in case viral isolation is impossible or for validation of virus isolation negative results (Mebus *et al.*, 1993, 1997; Petrini *et al.*, 2019).

5.7 Biosafety aspects of sampling African swine fever virus-infected material

5.7.1 Use of protective equipment

Although ASFV poses no threat to human health, considering the risk posed to pigs and livelihoods by the ASFV, to prevent its further dissemination ASF-suspected specimen sampling should be performed using personal protective equipment (PPE). PPE recommended includes coveralls, rubber boots or covers for boots, gloves, protective masks and glasses. Depending on the conditions and purpose of sampling, the choice of PPE varies from all the listed items (sampling from animals during outbreak investigation) to only gloves (e.g. for feed sampling during surveillance study). All equipment that gets directly into contact with potentially infected material (e.g. spoon for soil, gloves, etc.) must be disposed of safely after a single use. All the materials used while sampling must be disposed of appropriately on the sampling site or transported to the laboratory for disinfection.

5.7.2 Labelling and supplementary documentation

All samples must be clearly labelled in a way to ensure labels will be readable after transport and storage. Detailed information about the samples should be mentioned on the tag, which should be placed in a plastic envelope outside the transport container for easy access without getting into touch with the samples. A duplicate of the information should also be placed within the shipping container between the secondary and the sample packaging (especially important for samples from animals).

Information to be sent with environmental samples comprises:
- location of sampling, including geolocation if possible;
- date of collection;
- type and description of submitted material.

Information to be sent with samples collected from animals comprises:
- contact information of the owner of the animal (if domestic) and geolocation of sampling site if possible;
- location of sampling, including geolocation if possible, in case of wild boar;
- information about the sampled animal (species, age, sex, etc.);
- date of sampling;
- type and description of samples;
- the history of the suspected case;
- epidemiological information.

5.7.3 Packaging and shipping

Either national regulations or United Nations (UN) Recommendations on the Transport of Dangerous Goods must be followed for land transport. In case of transport by air, the Dangerous Goods Regulation of the International Air Transport Association (IATA) must be followed (OIE, 2018; United Nations, 2019a,b).

Appropriate packaging and transport are crucial for both safety and maintaining appropriate quality for further analysis. Packaging should prevent infecting other animals and cross-contamination of collected samples. According to UN regulations, environmental samples, faeces, dry blood spots and other samples that are unlikely to cause the disease are not subject to the Dangerous Goods Regulations if appropriately packaged and marked (OIE, 2018; United Nations, 2019a). Samples of organs and tissues belong to Category B (UN Dangerous Goods number 3373) and must be stored according to the Packaging instruction P650 (OIE, 2018; United Nations, 2019b).

Only general principles of packaging for listed materials are described here. A triple packaging system including a primary receptacle, a secondary packaging that might contain multiple primary receptacles, and an outer packaging that might contain several secondary packaged items should be used. Either secondary or outer packaging must be rigid. A primary receptacle is a sterile, sealable, leak-proof for liquids and sift-proof for solid materials tube or plastic bag, which contains one sample and is additionally sealed with Parafilm or adhesive tape. Each primary receptacle should be wrapped individually to prevent contact among samples placed in the same secondary packaging, which also should be leak- /sift-proof and hermetically sealable. Both, the primary receptacle and the secondary packaging should be capable of withstanding an internal pressure of 95 kPa. If samples contain any liquid, absorbing materials (cotton balls, paper towels, cellulose wadding) should be placed in the secondary packaging. The outer packaging should have passed a drop test and be clearly labelled. If ice, ice pack or dry ice is used for cooling the samples, it should be placed outside the secondary packaging. Samples transported within two or less days must be kept at 4 °C either in refrigerators or on ice. If sample shipping takes three or more days, they must be frozen at -20 or -70 °C (Beltrán-Alcrudo *et al.*, 2017; OIE, 2018; United Nations, 2019b). An example of the triple packaging system for the packing and labelling of Category B infectious substances is available in the OIE Manual (2018).

5.7.4 Waste management

Samples, all the packaging materials which possibly were or could be in direct contact with samples (including absorbing packaging) and materials used during the sampling process should be appropriately decontaminated before disposal. Either autoclaving (preferably) or chemical inactivation of this waste material must be carried out. ASFV is inactivated by 8/1000 sodium hydroxide (30 minutes), hypochlorites – 2.3% chlorine (30 minutes), 3/1000 formalin (30 minutes), 3% ortho-phenylphenol (30 minutes) and iodine compounds (Beltrán-Alcrudo *et al.*, 2017; OIE, 2019) (see also Chapter 11 for cleaning and disinfection procedures).

References

Aira, C., Ruiz, T., Dixon, L.K., Blome, S., Rueda, P. and Sastre, P., 2019. Bead-based multiplex assay for the simultaneous detection of antibodies to African swine fever virus and classical swine fever virus. Frontiers in Veterinary Science 6: 306: https://doi.org/10.3389/fvets.2019.00306

Arias, M., Jurado, C., Gallardo, C., Fernández-Pinero, J. and Sánchez-Vizcaíno, J.M., 2018. Gaps in African swine fever: analysis and priorities. Transboundary and Emerging Diseases 65 (Suppl. 1): 235-247. https://doi.org/10.1111/tbed.12695.

Beltrán-Alcrudo, D., Arias, M., Gallardo, C., Kramer, S. and Penrith, M.L., 2017. African swine fever: detection and diagnosis – A manual for veterinarians. FAO Animal Production and Health Manual No. 19. Food and Agriculture Organization of the United Nations (FAO), Rome, Italy. Available at: http://www.fao.org/3/a-i7228e.pdf

Boinas, F.S., Hutchings, G.H., Dixon, L.K. and Wilkinson, P.J., 2004. Characterization of pathogenic and non-pathogenic African swine fever virus isolates from *Ornithodoros erraticus* inhabiting pig premises in Portugal. Journal General Virology 85 (Part 8): 2177-2187. 10.1099/vir.0.80058-0

Cappai S, Loi F, Coccollone A, Cocco M, Falconi C, Dettori G, Feliziani F, Sanna ML, Oggiano A, Rolesu S.,2017. Evaluation of a commercial field test to detect African swine fever. Journal of Wild Life Diseases 53: 602-606. https://doi.org/10.7589/2016-05-112

Carlson, J., Zani, L., Schwaiger, T., Nurmoja, I., Viltrop, A., Vilem, A. and Blome, S., 2017. Simplifying sampling for African swine fever surveillance: Assessment of antibody and pathogen detection from blood swabs. Transboundary and Emerging Diseases 65: e165-e172. https://doi.org/10.1111/tbed.12706.

Carrascosa, A.L., Bustos, M.J., and de Leon, P., 2011. Methods for growing and titrating African swine fever virus: field and laboratory samples. Current Protocols in Cell Biology 53: 26.14.1-26.14.25. https://doi.org/10.1002/0471143030.cb2614s53

Cubillos, C., Gómez-Sebastian, S., Moreno, N., Nuñez, MC., Mulumba-Mfumu, L.K., Quembo, C.J., Heath, L., Etter, E.M., Jori, F., Escribano, J.M. and Blanco, E., 2013. African swine fever virus serodiagnosis: a general review with a focus on the analyses of African serum samples. Virus Research 173: 159-167. https://doi.org/10.1016/j.virusres.2012.10.021

Davies, K., Goatley, L.C., Guinat, C., Netherton, C.L., Gubbins, S., Dixon, L.K. and Reis, A.L., 2017. Survival of African swine fever virus in excretions from pigs experimentally infected with the Georgia 2007/1 isolate. Transboundary and Emerging Diseases 64: 425-431. https://doi.org/10.1111/tbed.12381

De Carvalho Ferreira, H.C., Weesendorp, E., Quak, S., Stegeman, J.A. and Loeffen, W.L., 2014. Suitability of faeces and tissue samples as a basis for non-invasive sampling for African swine fever in wild boar. Veterinary Microbiology 172: 449-454. https://doi.org/10.1016/j.vetmic.2014.06.016

Dixon, L.K., Stahl, K., Jori, F., Vial, L. and Pfeiffer, D.U., 2020. African Swine Fever Epidemiology and Control. Annual Review of Animal Biosciences 8: 221-246. https://doi.org/10.1146/annurev-animal-021419-083741

European Commission (EC), 2002. Council Directive 2002/60/EC of 27 June 2002 laying down specific provisions for the control of African swine fever and amending Directive 92/119/EEC as regards Teschen disease and African swine fever. Available at: https://eur-lex.europa.eu/legal-content/EN/TXT/?uri=CELEX:02002L0060-20080903

European Commission (EC), 2003. African swine fever diagnostic manual (notified under document number C (2003) 1696) 2003/422/EC). Available at https://eur-lex.europa.eu/legal-content/EN/TXT/PDF/?uri=CELEX:32003D0422&from=EN

European Food Safety Authority (EFSA) Panel on Animal Health and Welfare, 2010. Scientific opinion on African swine fever. EFSA Journal 8: 149. https://doi.org/10.2903/j.efsa.2010.1556

Fernández-Pinero, J., Gallardo, C., Elizalde, M., Robles, A., Gomez, C., Bishop, R., Heath, L., Couacy-Hymann, E., Fasina, F.O., Pelayo, V., Soler, A. and Arias, M., 2013. Molecular diagnosis of African swine fever by a new real-time PCR using universal probe library. Transboundary and Emerging Diseases 60: 48-58. https://doi.org/10.1111/j.1865-1682.2012.01317.x

Gallardo, C., Fernández-Pinero, J. and Arias, M., 2019b. African swine fever (ASF) diagnosis, an essential tool in the epidemiological investigation. Virus Research 271: 197676. https://doi.org/10.1016/j.virusres.2019.197676

Gallardo, C., Nieto, R., Soler, A., Pelayo, V., Fernández-Pinero, J., Markowska-Daniel, I., Pridotkas, G., Nurmoja, I., Granta, R., Simón, A., Pérez, C., Martín, E., Fernández-Pacheco, P. and Arias, M., 2015. Assessment of African swine fever diagnostic techniques as a response to the epidemic outbreaks in Eastern European Union countries: how to improve surveillance and control programs. Journal of Clinical Microbiology 53: 2555-2565. https://doi.org/10.1128/JCM.00857-15

Gallardo, C., Soler, A., Rodze, I., Nieto, R., Cano-Gómez, C., Fernandez-Pinero, J. and Arias, M., 2019a. Attenuated and non-haemadsorbing (non-HAD) genotype II African swine fever virus (ASFV) isolated in Europe, Latvia 2017. Transboundary and Emerging Diseases 66: 1399-1404. 10.1111/tbed.13132

Gaur, A., Umapathy, G., Vasudevan, K., Sontakke, S., Rao, S., Goel, S., Kumar, D., Gupta, B.J., Singh, D.N. and Kumar, A., 2017. Manual for biological sample collection and preservation for genetic, reproductive and disease analyses. New Delhi: Central Zoo Authority and Laboratory for the Conservation of Endangered Species (LaCONES) CSIR – Centre for Cellular and Molecular Biology. Available at: http://cza.nic.in/uploads/documents/publications/english/Fina%20A5%20Manual%20(1)%20(1).pdf

Herrman, T., 2001. Sampling: procedures for feed. Kansas State University Agricultural Experiment Station and Cooperative Extension Service. MF-2036. Available at: http://www.ksre.k-state.edu/bookstore/pubs/mf2036.pdf

Hurst, C.J. and Reynolds, K.A., 2007. Sampling viruses from soil. Available at https://www.asmscience.org/content/book/10.1128/9781555815882.ch50

James, H.E., Ebert, K., Mcgonigle, R., Reid, S.M., Boonham, N., Tomlinson, J.A. and King, D.P., 2010. Detection of African swine fever virus by loop-mediated isothermal amplification. Journal of Virological Methods 164: 68-74. https://doi.org/10.1016/j.jviromet.2009.11.034

Julian, T.R., Tamayo, F.J., Leckie, J.O. and Boehm, A.B., 2011. Comparison of surface sampling methods for virus recovery from fomites. Applied and Environmental Microbiology 77: 6918-6925. https://doi.org/10.1128/aem.05709-11

Khomenko, S., Alexandrov, T. and Sumption, K., 2013. Options for non-invasive collection of saliva from wild ungulates for disease surveillance. FAO EMPRES-Animal Health 360° Bulletin 42: 15-17. Available at: http://babh.government.bg/userfiles/files/ZJ/Publications/Khomenko%20Alexandrov%202013%20NI%20surv%202013.pdf

Kim, H.-J., Lee, M.-J., Lee, S.-K., Kim, D.-Y., Seo, S.-J., Kang, H.-E. and Nam, H.-M., 2019. African swine fever virus in pork brought into South Korea by travelers from China, August 2018. Emerging Infectious Diseases 25: 1231-1233. https://doi.org/10.3201/eid2506.181684

King, D.P., Reid, S.M., Hutchings, G.H., Grierson, S.S., Wilkinson, P.J., Dixon, L.K., Bastos, A.D. and Drew, T.W., 2003. Development of a TaqMan PCR assay with internal amplification control for the detection of African swine fever virus. Journal of Virological Methods 107: 53-61. https://doi.org/10.1016/S0166-0934(02)00189-1

Kumar, Y., Bansal, S. and Jaiswal, P., 2017. Loop-mediated isothermal amplification (LAMP): a rapid and sensitive tool for quality assessment of meat products. Comprehensive Reviews in Food Science and Food Safety 16: 1359-1378. https://doi.org/10.1111/1541-4337.12309

Leitão, A., Cartaxeiro, C., Coelho, R., Cruz, B., Parkhouse, R.M., Portugal, F., Vigário, J.D. and Martins, C.L., 2001. The non-haemadsorbing African swine fever virus isolate ASFV/NH/P68 provides a model for defining the protective anti-virus immune response. Journal of General Virology 82: 513-523. https://doi.org/10.1099/0022-1317-82-3-513

MacLachlan, N.J. and Dubovi, E.J., 2017. Asfarviridae and Iridoviridae. In: Fenner's Veterinary virology (5th ed.). Academic Press, Cambridge, MA, USA, pp. 175-188, https://doi.org/10.1016/B978-0-12-800946-8.00008-8

Majumdar, T.K. and Howard, D.R., 2011. The use of dried blood spots for concentration assessment in pharmacokinetic evaluations. In: Bonate P. and Howard D. (eds.) Pharmacokinetics in drug development. Springer, Boston, MA, USA, pp. 91-114. https://doi.org/10.1007/978-1-4419-7937-7_4

Männistö, H.E., 2018. Collection of oral fluid samples from wild boar in the field conditions to detect African swine fever virus (ASFV). Master's thesis, Estonian University of Life Sciences, Tartu, Estonia. Available at: https://dspace.emu.ee/xmlui/bitstream/handle/10492/4270/Hanna_M%C3%A4nnist%C3%B6_2018LA_VM.pdf?sequence=1&isAllowed=y

Mazur-Panasiuk, N., Żmudzki, J. and Woźniakowski, G., 2019. African swine fever virus – persistence in different environmental conditions and the possibility of its indirect transmission. Journal of Veterinary Research 63: 303-310. https://doi.org/10.2478/jvetres-2019-0058

Mebus, C., Arias, M., Pineda, J., Tapiador, J., House, C. and Sánchez-Vizcaíno, J., 1997. Survival of several porcine viruses in different Spanish dry-cured meat products. Food Chemistry 59: 555-559. https://doi.org/10.1016/s0308-8146(97)00006-x

Mebus, C., House, C., Gonzalvo, F., Pineda, J., Tapiador, J., Pire, J. and Sanchez-Vizcaino, J., 1993. Survival of foot-and-mouth disease, African swine fever, and hog cholera viruses in Spanish serrano cured hams and Iberian cured hams, shoulders and loins. Food Microbiology 10: 133-143. https://doi.org/10.1006/fmic.1993.1014

Menegat, M.B., Goodband, R.D., DeRouchey, J.M., Tokach, M.D., Woodworth, J.C. and Dritz, S.S., 2019. Kansas State University swine nutrition guide: feed sampling and analysis. Available at: https://www.asi.k-state.edu/research-and-extension/swine/swinenutritionguide/pdf/KSU%20Feed%20Sampling%20and%20Analysis%20fact%20sheet.pdf

Mouchantat, S., Haas, B., Böhle, W., Globig, A., Lange, E., Mettenleiter, T.C. and Depner, K., 2014. Proof of principle: non-invasive sampling for early detection of foot-and-mouth disease virus infection in wild boar using a rope-in-a-bait sampling technique. Veterinary Microbiology 172: 329-333. https://doi.org/10.1016/j.vetmic.2014.05.021

Nagamine, K., Hase, T. and Notomi, T., 2002. Accelerated reaction by loop-mediated isothermal amplification using loop primers. Molecular and Cellular Probes 16: 223-229. https://doi.org/10.1006/mcpr.2002.0415

Notomi, T., Okayama, H., Masubuchi, H., Yonekawa, T.,Watanabe, K., Amino, N. and Hase, T., 2000. Loop-mediated isothermal amplification of DNA. Nucleic Acids Research 28: e63. https://doi.org/10.1093/nar/28.12.e63

OIE (World Organisation for Animal Health), 2018. Transport of biological materials. In: Manual of diagnostic tests and vaccines for terrestrial animals 2018, Vol 1, Chapter 1.1.3 Available at: https://www.oie.int/fileadmin/Home/eng/Health_standards/tahm/1.01.03_TRANSPORT.pdf

OIE (World Organisation for Animal Health), 2019. African swine fever. In: Manual of diagnostic tests and vaccines for terrestrial animals 2019; Vol 2, Chapter 3.8.1. Available at: http://www.oie.int/fileadmin/Home/eng/Health_standards/tahm/3.08.01_ASF.pdf

Oura, C.A., Edwards, L. and Batten, C.A., 2013. Virological diagnosis of African swine fever--comparative study of available tests. Virus Research 173: 150-158. https://doi.org/10.1016/j.virusres.2012.10.022

Penrith, M.L., Thomson, G.R., Bastos, A.D.S., Phiri, O.C., Lubisi, B.A., Du Plessis, E.C., Macome, F., Pinto, F., Botha, B. and Esterhuysen, J.J., 2004. An investigation into natural resistance to African swine fever in domestic pigs from an endemic area in southern Africa. Revue Scientifique et Technique 23: 965-977. https://doi.org/10.20506/rst.23.3.1533

Petrini, S., Feliziani, F., Casciari, C., Giammarioli, M., Torresi, C. and Mia, G.M.D., 2019. Survival of African swine fever virus (ASFV) in various traditional Italian dry-cured meat products. Preventive Veterinary Medicine 162: 126-130. https://doi.org/10.1016/j.prevetmed.2018.11.013

Randriamparany, T., Kouakou, K.V., Michaud, V., Fernandez-Pinero, J., Gallardo, C., Le Potier, M.F., Rabenarivahiny, R., Couacy-Hymann, E., Raherimandimby, M. and Albina, E., 2016. African swine fever diagnosis adapted to tropical conditions by the use of dried-blood filter papers. Transboundary and Emerging Diseases 63: 379-388. https://doi.org/10.1111/tbed.12295

Rowlands, R.J., Michaud, V., Heath, L., Hutchings, G., Oura, C., Vosloo, W. and Dixon, L.K., 2008. African swine fever virus isolate, Georgia, 2007. Emerging Infectious Diseases 14: 1870-1874. https://doi.org/10.3201/eid1412.080591

Sánchez-Cordón, P.J., Jabbar, T., Berrezaie, M., Chapman, D., Reis, A., Sastre, P., Rueda, P., Goatley, L. and Dixon, L.K., 2018. Evaluation of protection induced by immunisation of domestic pigs with deletion mutant African swine fever virus BeninΔMGF by different doses and routes. Vaccine 36: 707-715. https://doi.org/10.1016/j.vaccine.2017.12.030.

Sánchez-Vizcaíno, J.M., Laddomada, A. and Arias, M.L., 2019. African swine fever virus. In: Zimmerman, J.J., Karriker, L.A., Ramirez, A., Schwartz, K.J., Stevenson, G.W. and Zhang, J. (eds.) Diseases of swine (11th ed.). John Wiley and Sons Inc., Hoboken, NJ, USA, pp. 443-452. https://doi.org/10.1002/9781119350927.ch25

Sastre, P., Gallardo, C., Monedero, A., Ruiz, T., Arias, M., Sanz, A. and Rueda, P., 2016a. Development of a novel lateral flow assay for detection of African swine fever in blood. BMC Veterinary Research 12: 206. https://doi.org/10.1186/s12917-016-0831-4.

Sastre, P., Pérez, T., Costa, S., Yang, X., Räber, A., Blome, S., Goller, K.V., Gallardo, C., Tapia, I., García, J., Sanz, A. and Rueda, P., 2016b. Development of a duplex lateral flow assay for simultaneous detection of antibodies against African and classical swine fever viruses. Journal of Veterinary Diagnostic Investigation 28: 543-549. https://doi.org/10.1177/1040638716654942.

Schulz, K., Staubach, C., Blome, S., Nurmoja, I., Viltrop, A., Conraths, FJ., Kristian, M. and Sauter-Louis, C., 2020. How to demonstrate freedom from African swine fever in wild boar-Estonia as an example. Vaccines 8: 336 https://dx.doi.org/10.3390%2Fvaccines8020336

Sindryakova, I., Morgunov, Y., Chichikin, A., Gazaev, I., Kudryashov, D. and Tsybanov, S., 2016. The influence of temperature on the Russian isolate of African swine fever virus in pork products and feed with extrapolation to natural conditions. Selskokhozyaistvennaya Biologiya 51: 467-474. https://doi.org/10.15389/agrobiology.2016.4.467eng

Tignon, M., Gallardo, C., Iscaro, C., Hutet, E., Van der Stede, Y., Kolbasov, D., De Mia, G.M., Le Potier, M.F., Bishop, R.P., Arias, M. and Koenen, F., 2011. Development and inter-laboratory validation study of an improved new real-time PCR assay with internal control for detection and laboratory diagnosis of African swine fever virus. Journal of Virological Methods 178: 161-170. https://doi.org/10.1016/j.jviromet.2011.09.007

United Nations, 2019a. Recommendations on the transport of dangerous goods. Model regulations. Volume I. Available at: https://www.unece.org/fileadmin/DAM/trans/danger/publi/unrec/rev21/ST-SG-AC10-1r21e_Vol1_WEB.pdf

United Nations, 2019b. Recommendations on the transport of dangerous goods. Model regulations. Volume II. Available at: https://www.unece.org/fileadmin/DAM/trans/danger/publi/unrec/rev21/ST-SG-AC10-1r21e_Vol2_WEB.pdf

Wang, D., Yu, J., Wang, Y., Zhang, M., Li, P., Liu, M. and Liu, Y., 2020. Development of a real-time loop-mediated isothermal amplification (LAMP) assay and visual LAMP assay for detection of African swine fever virus (ASFV). Journal of Virological Methods 276: 113775. https://doi.org/10.1016/j.jviromet.2019.113775

Wang, W.-H., Lin, C.-Y., Ishcol, M.R.C., Urbina, A.N., Assavalapsakul, W., Thitithanyanont, A. and Wang, S.-F., 2019. Detection of African swine fever virus in pork products brought to Taiwan by travellers. Emerging Microbes and Infections 8: 1000-1002. https://doi.org/10.1080/22221751.2019.1636615

6. African swine fever vaccines

C.L. Netherton

The Pirbright Institute, Ash Road, Pirbright, Woking, Surrey, GU24 0NF, United Kingdom; christopher.netherton@pirbright.ac.uk

Abstract

African swine fever virus causes an acute haemorrhagic fever in domestic pigs and wild boar which is invariably fatal. Introduction of the disease into Georgia in 2007 has led to the deaths of tens of millions of animals across Eastern Europe, Asia and Oceania with serious effects on animal welfare and global food security. Control of the disease is impaired by the lack of an effective vaccine and is dependent on strict biosecurity at the farm gate, and rapid diagnosis, quarantine and slaughter of infected herds. The few pigs that do recover from disease are robustly protected from a subsequent encounter with the same virus isolate, showing that immunity is achievable. This review provides a historical perspective on the approaches that researchers have taken to develop African swine fever vaccines, as well as discussing promising modern techniques such as targeted gene deleted viruses and viral vectored vaccines.

Keywords: ASF vaccine, protective antigens, ASF, protective immunity, African swine fever

This publication is based upon work from COST Action CA15116, ASF-STOP, supported by COST (European Cooperation in Science and Technology).
www.cost.eu

EUROPEAN COOPERATION IN SCIENCE & TECHNOLOGY

Funded by the Horizon 2020 Framework Programme of the European Union

DOI 10.3920/978-90-8686-910-7_6, © C.L. Netherton, 2021

6.1 Introduction and classic approaches to vaccination

African swine fever (ASF) was first described as an acute, invariably fatal disease in European breeds of domestic pigs. However, it was quickly realised that the few animals that did recover from acute fever were protected from a subsequent encounter with the virus, demonstrating the prospect of a protective immune response (Montgomery, 1921). These early studies also showed that wild suids, particularly warthogs and bushpigs, were associated with outbreaks of disease and that bushpigs were resistant to an inoculum that would normally kill a domestic pig. Since these early beginnings a number of different approaches have been taken to develop an effective African swine fever vaccine, but at the time of writing in June 2020 no licensed, commercial African swine fever vaccine is being used in the field.

6.1.1 Immune serum

Based on the observation that hyperimmune serum from animals recovered from European swine fever (now called classical swine fever) was able to protect against subsequent encounters with the virus, sera from the few animals that recovered from natural infection with African swine fever virus (ASFV) were tested for their immunological properties. These types of experiment were also used to show that African and classical swine fever were caused by two different aetiological agents. However, the results from these tests were not particularly promising as only a few animals survived (Montgomery, 1921), although both Walker and DeTray reported some positive results when serum and virus were inoculated at the same time (DeTray, 1957). Later experiments showed that transfer of serum from animals recovered from a moderately virulent isolate of ASFV (Malta 1978) reduced viraemia and clinical signs after homologous challenge and allowed survival of pigs after heterologous challenge (Uganda), although viraemia was not significantly reduced (Wardley *et al.*, 1985). Experiments also showed that feeding neonatal pigs with colostrum from sows recovered from ASF or injecting them with hyperimmune serum protected them from severe disease and death. Viraemia and clinical signs were much reduced compared to control colostrum deprived piglets or those fed normal colostrum, and piglets delivered by ASF recovered sows were also protected from challenge with homologous virus (Schlafer *et al.*, 1984). Immunoglobulin G (IgG) concentrated from animals that had been immunised with an attenuated strain (E75-CV1) and then challenged with a homologous virulent isolate (E75) was passively transferred into naïve pigs. Titres of ASFV-specific antibodies in the transferred pigs were similar to those from the pigs from which the serum was obtained, and in two separate experiments twelve out of fourteen pigs that had serum transferred survived challenge with E75. Although most of the animals were viraemic, this was 10,000 times less than controls and the duration and magnitude of fever were also significantly less (Onisk *et al.*, 1994). These experiments showed the importance of the antibody response to the protection mediated by the E75-CV1 strain, but did suggest that a vaccine that induced an antibody response alone might not be sufficient to generate a truly effective protective response. Later experiments would demonstrate the importance of the cellular response to protection mediated by live attenuated viruses (Oura *et al.*, 2005).

6.1.2 Inactivated virus

African swine fever emerged just before development of the two principal techniques that would dominate vaccine development for around fifty years, inactivation and attenuation by repeated passage through either animals or tissue culture (Figure 6.1). In his initial description of the disease, Montgomery showed that heat treatment could inactivate the aetiological agent of ASF but with the exception of a single pig, did not protect animals from a subsequent inoculation with live virus (Montgomery, 1921). This approach was repeated using a number of different agents including toluene, crystal violet, formalin, β-propiolactone, glutaraldehyde and binary ethyleneimine, all without a clear indication of protection. Experiments with spleen extracts treated with the non-ionic detergent n-octylglucoside yielded better results, with 70% of animals surviving challenge with homologous virus, but these animals were persistently infected and transmitted disease to in contact animals. Similar experiments with n-octyl β-D-glucopyranoside showed reasonable levels of protection against challenge but, again, the recovered animals were

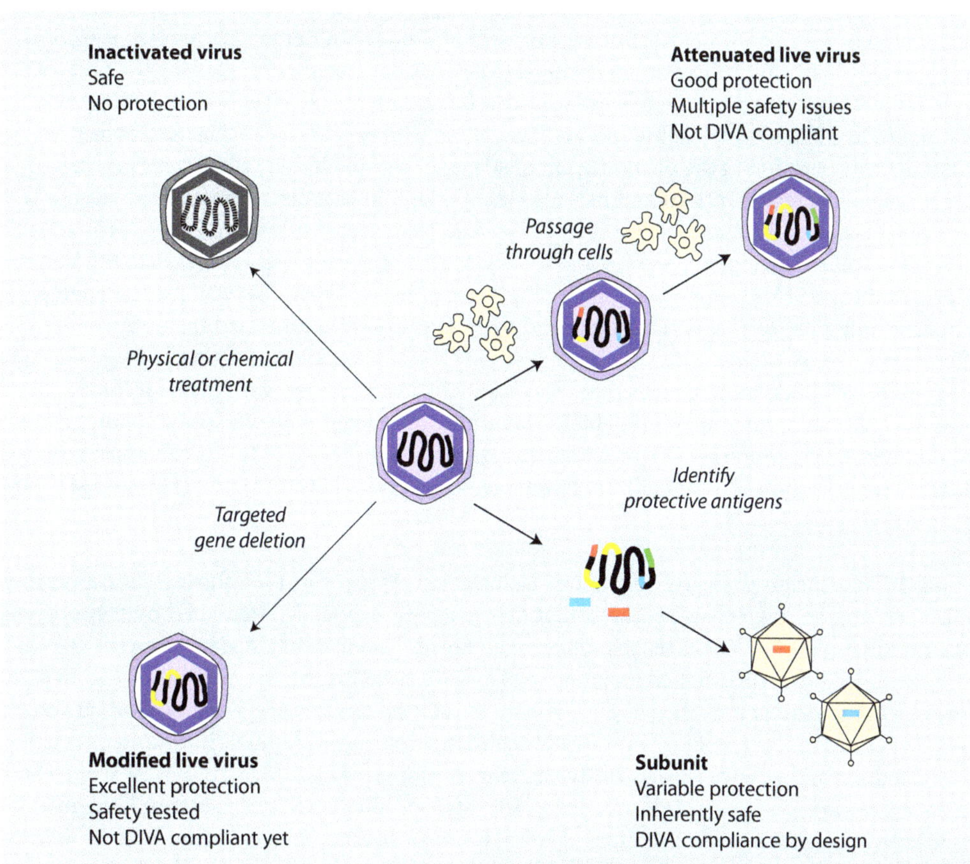

Figure 6.1. Approaches to African swine fever vaccine development. Schematic highlighting four approaches to generating an African swine fever virus vaccine. Inactivated virus, attenuated live virus (attenuation by passage through tissue culture), modified live virus (attenuation by targeted gene deletion) and subunit vaccines.

persistently infected, although they did not transmit to in contact pigs (Kihm *et al.*, 1987). Both approaches failed to identify antibody responses to their respective immunogens and therefore the mechanism of protection in each case is unclear. Moreover, attempts by another group to repeat the experiments that showed a protective effect of n-octylglucoside treated ASFV were not successful (Forman *et al.*, 1982). More recent work used binary ethyleneimine treated virus in conjunction with the adjuvants Polygen™ and Emulsigen®-D and, although both inactivated preparations induced ASF-specific antibodies, they did not prevent disease by subsequent challenge with homologous virus (Armenia 2008 isolate) (Blome *et al.*, 2014). As discussed in later sections, experiments with individual proteins were able to offer some protection, but the prospects of an effective vaccine based on inactivated virus material or particles seem low.

6.1.3 Attenuation by passage

Initial attempts to passage the virus involved both embryonic and adult rodents, eggs, rabbits and goats, sometimes in conjunction with alternate passage through pigs. Using these techniques infectious virus could be maintained for up to 80 passages in the case of rabbits, without significant attenuation of the virus. It was not until the work of Malmquist in Kenya (Malmquist, 1962; 1963) and Manso Ribeiro and colleagues in Portugal (Manso Ribeiro *et al.*, 1963) during the mid-1950s with tissue culture that reproducible progress towards an attenuated virus was made (Figure 6.1). Malmquist and Hay's work (Malmquist and Hay, 1960) with porcine leukocytes and bone marrow led to the identification of the phenomenon of adsorption of red blood cells to cells infected with ASFV (called haemadsorption) and its subsequent adoption as a discriminating laboratory test between African and classical swine fever. They also passaged virus through these primary cell cultures as well as through pig kidney and testis cell lines. The progressive passage of the Kenyan Hinde strain of ASFV through pig kidney cells 75 times led to a virus that was attenuated in pigs, inducing only mild or inapparent signs, and these animals were protected from severe disease after subsequent challenge with the parental strain (Malmquist, 1962). Further passage through the porcine cell lines led to a virus that was essentially avirulent and was unable to protect pigs from the parental strain, an effect which Malmquist associated with the loss of pathogenicity (Malmquist, 1963). Much later experiments adapting the Georgia strain to African green monkey kidney Vero cells showed a similar loss of pathogenicity between the 80th and 110th passage (Krug *et al.*, 2015).

Manso Ribeiro's team passaged Portuguese isolate 1455 obtained in 1961 through bone marrow cells derived from piglets and showed that the virus was gradually attenuated by this process (Manso Ribeiro *et al.*, 1963). The Portuguese team noted that virus obtained between the 40th and 60th passage induced unspecified locomotive disorders as well as the typical signs of ASF in pigs; however, as the virus approached its 60th passage the clinical signs of ASF were less evident and by the 70th passage the locomotive disorders were also much reduced. Animals immunised with this attenuated strain that were subsequently challenged with virus circulating in both Portugal and Spain at the time were protected from acute ASF. Manso Ribeiro and colleagues did not observe significant clinical signs in the vaccinated animals nor an effect on in contact animals. Immunity was evident 14 days post immunisation with strain 1455, but not at 7 days. Both Malmquist and Manso Ribeiro's team explored the idea of cross-protection between different strains of virus. However, strain 1455 protected against selected isolates circulating in Spain and Portugal in the 1960s, but did not protect against isolates obtained from the original outbreak in Lisbon in 1957 or from Katanga (now Democratic Republic of Congo). Similarly, the attenuated Hinde strain protected against homologous challenge, but not from viruses obtained from either Portugal or South Africa, although the numbers of animals used in these latter experiments were very low (Malmquist, 1963). Malmquist's work also provided the first hints of a potential mechanism to the observed protection, namely that serum from the animals that recovered from attenuated Hinde was capable of blocking the adsorption of red blood cells to cells infected with the parental Hinde isolate, but not to those infected with either of the Portuguese or South African isolates. While all of the viruses used in the Manso Ribeiro experiments were genetically related (p72 genotype I), they differed immunologically, as Katanga and Lisbon 1957 were haemadsorption inhibition assay serogroup 1 and Lisbon 60 was serogroup 4 (Malogolovkin *et al.*, 2015).

Although the performance of strain 1455 under experimental conditions appeared promising, unfortunately this was not replicated in field trials (Manso Ribeiro *et al.*, 1963). Between May and June 1962, 40,000 pigs were vaccinated, the number was increased to over half a million swine as the situation in Portugal worsened. During this time 128,684 animals across 1,247 herds were monitored closely and 3.4% of these suffered cases of acute ASF, out of which less than a third occurred more than three weeks after vaccination and could be assessed as vaccination failure (either through lack of protection or insufficient attenuation). However, there were significant complications in approximately 11% of vaccinated animals, principally concerning the respiratory, digestive and musculoskeletal systems as well as the skin. Between 7 and 10% of vaccinated animals died of these complications, depending on the age and breed, with Iberian pigs suffering the most. Pulmonary complications were particularly prevalent and were also seen in approximately 20% of the carcasses of pigs slaughtered in the area during early 1963. The vaccine also caused abortions, stillbirths and infertility in sows. The poorer performance of the vaccine in Iberian pigs was put down to the generally poorer health of these animals relative to other animals, rather than to the breed itself; however, Duroc-Jersey pigs were mentioned as being particularly sensitive to vaccine complications.

Around the same time researchers in Spain were also exploring the effect of passaging ASFV through pig kidney cells and leukocytes (Sánchez Botija, 1963). Isolate AL passaged through kidney cells 25 (AL-25), 30 or 60 times were compared to the same virus passaged 80 times (AL-80) through porcine leukocytes. AL-25 and AL-80 had some residual virulence, killing 3 to 6% of animals in the first 20 days after inoculation with the remainder recovering. However, between 10 and 50% of these recovered animals within a given experiment died either by a recurrence of acute disease or by chronic pulmonary disease. Sánchez Botija (1963) observed that the mortality from chronic pulmonary disease was much less in small scale laboratory animal experiments, an observation that he concluded was due to the better general health of the animals and the greater care that was taken with their husbandry. 'Hardy' pigs were also considered to be more susceptible to the vaccine strain. Recurrence of acute ASF and chronic forms of the disease was not seen in animals inoculated with the virus that had been passaged on kidney cells 60 times (AL-60), suggesting further passage through kidney cells had further attenuated the virus. All of the three attenuated viruses protected surviving animals from subsequent challenge with the parental strain. AL-25 and AL-60 were tested in field trials and natural cases of acute ASF were not seen in herds vaccinated with AL-25, whereas the protective efficacy of AL-60 showed more herd-to-herd variability. Sánchez Botija (1963) concluded that their attenuated strains required further analysis before wider usage and also highlighted that it was not possible to discriminate between vaccinated and naturally infected animals using the haemadsorption test.

The use of live attenuated ASF vaccines, particularly in Portugal where strain 1455 was used over much of the Southern part of the country, has often been linked with the endemicity of ASF in the Iberian Peninsula for 30 to 40 years (Costa, 1990). Although chronic forms of the disease were clearly associated with animals that were vaccinated during this time (Nunes Petisca, 1965), the presence of chronically affected animals had already been identified both in Spain (Sánchez Botija, 1962) and in Angola (Leite Velho, 1956), generally thought to be the origin of the European strains at that time. It also seems likely that at least two antigenically different strains of ASFV were circulating in Iberia at this point (Lacasta *et al.*, 2015; Malogolovkin *et al.*, 2015; Sereda *et al.*, 2020), which could explain some of the vaccine failures in the field. Additionally, the association of a suitable biological vector with the traditional farming practices in Iberia was something that no one could have predicted at the time. In short, ASF and Iberia was a case of the wrong virus in the wrong place and the unfortunate early use of vaccines is quite likely to have exacerbated the issue.

Researchers in the Soviet Union in the second half of the twentieth century used the haemadsorption inhibition assay as a tool to guide the generation of a panel of immunologically distinct attenuated strains of ASFV by passage through porcine bone marrow cultures (Sereda *et al.*, 2020). These viruses could then be used in the event of an outbreak of ASFV if the epidemiological situation permitted. Despite the outbreak of ASF in Odessa in 1977, this situation did not occur and the vaccines were never used outside of the laboratory. The process of building up the panel of attenuated strains also produced the most comprehensive map of ASFV immunotypes to date, with at least eleven different immunotypes identified based on the induction of antibodies capable of inhibiting haemadsorption to infected macrophages, in combination with immunisation and challenge studies in pigs (Malogolovkin *et al.*, 2015; Sereda *et al.*, 2020). An ASFV isolate from Rhodesia (now Zimbabwe) attenuated by serial passage was shown to have the same serogroup as the current Eurasian outbreak isolates and was capable of protecting pigs (Balyshev *et al.*, 2011) but it also induced chronic disease (Sereda *et al.*, 2020). Attenuated viruses were also used to test the effect of immune status of recipient animals and results supported earlier observations by Manso Ribeiro's team (1963) and Sánchez Botija (1962) that animals with meagre health had a poorer response to attenuated ASF vaccines (Budarkov *et al.*, 2017). Differences in the protection obtained with live attenuated viruses have been observed between farm animals and specific pathogen free animals (King *et al.*, 2011; Lacasta *et al.*, 2014). Host genetics are also likely to play a role, as differences in protection after OUR T88/3 immunisation were observed between different lines of minipigs inbred at the major histocompatibility locus (Oura *et al.*, 2005). Live attenuated viruses continued to be used extensively as tools to study the immune response to infection as discussed in Chapter 3.

Viruses with reduced virulence have been obtained from wild boar in the field during the present ASF panzootic. Notably one of them, a non-haemadsorbing strain from Latvia (Lv17/WB/Rie1), when delivered orally can protect wild boar from subsequent challenge with the virulent Armenia 2008 isolate (Barasona *et al.*, 2019). This strain is being considered as a potential vaccine; however, both intramuscular inoculation and natural transmission of this virus to domestic pigs causes non-specific clinical signs, including joint swelling, which are similar to those seen after infection with two Portuguese isolates NH/P68 and OUR T88/3, both of which cause chronic disease (Gallardo *et al.*, 2019). The differences between the Lv17/WB/Rie1 strain and the virulent parent are relatively minor compared to those seen in NH/P68 and OUR T88/3. Therefore, this naturally attenuated isolate from Latvia may represent an excellent candidate for further targeted genetic modification.

6.2 The recombinant revolution: targeted gene deletion

The emergence of the technology to genetically modify virus, along with the beginnings of the sequencing revolution of the late 1980s and early 1990s, greatly increased our ability to interrogate and understand a whole range of different aspects of ASFV biology, not least the identification of genetic factors related to virulence. The ability to excise genes from the ASFV genome meant that attenuated viruses could now be developed by targeted gene deletion rather than blind passage through tissue culture (Figure 6.1). The first targets for gene deletion were the very few ASFV genes which showed similarity to known cellular genes and/or genes from other viruses. The *DP71L* gene (also referred to as *NL*) is similar to the *ICP34.5* gene which is a known virulence factor from herpes simplex virus 1. Both *ICP34.5* and *DP71L* share similarity with a cellular protein called GADD34 that helps control a key step in the synthesis of proteins within the host. One of the host responses to cellular stress or viral infection is a decrease in protein synthesis within cells and GADD34 is part of the process that turns protein synthesis back on. By mimicking GADD34 activity *DP71L* can help maintain high levels of protein synthesis in the face of this host response and so ensure that ASFV can produce all of the proteins it requires to make new progeny (Barber *et al.*, 2017). Removal of the *DP71L* gene from the Spanish E70 ASFV isolate did not affect replication *in vitro*, but led to a significantly attenuated virus *in vivo*, with animals suffering a short fever and reduced viraemia. All of the animals survived the E70 deletion mutant and when they were subsequently challenged with the virulent parental virus all of them were protected without showing clinical signs or viraemia (Zsak *et al.*, 1996). This promising result led the authors to try and delete the *DP71L* gene from virulent isolates obtained from both Chalaswa in Malawi and Pretoriuskop in South Africa. Deletion of *DP71L* had no effect on the virulence of Malawi and only slightly altered disease progression of the Pretoriuskop strain (Afonso *et al.*, 1998). The observation that ASFV may encode multiple virulence factors and that different genes may be more or less important for virulence in a given isolate, led to a series of experiments where different genes were removed from these two African isolates. Individual ASFV genes or combinations of the genes *DP96R* (*UK*), *K196R* (*TK*), *B119L* (*9GL*) and a section of genome encoding six multigene family members were all shown to significantly attenuate different isolates in domestic swine (Lewis *et al.*, 2000; Moore *et al.*, 1998; Neilan *et al.*, 2002; Zsak *et al.*, 1998). A Malawi *B119L* deletion showed a significant growth defect in macrophages and in pigs viraemia was 100 to 10,000-fold lower than wild type virus, with a short fever duration and reduced clinical signs. All of the animals survived inoculation with the *B119L* deletion mutant and subsequent challenge with a lethal dose of the parental Malawi strain, although the dose of the deletion mutant affected the length of viraemia after challenge. The *B119L* gene encodes for a protein that is involved in formation of progeny viruses; the deletion of this gene did not block virus replication *per se*, but resulted in the assembly of aberrant capsids. This observation suggested that deletion of *B119L* might provide a route to attenuate any given strain of ASFV.

The first complete genome sequence of an attenuated strain obtained from the field (Chapman *et al.*, 2008) allowed the identification of potential virulence factors by comparative genomic analysis. The Portuguese OUR T88/3 had a large deletion that meant the virus lacked members of MGF360 and MGF505 similar to those that had previously been shown to be important for the virulence of the Pretoriuskop isolate (Neilan *et al.*, 2002). These multigene family members were then deleted from the Benin 1997/1 isolate and this virus was attenuated in pigs and could protect those pigs from subsequent challenge (Reis *et al.*, 2016). In another experiment the *DP148R* (*MGF360-18R*) gene was deleted from the Benin 1997/1 isolate and this virus was also significantly attenuated in swine. After a transient fever, animals recovered and were protected after challenge with parental virulent virus (Reis *et al.*, 2017). The level of protection achieved with the *DP148R* deleted virus was practically 100%, although vaccinated animals did have persistent viraemia that lasted for several weeks. However, the pigs were clinically normal and this suggested that further modification of the *DP148R* deleted virus may yield a vaccine with an attractive safety profile.

6.2.1 Genotype II vaccines

Research into ASF vaccines intensified as the situation in Europe deteriorated and the disease subsequently spread into China. What had been learnt from previous experiments with the genotype I viruses obtained from Europe in the second half of the last century, as well as those from other geographical regions, was applied to the current panzootic with mixed results. The *EP402R* gene encodes for a protein responsible for the adsorption of red blood cells to the surface of infected cells and although deletion of this gene from the Georgia isolate did not attenuate its virulence in pigs (Borca *et al.*, 2020), it did attenuate the Spanish Badajoz 1971 strain (Monteagudo *et al.*, 2017). Interestingly, this deletion mutant was capable of protecting pigs from the genotype II Georgia strain. This protection was linked to the ability of CD8 T-cells to recognise heterologous virus, as the antibody response induced by the *EP402R* deletion mutant was highly variable. Nevertheless, some of the animals immunised with the Badajoz *EP402R* deletion mutant did become sick after challenge and ASFV was found in some of the animal's secretions, suggesting further optimisation was required. The Badajoz *EP402R* deletion mutant was also successfully maintained on a continuous African green monkey kidney cell line (COS-1) for twenty passages without significant genetic modification. Generation and culture of recombinant deletion mutants of ASFV field isolates are for the most part still carried out on primary macrophages derived from porcine bone marrow, leukocytes or lung washes. Current EU guidelines for veterinary medicinal products require freedom from 23 different extraneous agents and an ASF vaccine derived from primary cells may require a herd of specific pathogen free animals. A prospective ASF vaccine that could be propagated on a defined cell line would therefore have significant advantage over one that required primary cell culture.

In contrast to the experiments with the Malawi isolate, a Georgia *B119L* deletion was not completely attenuated in swine (Lewis *et al.*, 2000; O'Donnell *et al.*, 2015b) although animals survived a low dose and were protected from challenge with the virulent parental strain. A multigene family deletion did attenuate the Georgia strain (O'Donnell *et al.*, 2015a), but surviving animals challenged with parental virus still harboured the challenge virus despite appearing clinically normal. Different combinations of *B119L*, the multigene family region, *DP71L* and the *DP96R* genes were then deleted from the Georgia backbone (O'Donnell *et al.*, 2016, 2017;

Ramirez-Medina *et al.*, 2019). A double deletion of *B119L* and *DP96R* was attenuated in swine, even at a dose 100 times greater than that which would have caused disease after infection with the single deletion of *B119L*. The double deletion also protected some pigs from challenge as early as 14 days post vaccination, but there was a marked difference in the response after challenge depending on the dose used for immunisation. The optimal dose protected all of the animals from severe disease with many of them showing minimal clinical signs, unfortunately a number of the animals still harboured virus three weeks later (O'Donnell *et al.*, 2017). However, a recent study has identified the *I177L* gene as a target for attenuating ASFV by gene deletion. This gene encodes for a highly conserved protein of unknown function that is a component of the virion. Interruption of this gene in the Georgia strain led to an interesting phenotype where viraemia was delayed by several days and the peak of circulating virus varied significantly from pig to pig. Crucially, despite some animals having as much virus circulating the blood as those infected with the wild-type virulent Georgia isolate, none of the pigs developed a temperature or exhibited any clinical signs (Borca *et al.*, 2020). All pigs were robustly protected when they were infected with the parental strain. Similarly, to what was seen in pigs infected with different Benin deletion mutants (Reis *et al.*, 2016, 2017), there was persistence of the virus in the blood stream; however, in all but one of the animals this was the vaccine and not the virulent parental virus. Preliminary data suggested that neither the vaccine nor the challenge virus was transmitted to sentinel pigs and even low doses of the *I177L* deletion mutant were capable of inducing robust protection, making this virus a potential vaccine candidate.

A number of deletion mutants were evaluated on a larger scale with a genotype II virus isolated from Heilongjiang in China in 2018 (Chen *et al.*, 2020) that is >99% identical to the Georgian strain used by other researchers. These experiments confirmed that deletion of *EP402R* did not attenuate Eurasian genotype II ASFV and also showed that the additional deletion of *DP96R* (*UK*) did not improve this attenuation. Furthermore, deletion of *DP148R* that attenuated the virulent Benin 1997/1 strain was not sufficient to attenuate the Chinese strain. However, deletions of the multigene family region and a double deletion of *B119L* and *DP96R* were attenuated, as was a combination of the multigene family region and *EP402R*. These three deletions were then tested for the ability to protect the immunised pigs against challenge with the virulent parental virus. Both of the multigene family-based deletions protected domestic pigs, but in contrast to the analogous virus based on the Georgia strain (O'Donnell *et al.*, 2017) the double deletion of *B119L* and *DP96R* did not induce a protective immune response. Pigs immunised with a high dose of the deletion mutant lacking the multigene family region and the *EP402R* gene only suffered a brief fever for one day after challenge, although ASFV was found in the tissues of animals three weeks later. Next, the multigene family deleted viruses were safety tested for reversion to virulence by passaging the viruses five times through pigs. Interestingly, the virus that did not have an *EP402R* deletion progressively replicated more efficiently as it was repeatedly passaged through pigs and by the fifth passage one animal died. This apparent reversion to virulence did not occur with the virus that had both the multigene family region and *EP402R* deleted and therefore the reason for the reversion is unknown. However, vaccinia virus can rapidly duplicate sections of its genome related to host immune evasion (Elde *et al.*, 2012). The *EP402R* deletion is not sufficient to attenuate genotype II ASFV on its own and therefore does not provide an additional safety measure if the attenuation conferred by the multigene family deletion can be reversed. A single dose of the multigene family and *EP402R* deletion mutant protected pigs from severe disease, but not infection, five months after immunisation, and this virus did not affect the health of a smaller number of pregnant sows or their piglets. The Chinese double deletion mutant is moving into field trials that may involve between 10,000 to 20,000 pigs and the results of these experiments will be of huge interest (Mallapaty, 2019). Previous experience in Spain and Portugal fifty years ago suggested that results in well husbanded pigs in clean animal isolation units did not translate into positive results in the field. However, none of the laboratories working with genotype II deletion mutants have reported any clinical signs that could be consistent with reactivation of the disease or chronic ASF, therefore it is to be hoped that second time round field trials of a live attenuated ASF vaccine will be successful.

6.3 Subunit vaccines: unfulfilled early promise

6.3.1 Protein vaccines

Antibody responses to specific ASFV proteins were first identified in the late 1970s and early 1980s (Black and Brown, 1976; Tabarés *et al.*, 1980). As the sequences of the genes encoding for these different antigens were deciphered, the possibility of generating immune responses in pigs without the use of ASFV became a reality (Figure 6.1). However, some of the first experiments of this type used protein purified directly from the supernatants of ASFV infected cells. Most ASFV isolates haemadsorb red blood cells to the surface of infected cells. However, Ruiz-Gonzalvo and Coll (Ruíz-Gonzalvo and Coll, 1993) identified an unusual strain which caused red blood cells to clump together (called haemagglutination) without needing to adhere to cells. In a series of experiments the haemagglutinating activity was separated, concentrated and used to immunise pigs. The immunisation induced antibodies in the serum of the animals that were capable of blocking the haemadsorption to infected cells *in vitro*. In a subsequent pilot study immunisation with the haemagglutinating activity protected one of two pigs from severe disease after contact challenge with a pig infected with the parental strain. Around the same time researchers in Russia had identified that the haemadsorption activity of virally infected cells was associated with glycoproteins (proteins modified by the addition of sugar residues). The haemadsorbing activity was then separated from other glycoproteins using anti-sera from recovered pigs. Similar to the results obtained by Ruiz-Gonzalvo and Coll (1993) this purified haemadsorbing activity could protect pigs from severe disease after infection with a related strain of ASFV. The ability of the protein to induce antibodies capable of blocking haemadsorption and protecting pigs from disease was also found to be dependent on the sugar residues attached to the protein (Sereda, 2013; Sereda *et al.*, 2018).

The identification of the gene encoding for the protein responsible for haemadsorption (Rodríguez *et al.*, 1993) enabled the generation of a recombinant baculovirus that could reproduce the haemadsorbing activity *in vitro*. The baculovirus was then used to immunise three pigs, which resulted in the animals having antibodies that could inhibit red blood cells sticking to infected macrophages and, in one of the pigs, antibodies were also capable of blocking infection of macrophages (Ruiz-Gonzalvo *et al.*, 1996). Strikingly, all three pigs survived challenge with the E75CV1 strain of ASF and the pig that had inhibiting antibodies had no circulating virus in the bloodstream throughout the course of the experiment. The cellular response to EP402R or ASFV was not tested in this study, but more recent work has identified T-cell epitopes in the EP402R protein (Argilaguet *et al.*, 2012; Burmakina *et al.*, 2019) and it is possible that immunisation with baculovirus infected cells also induced a cellular response against EP402R that could have contributed to protection.

As well as identifying the ASFV gene responsible for the haemadsorption phenomenon, significant progress has been made in identifying targets of neutralising antibodies against ASFV. In a series of experiments researchers showed that neutralising activity could be found in the sera of recovered pigs (Ruiz Gonzalvo *et al.*, 1986a,b) and that this activity targeted three different proteins that were part of the virus particle (Gómez-Puertas *et al.*, 1996). Antibodies against p72 and p54, encoded by the *B646L* and *E183L* genes respectively, inhibited the attachment of virus to the surface of susceptible cells, whereas antibodies against p30 (*CP204L* gene) were also capable of blocking the internalisation of virus that had already bound to cells (Gomez-Puertas *et al.*, 1996, 1998). Immunisation of pigs with cells infected with a baculovirus expressing p30 and p54, or a fusion of the two genes, induced antibodies against the two proteins that were capable of neutralising the virus (Barderas *et al.*, 2001; Gomez-Puertas *et al.*, 1998). Some of these pigs survived infection with virulent strain(s) and although the animals became sick and viraemic, both the clinical signs and levels of circulating virus in the bloodstream were lower than in control animals. All of the surviving vaccinated animals had cleared the virus by the end of the experiment. Unfortunately, these promising results did not translate to other strains of ASF. Baculoviruses expressing p30, p54, p72 and p22 (*KP177R* gene) derived from the South African Pretoriuskop isolate induced ASFV-specific antibodies capable of neutralising the virus. However, in contrast to the experiments with the E75 isolate the immunised pigs were not protected against challenge (Neilan *et al.*, 2004).

6.3.2 DNA vaccines

Immune responses can also be induced to individual proteins by immunising with the sequence of DNA that codes for that particular protein. Typically, genes in DNA vaccines are encoded by plasmids that can be directly injected into the target species or they can be introduced by specialist equipment, such as a gene gun. Manufacturing costs are typically a lot lower than producing the proteins themselves; lyophilised DNA is very stable and can be easily reconstituted prior to immunisation. DNA vaccination also lends itself to multi-valent vaccines or an approach called 'library' immunisation where many different genes or fragments of genes can be inoculated at the same time. This could be advantageous for complex pathogens like ASFV that have multiple genes and/or for which the antigenic determinants are unknown. DNA vaccines have been licensed for two different salmon viruses (APEX-IHN® and Clynav), avian influenza, West Nile virus in horses and canine melanoma (Oncept®).

Initial results with DNA vaccines against ASFV were not promising. A plasmid expressing the same p30 and p54 fusion protein described in the previous section was not immunogenic in pigs without the use of a molecular adjuvant (Argilaguet *et al.*, 2011). The p30 and p54 fusion protein were then fused to part of an antibody that specifically targeted ASFV proteins to pig macrophages. This led to the induction of immune responses to the two viral proteins and to the whole virus, but in contrast to the results with the baculovirus expressed proteins, the antibody response elicited did not inhibit the ability of the virus to infect macrophages, rather it enhanced it. Immunised pigs were not protected and had elevated viral titres three days after challenge, correlating with the infection enhancement seen *in vitro*. The difference in both the immune response and the clinical outcome with the same antigens after immunisation by DNA or baculovirus highlights the importance of the delivery system in the ultimate outcome of the vaccine. Disease and/or infection enhancement is an additional complication in designing ASFV subunit vaccines and it has been described by multiple groups using different antigens and vaccine platforms (Jancovich *et al.*, 2018; Netherton *et al.*, 2019; Sunwoo *et al.*, 2019).

The cellular immune response induced by the initial ASF DNA vaccine appeared to be biased toward a CD4 response (Argilaguet *et al.*, 2011); however, experiments with live attenuated ASFV had shown that a CD8 T-cell response was important for protection (Oura *et al.*, 2005). The protein responsible for ASFV haemadsorption encoded by the *EP402R* gene had already been shown as a protective antigen (Ruiz-Gonzalvo *et al.*, 1996) and so was also included in a follow up DNA vaccine study. A fusion between EP402R, p54 and p30 was created and combined with a different molecular adjuvant designed to direct the immune response toward a cellular response dominated by CD8 T-cells (Argilaguet *et al.*, 2012). Immunisation of pigs with the non-adjuvanted EP402R-p54-p30 fusion induced both antibody and cellular responses to the antigens, but again the antibody responses were not able to neutralise ASFV infectivity *in vitro* and were not able to protect pigs from challenge with virulent ASFV. Strikingly, the construct designed to target a cellular immune response did protect some of the pigs despite not inducing any detectable antibodies. This approach was expanded to include approximately three quarters of the entire ASFV genome by fusing a library of 4,000 separate fragments of ASFV genomic material to the molecular adjuvant (Lacasta *et al.*, 2014). Although there was no detectable ASF-specific immune response prior to infection 60% of the animals recovered from fever. An expansion of CD8 T-cells was observed after challenge suggesting that the cellular immune response played a role in the protection, although the exact nature of the protective response or the components of the ASFV genome that it targeted are unclear. This work suggested that with further improvement and optimisation a DNA vaccine approach might be feasible for ASF.

An alternative strategy was to combine DNA and protein immunisation. Plasmids and proteins for eight different ASFV proteins, including p30 and p54, were used to immunise pigs (Perez-Nunez *et al.*, 2019). Different combinations of antigens induced antibody responses that were capable of inhibiting or enhancing infection of macrophages in *in vitro* tests. Infection of pigs that had antibodies that enhanced infection also led to a more rapid appearance of viraemia and clinical signs compared to controls (Sunwoo *et al.*, 2019) and none of these immunised animals survived. Both studies where enhancement was observed included p54 and p30 in the pool, suggesting antibodies against these proteins might be responsible for infection enhancement (Argilaguet *et al.*, 2011; Perez-Nunez *et al.*, 2019). However, previous work with these two proteins showed that a protective response that included the induction of antibodies could be mounted when p54 and p30 were injected as a preparation of baculovirus infected cells (Barderas *et al.*, 2001; Gomez-Puertas *et al.*, 1998). This again highlights the complexities of developing subunit vaccines as individual ASFV proteins appear to induce immune responses that lead to diametrically opposed clinical outcomes depending on the delivery system.

6.3.3 Viral vectors

Many vaccination strategies require booster immunisations in order to generate effective immune responses. Normally a boost is identical to the original first immunisation, a procedure called homologous prime-boost immunisation. However, immunisation with antigens in one format followed by a boost with the same antigens in a different format, so called heterologous prime-boost, can prove advantageous. Heterologous prime-boost can avoid immune responses against the vaccine platform itself, preventing an effective recall response to the antigen of interest after boost and, in some cases, it can help shape the type of immune response. Viral vectors have been shown to be a potent way of generating both antibody and cellular responses in a variety of different species. Attenuated strains of vaccinia virus, the vaccine used to eradicate smallpox, and genetically engineered replication deficient adenoviruses have been used extensively over the last decade or so. A combination of DNA vaccination and vaccinia virus was used in a heterologous prime boost approach in the first instance to identify potentially immunogenic proteins and then to test their protective effects in pigs.

The work on ASF subunit vaccines described in Sections 6.3.1 and 6.3.2 focused on the few proteins that induced a clearly defined antibody response. However, as described in Chapter 2, ASFV encodes for more than 150 different proteins and the antigenicity and any potential role in protection had not been tested. Both the antibody and cellular responses to 41 different ASFV genes were determined in pigs after DNA plasmid prime followed by vaccinia virus boost. This approach identified thirteen different viral genes, eight of which were novel and capable of inducing consistent immune responses. However, immunised pigs were not protected from virulent ASFV in a subsequent challenge experiment and, as seen with other experiments, showed more rapid onset of clinical signs than the controls. In contrast to these other studies, however, viral load in the vaccinated animals was less than in the controls, suggesting there was also a protective effect. Increased clinical signs against the background of reduced viral replication has been observed in vaccination models with other viruses.

Live attenuated strains of ASFV were used in a series of experimental screens to understand the breadth and specificity of both the humoral and cellular immune response to natural infection (Jenson *et al.*, 2000; Kollnberger *et al.*, 2002; Netherton *et al.*, 2019; Reis *et al.*, 2007). Viral vectors were used to induce specific immune responses to these proteins in pigs and test for their protective efficacy. A range of immunogenic ASFV proteins were incorporated into adenovirus vectors and the majority of them proved immunogenic in swine (Goatley *et al.*, 2020; Lokhandwala *et al.*, 2016, 2017, 2019; Netherton *et al.*, 2019). However, the protective efficacy of these different combinations of viral vector antigens has been mixed. A pool of eleven different adenoviruses, containing 35 different ASFV proteins, did not protect wild boar from challenge with a virulent strain from Armenia (Cadenas-Fernandez *et al.*, 2020). Nine adenovirus vectored ASFV antigens induced enhanced disease in pigs after challenge with the virulent Georgia strain (very similar to the Armenia strain). However, immunisation with seven adenoviruses expressing eight different structural proteins showed more promise (Lokhandwala *et al.*, 2019). This second cocktail of adenoviruses induced antigen specific responses that were influenced by the choice of adjuvant, with the adjuvant BioMise inducing higher antibody responses than ZTS-01. However, the pigs that were immunised in combination with ZTS-01 had better clinical outcomes, with five of the nine pigs in the group surviving to the end of the study compared to two in the group immunised in combination with BioMise. Interpretation of this result is complicated, however, by the fact that not all of the control animals suffered acute disease after challenge despite some pigs having low level viraemia. Interestingly, all of the ASFV genes that were potentially protective were included in the 35 genes that were combined and trialled in wild boar. In the experiment that showed some promise the antigens were all delivered as individual genes, whereas in the wild boar study multiple fusions of the ASFV genes were used, but it is difficult to draw comparisons as these two studies used different challenge models as well as vector design (Cadenas-Fernandez *et al.*, 2020; Lokhandwala *et al.*, 2019).

A number of ASFV genes were identified as targets of the cellular immune response and these were also vectored using adenovirus in combination with a highly attenuated vaccinia virus in a heterologous prime-boost strategy (Netherton *et al.*, 2019). Two different pools were trialled and a combination of 12 ASFV proteins reduced viraemia and clinical signs in some pigs, but ultimately did not protect them from severe disease. Animals with reduced viraemia had higher ASF-specific cellular immune responses than those that did not, highlighting the potential role of the T-cell response in protection. Antibody responses, but not cellular responses, were observed in pigs immunised with a pool of nine ASFV antigens; however, these pigs suffered enhanced clinical signs compared to the controls. In a follow-up study eight antigens that were identified using the DNA-prime vaccinia virus-boost screen described above (Jancovich *et al.*, 2018) were also trialled using adenovirus vectors. The animals were then boosted with a combination of adenoviruses and highly attenuated vaccinia viruses and in one experiment all of the vaccinated animals survived challenge with a virulent strain of ASFV (Goatley *et al.*, 2020). The *caveat* to this positive result is the observation that all of the pigs got sick and although clinically normal at the end of the study were still viraemic. The strain of ASFV used as the challenge material in these studies was a virulent strain obtained from Portugal in 1988 and it will be interesting to see if these eight antigens are effective against the genotype II strains circulating in Europe and Asia.

6.4 Disabled infectious single cycle African swine fever vaccines

An alternative approach to vaccine development is to try and convert ASFV itself into a viral vector, also referred to as a disabled infectious single cycle (DISC) vaccine. The replication deficient adenoviruses described in the previous section typically lack the *E1* genes which are involved in early steps of adenovirus replication. Replication deficient adenoviruses are made with cell lines that contain these crucial E1 proteins, allowing the generation of progeny viruses that lack the *E1* gene. Therefore, an adenovirus vector can infect a cell in a pig and express the ASF gene of interest, but the adenovirus cannot make any more copies of itself. This type of strategy has been used with other large DNA viruses such as herpesvirus and therefore it may be possible to take a similar approach with ASFV. A replication deficient ASFV would have the potential to express all but one of the 150 or more genes that the virus encodes, allowing a full range of T-cell antigens to be displayed to the cellular immune system. Although the use of inactivated ASFV has not been successful as a vaccine, it is conceivable that inactivation methods may have induced structural changes to the virus particle that could have led to a loss of antigenicity (Schloer, 1980). Therefore, a replication deficient ASFV could conceivably induce both cellular and humoral immune responses that would be representative of infection with an attenuated strain of the virus without the attendant risk of causing disease. However, there are numerous technological roadblocks that need to be overcome before this approach can be brought to fruition (Figure 6.2). In the first instance one is the availability of suitable cell lines in which to grow the virus. As discussed above, field isolates of ASFV are typically grown in primary macrophages, which are not suitable for the generation of long-term cultures that express the ASFV genes required

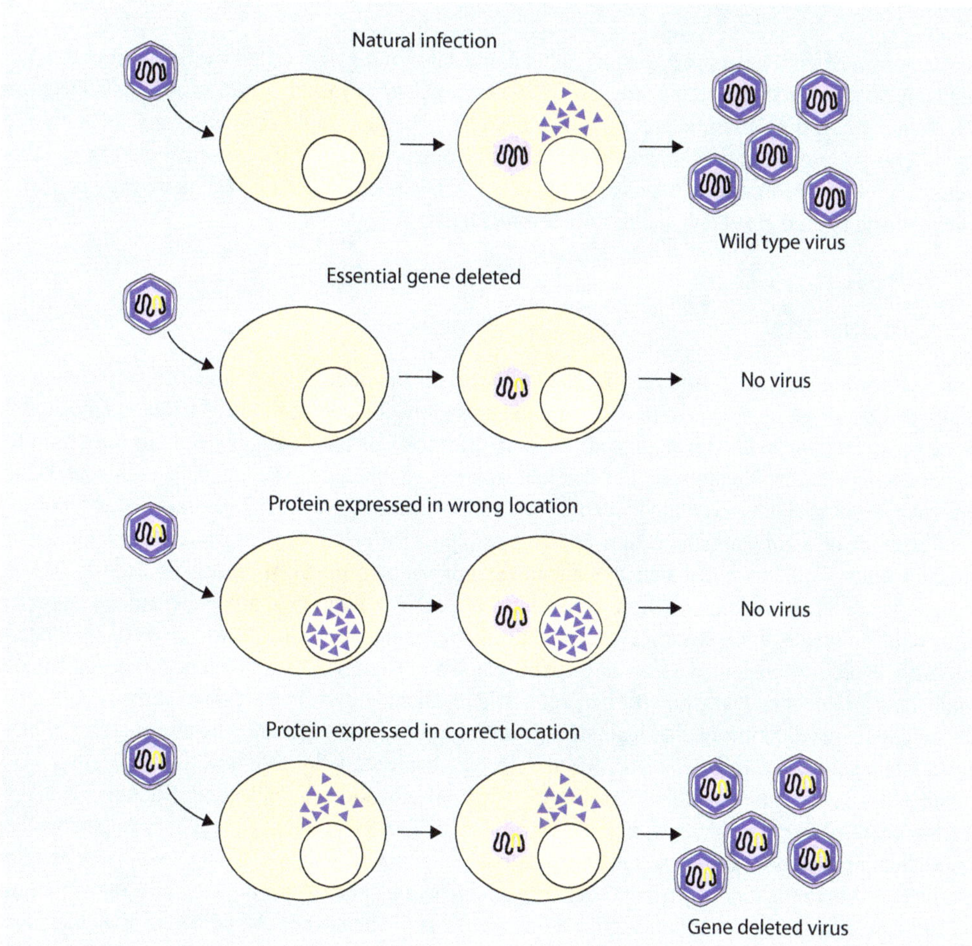

Figure 6.2. Challenges in developing defective infectious single cycle vaccines. Schematic highlighting of the principal challenges for developing a DISC vaccine for a complex pathogen like ASFV. Virus replication is compartmentalised within an infected cell with different viral proteins found in different locations. The essential ASFV gene selected as the target for the DISC vaccine therefore needs to be available in the correct location at the correct time in order to make gene deleted virus.

to support an ASFV DISC vaccine. Some field isolates can be grown in standard cell cultures or macrophage-like cell lines, but this could lead to adaptation of the virus to culture and loss of immunogenicity. The other problem is the choice of essential gene or genes to delete in order to block the ability of ASFV to replicate. Although a number of elegant studies have demonstrated the dependence of viral replication on particular genes, in order for a DISC vaccine to work the gene in question also has to be suitable for efficient expression in mammalian cells and crucially needs to be in the right place at the right time in the virus replication cycle. ASFV replication is compartmentalised both in space and time and viral structural proteins typically misfold or localise to cellular structures when expressed outside the context of an ASFV infected cell. However, recent work has suggested that the *A104R* gene that encodes for a protein involved in transcription of viral genes may be a suitable candidate for an ASF DISC vaccine (Frouco *et al.*, 2017). A proof of principle study showed that a genetically modified strain of ASFV lacking the *A104R* gene could be maintained on tissue culture cells that stably expressed the *A104R* gene for several passages (Freitas *et al.*, 2019). Further optimisation will be required to bring this technology to fruition and it is possible the other genes such as the *P1192R* gene may be more suitable targets than *A104R* (Coelho and Leitão, 2020).

6.5 Conclusions

ASF vaccine research did not get off to a particularly auspicious start and the experience in Spain and Portugal in the early 1960s cast a long shadow. Results of the field trials in China will be of great interest, with the hope that modern methods of targeted gene deletion are able to succeed where the classic approach of attenuation through passage in tissue culture did not. If the trials are successful, international acceptance of vaccination against ASF will most likely require development of a robust companion test to be able to differentiate vaccinated from naturally infected animals. Such a test will be crucial for allowing trade from countries that decide to include vaccination as part of their control strategies. A vaccine capable of eradicating virus from wildlife reservoirs is another complication, particularly relevant to the current European situation. Vaccination by injection is impractical when the virus is so widely dispersed in the wild, therefore an oral vaccine will be preferable and has shown to be particularly effective in the successful eradication of classical swine fever from wild boar. Of note, the attenuated Latvian strain has been shown to be effective when delivered orally to wild boar and therefore it may be possible to use live attenuated strains of ASFV in baited food for deployment in the field. An oral vaccine based on a poxvirus, a double stranded DNA virus related to ASFV, was used in the successful campaign to eradicate rabies from foxes in parts of Western Europe. As ASFV is also relatively stable in the environment it may be possible to carry out a similar vaccination campaign to eradicate ASF from European wild boar. Regardless of the success or failure of the scientific community to generate a safe and effective ASF vaccine, control of ASF will always require a multi-factorial approach based on the principles of biosecurity and rapid diagnosis in the case of breaches of that biosecurity. As Donald E. DeTray wrote of ASF nearly fifty years ago 'We should not be lulled into a false sense of security in the belief that a vaccine will solve the problem'.

Acknowledgements

This publication is based on work from 'Understanding and combating African swine fever in Europe (ASF-STOP COST action 15116)' supported by COST (European Cooperation in Science and Technology). The author would like to thank Ana-Luisa Reis for critical reading of the manuscript. The author's work on ASFV is supported by the Department for Environment, Food and Rural Affairs and the Biotechnology and Biological Sciences Research Council grants BBS/E/I/00002120, BBS/E/I/00007030, BBS/E/I/00007031, BBS/E/I/00007037 and BBS/E/I/00007039. This project has received funding from the European Union's Horizon 2020 research and innovation programme under grant agreement No 773701.

References

Afonso, C.L., Zsak, L., Carrillo, C., Borca, M.V. and Rock, D.L., 1998. African swine fever virus NL gene is not required for virus virulence. Journal of General Virology 79: 2543-2547. https://doi.org/10.1099/0022-1317-79-10-2543

Argilaguet, J.M., Perez-Martin, E., Gallardo, C., Salguero, F.J., Borrego, B., Lacasta, A., Accensi, F., Diaz, I., Nofrarias, M., Pujols, J., Blanco, E., Perez-Filgueira, M., Escribano, J.M. and Rodriguez, F., 2011. Enhancing DNA immunization by targeting ASFV antigens to SLA-II bearing cells. Vaccine 29: 5379-5385. https://doi.org/10.1016/j.vaccine.2011.05.084

Argilaguet, J.M., Perez-Martin, E., Nofrarias, M., Gallardo, C., Accensi, F., Lacasta, A., Mora, M., Ballester, M., Galindo-Cardiel, I., Lopez-Soria, S., Escribano, J.M., Reche, P.A. and Rodriguez, F., 2012. DNA vaccination partially protects against African swine fever virus lethal challenge in the absence of antibodies. PloS One 7: e40942. https://doi.org/10.1371/journal.pone.0040942

Balyshev, V.M., Kalantaenko, Y.F., Bolgova, M.V. and Prodnikova, E.Y., 2011. Seroimmunological affiliation of African swine fever virus isolated in the Russian Federation. Russian Agricultural Sciences 37: 427-429.

Barasona, J.A., Gallardo, C., Cadenas-Fernandez, E., Jurado, C., Rivera, B., Rodriguez-Bertos, A., Arias, M. and Sanchez-Vizcaino, J.M., 2019. First oral vaccination of Eurasian wild boar against african swine fever virus genotype II. Frontiers in Veterinary Science 6: 137. https://doi.org/10.3389/fvets.2019.00137

Barber, C., Netherton, C., Goatley, L., Moon, A., Goodbourn, S. and Dixon, L., 2017. Identification of residues within the African swine fever virus DP71L protein required for dephosphorylation of translation initiation factor eIF2alpha and inhibiting activation of pro-apoptotic CHOP. Virology 504: 107-113. https://doi.org/10.1016/j.virol.2017.02.002

Barderas, M.G., Rodriguez, F., Gomez-Puertas, P., Aviles, M., Beitia, F., Alonso, C. and Escribano, J.M., 2001. Antigenic and immunogenic properties of a chimera of two immunodominant African swine fever virus proteins. Archives of Virology 146: 1681-1691. https://doi.org/10.1007/s007050170056

Black, D.N. and Brown, F., 1976. Purification and physicochemical characteristics of African swine fever virus. Journal of General Virology 32: 509-518. https://doi.org/10.1099/0022-1317-32-3-509

Blome, S., Gabriel, C. and Beer, M., 2014. Modern adjuvants do not enhance the efficacy of an inactivated African swine fever virus vaccine preparation. Vaccine 32: 3879-3882. https://doi.org/10.1016/j.vaccine.2014.05.051

Borca, M.V., O'Donnell, V., Holinka, L.G., Risatti, G.R., Ramirez-Medina, E., Vuono, E.A., Shi, J., Pruitt, S., Rai, A., Silva, E., Velazquez-Salinas, L. and Gladue, D.P., 2020. Deletion of CD2-like gene from the genome of African swine fever virus strain Georgia does not attenuate virulence in swine. Science Reports 10: 494. https://doi.org/10.1038/s41598-020-57455-3

Budarkov, V.A., Sereda, A.D. and Balyshev, V.M., 2017. Protective properties of attenuated strains of the african swine fever virus in the course of immunodeficiency induced by radiation. Russian Agricultural Sciences 43: 432-436. https://doi.org/10.3103/S1068367417050044

Burmakina, G., Malogolovkin, A., Tulman, E.R., Xu, W., Delhon, G., Kolbasov, D. and Rock, D.L., 2019. Identification of T-cell epitopes in African swine fever virus CD2v and C-type lectin proteins. Journal of General Virology 100: 259-265. https://doi.org/10.1099/jgv.0.001195

Cadenas-Fernandez, E., Sanchez-Vizcaino, J.M., Kosowska, A., Rivera, B., Mayoral-Alegre, F., Rodriguez-Bertos, A., Yao, J., Bray, J., Lokhandwala, S., Mwangi, W. and Barasona, J.A., 2020. Adenovirus-vectored African swine fever virus antigens cocktail is not protective against virulent Arm07 isolate in Eurasian wild boar. Pathogens 9: 171. https://doi.org/10.3390/pathogens9030171

Chapman, D.A., Tcherepanov, V., Upton, C. and Dixon, L.K., 2008. Comparison of the genome sequences of non-pathogenic and pathogenic African swine fever virus isolates. Journal of General Virology 89: 397-408. https://doi.org/10.1099/vir.0.83343-0

Chen, W., Zhao, D., He, X., Liu, R., Wang, Z., Zhang, X., Li, F., Shan, D., Chen, H., Zhang, J., Wang, L., Wen, Z., Wang, X., Guan, Y., Liu, J. and Bu, Z., 2020. A seven-gene-deleted African swine fever virus is safe and effective as a live attenuated vaccine in pigs. Science China Life Sciences 63: 623-634. https://doi.org/10.1007/s11427-020-1657-9

Coelho, J. and Leitão, A., 2020. The African swine fever virus (ASFV) topoisomerase II as a target for viral prevention and control. Vaccines 8: 312. https://doi.org/https://doi.org/10.3390/vaccines8020312

Costa, J.V., 1990. African swine fever virus. In: G. Darai (ed.) Molecular biology of Iridoviruses. Kluwer Academic Publishers, Dordrecht, the Netherlands, pp. 247-270.

DeTray, D.E., 1957. Persistence of viremia and immunity in African swine fever. American Journal of Veterinary Research 18: 811-816.

Elde, N.C., Child, S.J., Eickbush, M.T., Kitzman, J.O., Rogers, K.S., Shendure, J., Geballe, A.P. and Malik, H.S., 2012. Poxviruses deploy genomic accordions to adapt rapidly against host antiviral defenses. Cell 150: 831-841. https://doi.org/10.1016/j.cell.2012.05.049

Forman, A.J., Wardley, R.C. and Wilkinson, P.J., 1982. The immunological response of pigs and guinea pigs to antigens of African swine fever virus. Archives of Virology 74: 91-100. https://doi.org/10.1007/BF01314703

Freitas, F.B., Simoes, M., Frouco, G., Martins, C. and Ferreira, F., 2019. Towards the generation of an ASFV-pA104R DISC mutant and a complementary cell line – a potential methodology for the production of a vaccine candidate. Vaccines 7: 68. https://doi.org/10.3390/vaccines7030068

Frouco, G., Freitas, F.B., Coelho, J., Leitao, A., Martins, C. and Ferreira, F., 2017. DNA-binding properties of African swine fever virus pA104R, a histone-like protein involved in viral replication and transcription. Journal of Virology 91: e02498-16. https://doi.org/10.1128/jvi.02498-16

Gallardo, C., Soler, A., Rodze, I., Nieto, R., Cano-Gomez, C., Fernandez-Pinero, J. and Arias, M., 2019. Attenuated and non-haemadsorbing (non-HAD) genotype II African swine fever virus (ASFV) isolated in Europe, Latvia 2017. Transboundary and Emerging Diseases 66: 1399-1404. https://doi.org/10.1111/tbed.13132

Goatley, L., Reis, A., Portugal, R., Goldswain, H., Shimmon, G., Hargreaves, Z., Ho, C.-S., Montoya, M., Sánchez-Cordón, P., Taylor, G., Dixon, L. and Netherton, C., 2020. A pool of eight virally vectored African swine fever antigens protect pigs against fatal disease. Vaccines 8: 234. https://doi.org/10.3390/vaccines8020234

Gomez-Puertas, P., Rodriguez, F., Oviedo, J.M., Brun, A., Alonso, C. and Escribano, J.M., 1998. The African swine fever virus proteins p54 and p30 are involved in two distinct steps of virus attachment and both contribute to the antibody-mediated protective immune response. Virology 243: 461-471. https://doi.org/10.1006/viro.1998.9068

Gómez-Puertas, P., Rodríguez, F., Oviedo, J.M., Ramiro-Ibáñez, F., Ruiz-Gonzalvo, F., Alonso, C. and Escribano, J.M., 1996. Neutralizing antibodies to different proteins of African swine fever virus inhibit both virus attachment and internalization. Journal of Virology 70: 5689-5694. https://doi.org/10.1128/JVI.70.8.5689-5694.1996

Jancovich, J.K., Chapman, D., Hansen, D.T., Robida, M.D., Loskutov, A., Craciunescu, F., Borovkov, A., Kibler, K., Goatley, L., King, K., Netherton, C.L., Taylor, G., Jacobs, B., Sykes, K. and Dixon, L.K., 2018. Immunisation of pigs by DNA prime and recombinant vaccinia virus boost to identify and rank African swine fever virus immunogenic and protective proteins. Journal of Virology 92: e02219-17. https://doi.org/10.1128/jvi.02219-17

Jenson, J.S., Childerstone, A., Takamatsu, H., Dixon, L.K. and Parkhouse, R.M., 2000. The cellular immune recognition of proteins expressed by an African swine fever virus random genomic library. Journal of Immunological Methods 242: 33-42. https://doi.org/10.1016/s0022-1759(00)00222-2

Kihm, U., Ackermann, M., Mueller, H. and Pool, R., 1987. Approaches to vaccination. In: Y. Becker (ed.), African swine fever. Developments in veterinary virology. Martinus Nijhoff Publishing, Boston, MA, USA, pp. 127-144.

King, K., Chapman, D., Argilaguet, J.M., Fishbourne, E., Hutet, E., Cariolet, R., Hutchings, G., Oura, C.A., Netherton, C.L., Moffat, K., Taylor, G., Le Potier, M.F., Dixon, L.K. and Takamatsu, H.H., 2011. Protection of European domestic pigs from virulent African isolates of African swine fever virus by experimental immunisation. Vaccine 29: 4593-4600. https://doi.org/10.1016/j.vaccine.2011.04.052

Kollnberger, S.D., Gutierrez-Castaneda, B., Foster-Cuevas, M., Corteyn, A. and Parkhouse, R.M., 2002. Identification of the principal serological immunodeterminants of African swine fever virus by screening a virus cDNA library with antibody. Journal of General Virology 83: 1331-1342. https://doi.org/10.1099/0022-1317-83-6-1331

Krug, P.W., Holinka, L.G., O'Donnell, V., Reese, B., Sanford, B., Fernandez-Sainz, I., Gladue, D.P., Arzt, J., Rodriguez, L., Risatti, G.R. and Borca, M.V., 2015. The progressive adaptation of a georgian isolate of African swine fever virus to Vero cells leads to a gradual attenuation of virulence in swine corresponding to major modifications of the viral genome. Journal of Virology 89: 2324-2332. https://doi.org/10.1128/jvi.03250-14

Lacasta, A., Ballester, M., Monteagudo, P.L., Rodriguez, J.M., Salas, M.L., Accensi, F., Pina-Pedrero, S., Bensaid, A., Argilaguet, J., Lopez-Soria, S., Hutet, E., Le Potier, M.F. and Rodriguez, F., 2014. Expression library immunization can confer protection against lethal challenge with African swine fever virus. Journal of Virology 88: 13322-13332. https://doi.org/10.1128/jvi.01893-14

Lacasta, A., Monteagudo, P.L., Jimenez-Marin, A., Accensi, F., Ballester, M., Argilaguet, J., Galindo-Cardiel, I., Segales, J., Salas, M.L., Dominguez, J., Moreno, A., Garrido, J.J. and Rodriguez, F., 2015. Live attenuated African swine fever viruses as ideal tools to dissect the mechanisms involved in viral pathogenesis and immune protection. Veterinary Research 46: 135. https://doi.org/10.1186/s13567-015-0275-z

Leite Velho, E., 1956. Observations sur la peste porcine en Angola. Bulletin de l'Office International des Epizooties 46: 335-340.

Lewis, T., Zsak, L., Burrage, T.G., Lu, Z., Kutish, G.F., Neilan, J.G. and Rock, D.L., 2000. An African swine fever virus ERV1-ALR homologue, 9GL, affects virion maturation and viral growth in macrophages and viral virulence in swine. Journal of Virology 74: 1275-1285. https://doi.org/10.1128/jvi.74.3.1275-1285.2000

Lokhandwala, S., Petrovan, V., Popescu, L., Sangewar, N., Elijah, C., Stoian, A., Olcha, M., Ennen, L., Bray, J., Bishop, R.P., Waghela, S.D., Sheahan, M., Rowland, R.R.R. and Mwangi, W., 2019. Adenovirus-vectored African swine fever virus antigen cocktails are immunogenic but not protective against intranasal challenge with Georgia 2007/1 isolate. Veterinary Microbiology 235: 10-20. https://doi.org/10.1016/j.vetmic.2019.06.006

Lokhandwala, S., Waghela, S.D., Bray, J., Martin, C.L., Sangewar, N., Charendoff, C., Shetti, R., Ashley, C., Chen, C.H., Berghman, L.R., Mwangi, D., Dominowski, P.J., Foss, D.L., Rai, S., Vora, S., Gabbert, L., Burrage, T.G., Brake, D., Neilan, J. and Mwangi, W., 2016. Induction of robust immune responses in swine by using a cocktail of adenovirus-vectored African swine fever virus antigens. Clinical and Vaccine Immunology 23: 888-900. https://doi.org/10.1128/cvi.00395-16

Lokhandwala, S., Waghela, S.D., Bray, J., Sangewar, N., Charendoff, C., Martin, C.L., Hassan, W.S., Koynarski, T., Gabbert, L., Burrage, T.G., Brake, D., Neilan, J. and Mwangi, W., 2017. Adenovirus-vectored novel African swine fever virus antigens elicit robust immune responses in swine. PloS One 12: e0177007. https://doi.org/10.1371/journal.pone.0177007

Mallapaty, S., 2019. Spread of deadly pig virus in China hastens vaccine research. Nature 569: 13-14. https://doi.org/10.1038/d41586-019-01269-5

Malmquist, W.A., 1962. Propagation, modification, and hemadsorption of African swine fever virus in cell cultures. American Journal of Veterinary Research 23: 241-247.

Malmquist, W.A., 1963. Serologic and immunologic studies with African swine fever virus. American Journal of Veterinary Research 24: 450-459.

Malmquist, W.A. and Hay, D., 1960. Hemadsorption and cytopathic effect produced by ASFV in swine bone marrow and buffy coat cultures. American Journal of Veterinary Research 21: 104-108.

Malogolovkin, A., Burmakina, G., Tulman, E.R., Delhon, G., Diel, D.G., Salnikov, N., Kutish, G.F., Kolbasov, D. and Rock, D.L., 2015. African swine fever virus CD2v and C-type lectin gene loci mediate serological specificity. Journal of General Virology 96: 866-873. https://doi.org/10.1099/jgv.0.000024

Manso Ribeiro, J., Nunes Petisca, J.L., Lopes Frazao, F. and Sobral, M., 1963. Vaccination contre la peste porcine africaine. Bulletin de L'Office International des Épizooties 60: 921-937.

Monteagudo, P.L., Lacasta, A., Lopez, E., Bosch, L., Collado, J., Pina-Pedrero, S., Correa-Fiz, F., Accensi, F., Navas, M.J., Vidal, E., Bustos, M.J., Rodriguez, J.M., Gallei, A., Nikolin, V., Salas, M.L. and Rodriguez, F., 2017. BA71DeltaCD2: a new recombinant live attenuated African swine fever virus with cross-protective capabilities. Journal of Virology 91: e01058-1. https://doi.org/10.1128/jvi.01058-17

Montgomery, R.E., 1921. On a form of swine fever occurring in British East Africa (Kenya colony). Journal of Comparative Pathology and Therapeutics 34: 159-191 (part I), 243-269 (part II).

Moore, D.M., Zsak, L., Neilan, J.G., Lu, Z. and Rock, D.L., 1998. The African swine fever virus thymidine kinase is required for efficient replication in swine macrophages and for virulence in swine. Journal of Virology 72: 10310-10315. https://doi.org/10.1128/JVI.72.12.10310-10315.1998

Neilan, J.G., Zsak, L., Lu, Z., Burrage, T.G., Kutish, G.F. and Rock, D.L., 2004. Neutralizing antibodies to African swine fever virus proteins p30, p54, and p72 are not sufficient for antibody-mediated protection. Virology 319: 337-342. https://doi.org/10.1016/j.virol.2003.11.011

Neilan, J.G., Zsak, L., Lu, Z., Kutish, G.F., Afonso, C.L. and Rock, D.L., 2002. Novel swine virulence determinant in the left variable region of the African swine fever virus genome. Journal of Virology 76: 3095-3104. https://doi.org/10.1128/jvi.76.7.3095-3104.2002.

Netherton, C.L., Goatley, L.C., Reis, A.L., Portugal, R., Nash, R.H., Morgan, S.B., Gault, L., Nieto, R., Norlin, V., Gallardo, C., Ho, C.-S., Sánchez-Cordón, P.J., Taylor, G. and Dixon, L.K., 2019. Identification and immunogenicity of African swine fever virus antigens. Frontiers in Immunology 10: 1318. https://doi.org/10.3389/fimmu.2019.01318

Nunes Petisca, J.L., 1965. Quelques aspects morphologiques des suites de la vaccination contre la peste porcine africaine (Virose L) au Portugal. Bulletin de l'Office International des Epizooties 63: 199-237.

O'Donnell, V., Holinka, L.G., Gladue, D.P., Sanford, B., Krug, P.W., Lu, X., Arzt, J., Reese, B., Carrillo, C., Risatti, G.R. and Borca, M.V., 2015a. African swine fever virus Georgia isolate harboring deletions of MGF360 and MGF505 genes is attenuated in swine and confers protection against challenge with virulent parental virus. Journal of Virology 89: 6048-6056. https://doi.org/10.1128/jvi.00554-15

O'Donnell, V., Holinka, L.G., Krug, P.W., Gladue, D.P., Carlson, J., Sanford, B., Alfano, M., Kramer, E., Lu, Z., Arzt, J., Reese, B., Carrillo, C., Risatti, G.R. and Borca, M.V., 2015b. African swine fever virus Georgia 2007 with a deletion of virulence-associated gene 9GL (B119L), when administered at low doses, leads to virus attenuation in swine and induces an effective protection against homologous challenge. Journal of Virology 89: 8556-8566. https://doi.org/10.1128/jvi.00969-15

O'Donnell, V., Holinka, L.G., Sanford, B., Krug, P.W., Carlson, J., Pacheco, J.M., Reese, B., Risatti, G.R., Gladue, D.P. and Borca, M.V., 2016. African swine fever virus Georgia isolate harboring deletions of 9GL and MGF360/505 genes is highly attenuated in swine but does not confer protection against parental virus challenge. Virus Research 221: 8-14. https://doi.org/10.1016/j.virusres.2016.05.014

O'Donnell, V., Risatti, G.R., Holinka, L.G., Krug, P.W., Carlson, J., Velazquez-Salinas, L., Azzinaro, P.A., Gladue, D.P. and Borca, M.V., 2017. Simultaneous deletion of the 9GL and UK genes from the African swine fever virus Georgia 2007 isolate offers increased safety and protection against homologous challenge. Journal of Virology 91: e01760-16. https://doi.org/10.1128/jvi.01760-16

Onisk, D.V., Borca, M.V., Kutish, G., Kramer, E., Irusta, P. and Rock, D.L., 1994. Passively transferred African swine fever virus antibodies protect swine against lethal infection. Virology 198: 350-354. https://doi.org/10.1006/viro.1994.1040

Oura, C.A., Denyer, M.S., Takamatsu, H. and Parkhouse, R.M., 2005. *In vivo* depletion of CD8+ T lymphocytes abrogates protective immunity to African swine fever virus. Journal of General Virology 86: 2445-2450. https://doi.org/10.1099/vir.0.81038-0

Perez-Nunez, D., Sunwoo, S.Y., Sanchez, E.G. Haley, N., Garcia-Belmonte, R., Nogal, M., Morozov, I., Madden, D., Gaudreault, N.N., Mur, L., Shivanna, V., Richt, J.A. and Revilla, Y., 2019. Evaluation of a viral DNA-protein immunization strategy against African swine fever in domestic pigs. Veterinary Immunology and Immunopathology 208: 34-43. https://doi.org/10.1016/j.vetimm.2018.11.018

Ramirez-Medina, E., Vuono, E., O'Donnell, V., Holinka, L.G., Silva, E., Rai, A., Pruitt, S., Carrillo, C., Gladue, D.P. and Borca, M.V., 2019. Differential effect of the deletion of African swine fever virus virulence-associated genes in the induction of attenuation of the highly virulent Georgia strain. Viruses 11: 599. https://doi.org/10.3390/v11070599

Reis, A.L., Abrams, C.C., Goatley, L.C., Netherton, C., Chapman, D.G., Sanchez-Cordon, P. and Dixon, L.K., 2016. Deletion of African swine fever virus interferon inhibitors from the genome of a virulent isolate reduces virulence in domestic pigs and induces a protective response. Vaccine 34: 4698-4705. https://doi.org/10.1016/j.vaccine.2016.08.011

Reis, A.L., Goatley, L.C., Jabbar, T., Sanchez-Cordon, P.J., Netherton, C.L., Chapman, D.G. and Dixon, L.K., 2017. Deletion of the African swine fever virus gene DP148R does not reduce virus replication in culture but reduces virus virulence in pigs and induces high levels of protection against challenge. Journal of Virology 91: e01428-17. https://doi.org/10.1128/jvi.01428-17

Reis, A.L., Parkhouse, R.M., Penedos, A.R., Martins, C. and Leitao, A., 2007. Systematic analysis of longitudinal serological responses of pigs infected experimentally with African swine fever virus. Journal of General Virology 88: 2426-2434. https://doi.org/10.1099/vir.0.82857-0

Rodríguez, J.M., Yáñez, R.J., Almazán, F., Viñuela, E. and Rodriguez, J.F., 1993. African swine fever virus encodes a CD2 homolog responsible for the adhesion of erythrocytes to infected cells. Journal of Virology 67: 5312-5320.

Ruíz-Gonzalvo, F. and Coll, J.M., 1993. Characterization of a soluble hemagglutinin induced in African swine fever virus-infected cells. Virology 196: 769-777.

Ruiz Gonzalvo, F., Caballero, C., Martinez, J. and Carnero, M.E., 1986a. Neutralization of African swine fever virus by sera from African swine fever-resistant pigs. American Journal of Veterinary Research 47: 1858-1862.

Ruiz Gonzalvo, F., Carnero, M.E., Caballero, C. and Martinez, J., 1986b. Inhibition of African swine fever infection in the presence of immune sera *in vivo* and *in vitro*. American Journal of Veterinary Research 47: 1249-1252.

Ruiz-Gonzalvo, F., Rodriguez, F. and Escribano, J.M., 1996. Functional and immunological properties of the baculovirus-expressed hemagglutinin of African swine fever virus. Virology 218: 285-289. https://doi.org/10.1006/viro.1996.0193

Sánchez Botija, C., 1962. Estudios sobre la peste porcina Africana en España. Bulletin de l'Office International des Epizooties 58: 707-727.

Sánchez Botija, C., 1963. Modificatión del virus de la peste porcina Africana en cultivos celulares. Bulletin de l'Office International des Epizooties 60: 901-919.

Schlafer, D.H., McVicar, J.W. and Mebus, C.A., 1984. African swine fever convalescent sows: subsequent pregnancy and the effect of colostral antibody on challenge inoculation of their pigs. American Journal of Veterinary Research 45: 1361-1366.

Schloer, G.M., 1980. Effect of storage at different temperatures on African swine fever (ASF) antigen treated with 10 mM binary ethylenimine. In: Proceedings of the Annual Meeting of the American Association of Veterinary Laboratory Diagnosticians vol. 23, pp. 351-363.

Sereda, A.D., 2013. Immunogenic and protective characteristics of African swine fever virus glycoproteins Actual'nye Voprosy Veterinarnoi Biologii 4: 31-35.

Sereda, A.D., Balyshev, V.M., Kazakova, A.S., Imatdinov, A.R. and Kolbasov, D.V., 2020. Protective properties of attenuated strains of African swine fever virus belonging to seroimmunotypes I-VIII. Pathogens 9: 274. https://doi.org/10.3390/pathogens9040274

Sereda, A.D., Kazakova, A.S., Imatdinov, I.R. and Kolbasov, D.V., 2018. Serotype-specific and haemadsorption protein of the African swine fever virus. Slovenian Veterinary Research 55: 141-150. https://doi.org/http://dx.doi.org/10.26873/SVR-454-2018

Sunwoo, S.Y., Perez-Nunez, D., Morozov, I., Sanchez, E.G. Gaudreault, N.N., Trujillo, J.D., Mur, L., Nogal, M., Madden, D., Urbaniak, K., Kim, I.J., Ma, W., Revilla, Y. and Richt, J.A., 2019. DNA-protein vaccination strategy does not protect from challenge with African swine fever virus Armenia 2007 strain. Vaccines 7: 12. https://doi.org/10.3390/vaccines7010012

Tabarés, E., Marcotegui, M.A., Fernández, M. and Sánchez-Botija, C., 1980. Proteins specified by African swine fever virus I. Analysis of viral structural proteins and antigenic properties. Archives of Virology 66: 107-117. https://doi.org/10.1007/BF01314979

Wardley, R.C., Norley, S.G., Wilkinson, P.J. and Williams, S., 1985. The role of antibody in protection against African swine fever virus. Veterinary Immunology and Immunopathology 9: 201-212. https://doi.org/10.1016/0165-2427(85)90071-6

Zsak, L., Caler, E., Lu, Z., Kutish, G.F., Neilan, J.G. and Rock, D.L., 1998. A nonessential African swine fever virus gene UK is a significant virulence determinant in domestic swine. Journal of Virology 72: 1028-1035. https://doi.org/10.1128/JVI.72.2.1028-1035.1998

Zsak, L., Lu, Z., Kutish, G.F., Neilan, J.G. and Rock, D.L., 1996. An African swine fever virus virulence-associated gene NL-S with similarity to the herpes simplex virus ICP34.5 gene. Journal of Virology 70: 8865-8871. https://doi.org/10.1128/JVI.70.12.8865-8871.1996

7. The pig sector in the European Union

S. Bellini
Istituto Zooprofilattico Sperimentale della Lombardia e dell'Emilia Romagna (IZSLER), via Bianchi 9, 25124 Brescia, Italy; silvia.bellini@izsler.it

Abstract

The pig sector is one of the most economically significant farming sectors in the European Union (EU) and pork is the most consumed meat. The EU is the world's second biggest producer of pork, after China, and the biggest exporter of pork and pork products. The main producer countries, Germany, Spain and France, represent about half of the EU's total production. A significant number of animals are kept in backyard systems to provide pork for the family and the local market. The major production basin extends from Germany to Belgium and accounts for 30% of the sows in the EU. Other important pig producing regions are Catalonia, Murcia, Lombardy, Bretagne and some areas of central Poland and Northern Croatia. Small-scale pig producers are mostly found in Eastern Europe where small units rearing 3.8% of pigs account for 73.3% of the pig farms. Following the current African swine fever (ASF) epidemic in the Eastern EU countries since 2014, the pig sector has greatly changed, with a decrease in the number of small sized pig holdings. An in-depth analysis of the impact of ASF on the structure of the pig farming system in the EU has not yet been conducted. The Food and Agriculture Organization of the United Nations (FAO) in the last report on Global Food Markets reported that there has been a contraction in the meat markets, due to both the COVID-19 pandemic and losses linked to the occurrence of serious epidemic diseases in livestock. While the effects of ASF and other animal diseases are confined to the countries affected, COVID-19 is having widespread impacts on all types of meat production systems.

Keywords: pigs, pork market, pork producers

This publication is based upon work from COST Action CA15116, ASF-STOP, supported by COST (European Cooperation in Science and Technology).
www.cost.eu

EUROPEAN COOPERATION IN SCIENCE & TECHNOLOGY

Funded by the Horizon 2020 Framework Programme of the European Union

DOI 10.3920/978-90-8686-910-7_7,

7.1 Introduction

Over the past decade, African swine fever (ASF) spilled out of its original area of endemicity and spread widely in Europe and Asia. The evolution of the ASF situation is causing serious economic damage in the affected countries, reshaping the pig farming sector and the commercial pork network around the world. Globally, the main economic impact of ASF is occurring in China where, since 2010 the annual pork production increased by over 50 million tons. Indeed, prior to ASF, half of the world's total output of pork was produced in China whereas, by the end of 2019, the Chinese national pig herd fell by half, and it is expected that production will continue to fall by 10-15% in 2020 (FAO, 2020). Furthermore, in 2020, in a few months the COVID-19 pandemic disrupted communities and lifestyles, also affecting livestock production, with consequent market disruptions.

The world population of pigs in 2013 was ~977 million heads and in the period between 2005 and 2013 the pig population increased by 8% to a total of over 970 million pigs. A significant number of animals are kept in backyard systems to provide pork meat for the family and the local market. In 2013, nearly 113 million tons of pig-meat carcasses were produced globally and almost 98% of the total pig-meat production originated from 3 regions: Asia, Europe and the Americas, with China contributing about 48% to the total global pig-meat production (FAO, 2016).

The Food and Agriculture Organization of the United Nations (FAO) in the last biannual report on Global Food Markets reported that for the second year in a row, world meat output is forecast to fall in 2020 to 333 million tons (in carcass weight equivalent), 1.7% less than in 2019 (FAO, 2020). Much of the reduction would again reflect a sharp drop in global production of pig meat, largely concentrated in Asian countries affected by ASF; conversely, global production of poultry meat is forecast to expand. In general, there has been a contraction in the meat markets, due to both the COVID-19 and losses linked to the occurrence of serious epidemic diseases in livestock, such as ASF and highly pathogenic avian influenza. While the effects of ASF and other animal diseases are confined to the countries affected, COVID-19 is having widespread impacts on all types of meat production systems. Social distancing measures have resulted in labour shortages in slaughterhouses, meat processing and packing plants, forcing some to shut down, creating disruptions through the whole supply chain. Indeed, in 2020 in key exporting countries COVID-19 lockdowns and restrictions on movement have led to loss of food service sales and substantial volumes of unsold meat products. Prices in the pork sector were also negatively influenced by the closure of restaurants, fast food and other outlets where a significant proportion of pork meat is sold (FAO, 2020).

7.2 The pig sector in the European Union

The pig sector is one of the most economically significant farming sectors in the European Union (EU) and, among terrestrial animals, pork is the most consumed meat followed by chicken and beef (Figure 7.1) (EC, 2019).

The EU is the world's second biggest producer of pork, after China, and the biggest exporter of pork and pork products (EC, 2019) (Figure 7.2). The EU exports about 13% of its total production

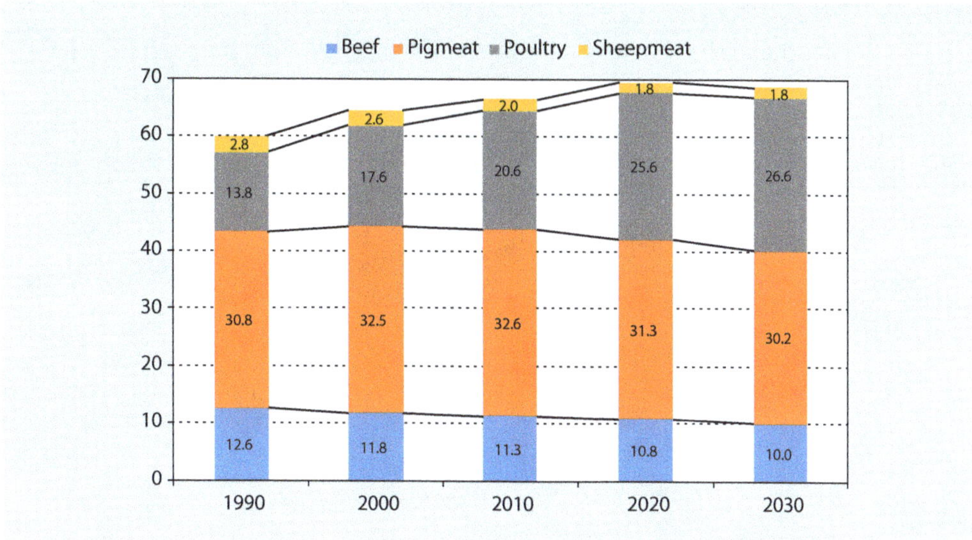

Figure 7.1. EU meat consumption by livestock meat type (kg per capita) (EC, 2019).

PIGMEAT		2019		2020
Production	⇩	-0.90%	⇨	0.50%
Exports	⇧	8.90%	⇧	10%
Consumption	⇩	-3.60%	⇩	-2.70%

Figure 7.2. EU market developments – pig meat (EC, 2020).

and most of these exports go to East Asia, in particular China. In 2019, despite the presence of ASF, the EU pig meat exports have risen significantly, driven by Chinese demand. Based on the last outlook performed by the Directorate-General for Agriculture and Rural Development (DGAGRI), pre-COVID-19 predictions forecast that pig shipments should remain high over the next years, peaking around 2022 and falling as Chinese supply recovers (EC, 2019).

Pork prices in the EU rose in 2019 because of Chinese demand and are expected to continue rising in 2020. China's imports remain the main driver of global meat trade in 2020. Based on the projection carried out by DGAGRI, prices should remain high until Chinese production recovers, and may fall sharply depending on the speed of the recovery and how much the production of EU competitors (the US, Brazil and Canada) grows (EC, 2019).

The EU's main producing countries are Germany, Spain and France, representing about half of the EU's total production (Figure 7.3). Germany and Spain are the leading producers of pork in the EU, with respectively 59.4 million (23% of the EU total) and 47.7 million (19%) pigs slaughtered in 2016, followed by France (23.8 million, 9%), Poland (21.8 million, 8%), Denmark (18.2 million, 7%), the Netherlands (15.4 million, 6%), Italy (11.8 million, 5%), Belgium (11.2

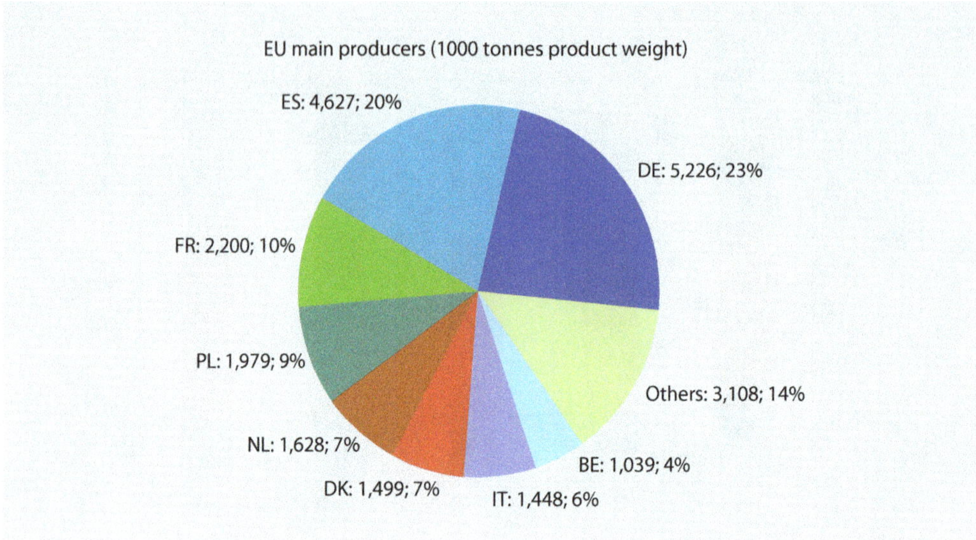

Figure 7.3. EU pork producers 2016 – FAO animal production and health (EUROSTAT, 2016).

million, 4%) and the United Kingdom (11.0 million, 4%) (EUROSTAT, 2016). However, EU pig-meat production is limited by public policy choices stemming in particular from environmental concerns in several EU Member States (e.g. Germany and the Netherlands), and by the ASF risk in central and eastern Europe.

7.3 Characteristics of pig farming in the European Union

Historically, pigs were kept in small numbers and were closely associated with the residence of the owner. They were valued as a source of meat and fat, for their ability to convert inedible food into meat and were often fed household food waste when kept on a homestead. The majority of pigs are used for human food but also supply skin, fat and other materials for use as clothing, ingredients for processed foods, cosmetics and medical use. Since the industrialisation of swine production in the late 1900s, swine farms have largely become consolidated in the EU. Pig farms became larger in size but fewer in number and pork production shifted to the intensive indoor confinement system, due to its potential to raise a large amount of pigs in a very cost-efficient manner. In general, the pig farming is characterised by three main different systems: the large scale industrialised pig farming system with increasing vertical integration, the traditional small scale production system, frequently found in households and the outdoor system, which is a traditional way of raising local pig breeds, originally practised in the southern European area. In most countries, the different pig production systems cohabit, from the simplest system with minimal investment, to large-scale market-oriented enterprises. Backyard farming remains the basic traditional system of keeping pigs and the most common in developing countries. Backyards supply mainly the rural population, whereas intensive holdings focus on satisfying the needs of urban areas, the export market and the pork processing sector. Smallholders normally raise pigs for both subsistence and commercial reasons. Pork is supplied to local markets and to more distant urban markets through a complex transport and marketing system. The backyard production is the farming system mostly affected during the last ASF epidemic in Europe.

Pig production in the EU is concentrated in a number of countries, with Denmark, Germany, Spain, France, the Netherlands and Poland having more than two thirds of the European breeding pigs. At regional level, more than half of the breeding pigs are concentrated in eleven regions, all of which are located in these six countries. Naturally, the size of the countries and regions plays a role in this ranking. In relative terms, the average share of pig production in agricultural output is higher in Denmark (29%), followed by Belgium (20%), Spain (14.7%) and Germany (14.5%).

The major production basin extends from Germany (namely from Nordhein-Westfalen and Niedersachen), to Belgium (Vlaams Gewest) and accounts for 30% of EU sows. However, there are other important regions, such as Catalonia, or Murcia (Spain), Lombardy (Italy), Bretagne (France) and some areas of central Poland and Northern Croatia (Figure 7.4). Small-scale pig producers are mostly found in the 13 Member States that joined the EU since 2004 (EUROSTAT, 2014).

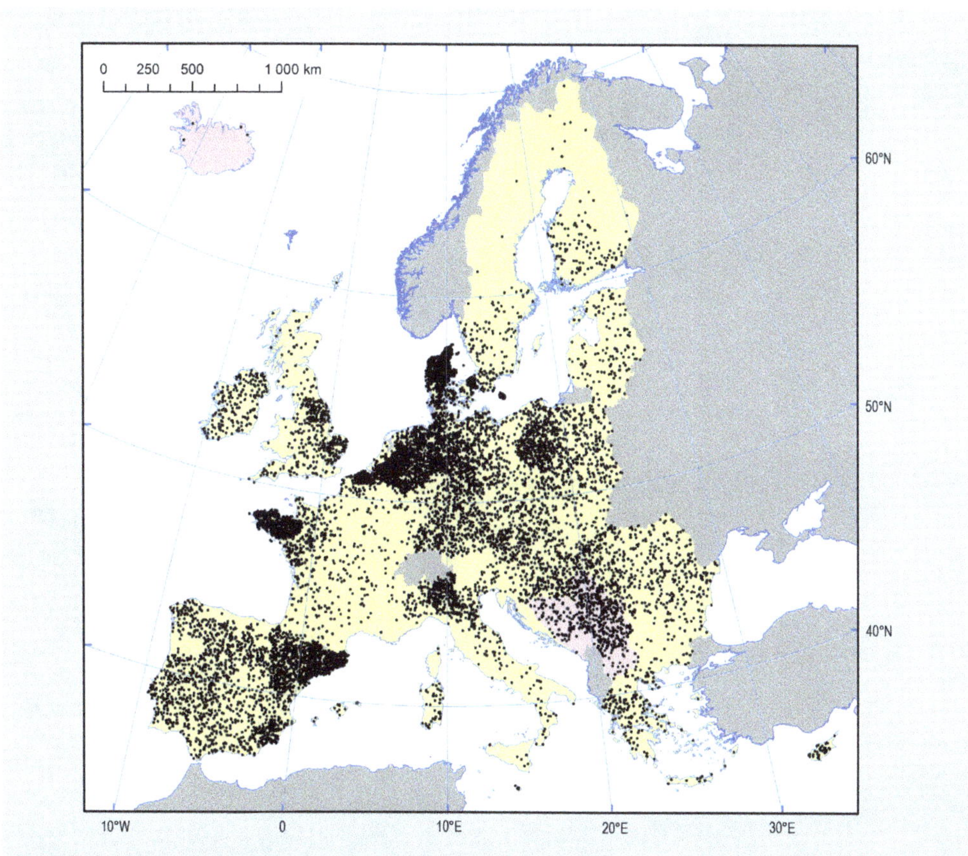

Figure 7.4. Number of sows by region (EUROSTAT, 2014).

The data reported in this chapter were collected from the 2014 EUROSTAT pig statistics, which is the latest EU consolidated document evaluating pig farming in the EU (EUROSTAT, 2014). This report includes data from the United Kingdom, which at the time was still part of the EU. It is worth mentioning that in recent years the pig sector has changed substantially, following the occurrence of the ASF epidemic in the eastern EU countries, and this phenomenon has particularly affected small sized pig holdings. An in-depth analysis of the impact of ASF on the structure of the pig farming system in the EU has not been conducted yet. This is why in this document for some considerations on the structure of the pig farming system reference is made to EUROSTAT 2014, whereas updated figures on the number of pigs and trends in the member states are shown in Table 7.1 and Figure 7.5 (EUROSTAT, 2020).

Table 7.1. Pig population in 28-EU in the period 2013-2019 (×1000; EUROSTAT, 2020).

GEO/TIME	2011	2012	2013	2014	2015	2016	2017	2018	2019
European Union – 28 countries	149,809	146,955	146,241	148,330	148,716	147,188	150,257	148,167	147,830
Belgium	6,328	6,448	6,351	6,350	6,364	6,177	6108	6209	6,085
Bulgaria	608	531	586	553	600	616	593	655	492
Czechia	1,487	1,534	1,548	1,607	1,555	1,479	1,532	1,508	1,509
Denmark	12,348	12,281	12,402	12,709	12,702	12,281	12,832	12,642	12,728
Germany	27,402	28,331	28,133	28,339	27,652	27,376	27,578	26,445	26,053
Estonia	365	375	359	358	305	266	289	290	302
Ireland	1,553	1,493	1,469	1,506	1,475	1,528	1,616	1,572	1,613
Greece	1,120	1,044	1,031	1,046	877	743	744	721	733
Spain	25,635	25,250	25,495	26,568	28,367	29,232	29,971	30,804	31,246
France	13,967	13,778	13,428	13,300	13,307	12,791	13,353	13,713	13,510
Croatia	1,233	1,182	1,110	1,156	1,167	1,163	1,121	1,049	1,022
Italy	9,351	8,662	8,561	8,676	8,675	8,478	8,570	8,492	8,510
Cyprus	439	395	358	342	328	353	350	362	352
Latvia	375	355	368	349	334	336	321	305	314
Lithuania	790	808	755	714	688	664	612	572	551
Luxembourg	91	89	90	93	89	95	91	83	84
Hungary	3,044	2,989	3,004	3,136	3,124	2,907	2,870	2,872	2,634
Malta	46	45	49	47	44	41	34	36	35
The Netherlands	12,103	12,104	12,013	12,065	12,453	11,881	12,296	11,909	11921
Austria	3,005	2,983	2,896	2,868	2,845	2,793	2,820	2,777	2,773
Poland	13,056	11,132	10,994	11,266	10,590	11,107	11,908	11,028	11,216
Portugal	1,985	2,024	2,014	2,127	2,247	2,151	2,165	2,205	2,216
Romania	5,364	5,234	5,180	5,042	4,926	4,708	4,406	3,925	3,816
Slovenia	347	296	2,885	282	271	266	257	259	240
Slovakia	580	631	637	642	633	586	614	627	589
Finland	1,290	1,271	1,258	1,223	1,239	1,197	1,108	1,041	1,062
Sweden	1,568	1,474	1,480	1,458	1,435	1,436	1,382	1,417	1,481
United Kingdom	4,326	4,216	4,383	4,510	4,422	4,538	4,713	4,648	4,741
Iceland	38	36	36	36	36	38	37	36	:

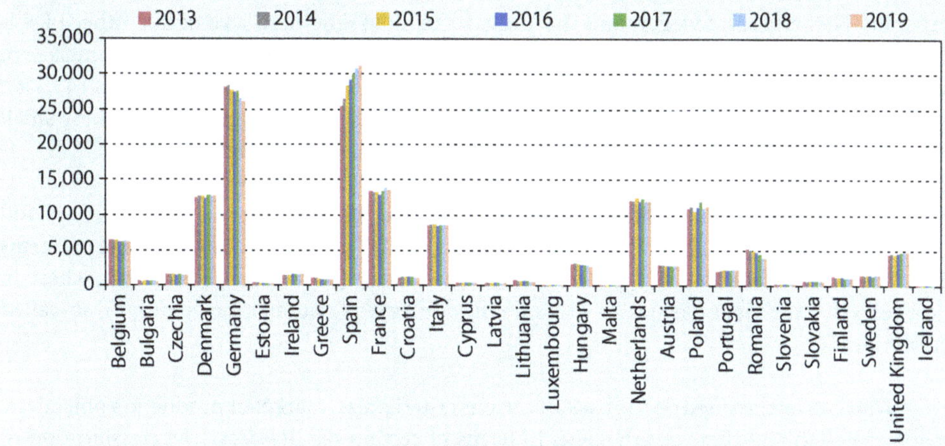

Figure 7.5. Pig population in the 28-EU countries in the period 2013-2019 (EUROSTAT, 2020).

In the EUROSTAT survey pigs are recorded in three categories, i.e. piglets with a live weight of less than 20 kg, breeding sows (weighing 50 kg and over) and other pigs. This latter category covers fattening pigs and also boars, cull sows, gilts, and various other pigs of at least 20 kg. Most of these other pigs are fattening pigs. The sows reflect the permanent pig herd and the other pigs are the pigs fattened before slaughtering. In the 2019 survey, some changes in the types of pig farm were already observed, with the number of small fatteners following a general downward trend (DGSANTE, 2020). This trend is especially noticeable in Member States with more small farms. The overall number of pigs is also falling at a similar rate, as there is a strong link between these two, whereas the number of large fatteners is increasing compared to 2007 and, although the number of large farms is stable, their average size is continuing to increase.

The number of large breeders has increased in some Member States, such as Czech Republic, Estonia, Latvia, Hungary and Slovenia, and the average number of breeding sows per farm has increased by 14% at EU-28 (28 countries of the EU) level compared to 2007, this figure has dropped in six Member States (Bulgaria, Ireland, Cyprus, Slovenia, Slovakia and the United Kingdom).

Concerning the average number of other pigs per farm, the total increase has been around 22% except in Ireland, Cyprus, Malta and the United Kingdom. These data confirm the trend of concentrating the pig production sector in large farms where fixed costs are divided by a larger quantity of animals, increasing productivity and thus reducing the average cost of production.

In the 13 newest EU Member States, a high proportion of sows in small herds has been reported. More than 98.6% of the sows are in farms with at least 10 sows.

The distribution of the pig population by size of the pig herds (in numbers of other pigs) shows that 1.7% of pig farms have at least 400 other pigs and rear 77.9% of these and 48.6% of the sows. In 12 Member States (Belgium, Czech Republic, Denmark, Estonia, Ireland, Spain, France, Italy,

Cyprus, the Netherlands, Sweden and the United Kingdom) the herd size of 400 other pigs is more than 90%, while in Poland and Romania this category is approximately 33%. Animals kept in small units of less than 10 other pigs are important in Romania (62.8%), Croatia (45.3%), Slovenia (31.4%), Lithuania (28.8%) and Bulgaria (25.8%). At the EU level, although these small units rear 3.8% of other pigs, they account for 73.3% of the pig farms (Figure 7.6).

Two types of pig farms are classified based on the number of breeding sows: the fatteners and the breeder-fatteners. Almost half of the other pigs (41.6%) are kept as fatteners, i.e. on farms without sows. More than half of these numerous small farms (58.4%) are in Romania, where in the last years the ASF virus has spread widely in the domestic pig population, mainly on so-called backyard farms.

The pig farms can be classified on the basis of various criteria: size, market or production objectives, husbandry system and their health status in terms of certain pig diseases. The classification of pig farms according to their size class shows that, on average, the larger farms (more than 400 sows) are more technically efficient than medium- and small-sized farms. The size is a crucial element in the economic viability of pig farms. Smaller farms are impeded by greater technical inefficiencies, whereas the large farms achieve better performance benefiting from both increased technical efficiencies and greater scale of economies.

The breakdown of other pigs among the four types of pig farm is shown by country in Figure 7.6.

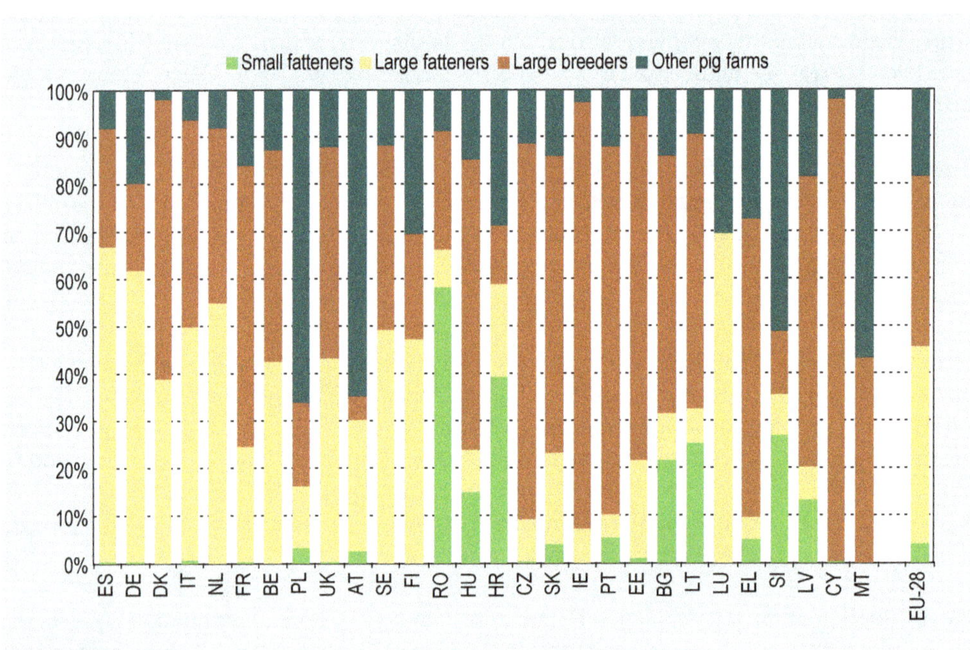

Figure 7.6. Distribution of other pigs by type of pig farm (EUROSTAT, 2014).

- The small fatteners (no sows and fewer than 10 other pigs) represent a significant share of pig production and at least 10% of other pigs are in seven of the newest Member States (Bulgaria, Croatia, Latvia, Lithuania, Hungary, Romania and Slovenia).
- The large fatteners (no sows and at least 400 other pigs) account for more than one third of other pigs in ten countries (Belgium, Denmark, Germany, Spain, Italy, Luxembourg, the Netherlands, Finland, Sweden and the United Kingdom). These 10 countries represent two thirds of the other pigs and three quarters of the EU pork-meat production. In France, the distribution is intermediate between typical fatteners and large breeders.
- The large breeders (at least 400 pigs and 100 sows) manage more than two thirds of the other pigs in six countries (Czech Republic, Estonia, Ireland, Greece, Cyprus and Portugal),
- The other pig farms manage more than two thirds of other pigs in Greece, Malta, Austria and Poland, which reflects a certain level of concentration, although limited by the farm size. Latvia, Hungary, Slovenia and Slovakia, with almost two thirds of the other pigs in such farms, can also to be included into this group.

7.4 Husbandry and management system in pig farms

Pigs throughout the EU are raised on different types of farms with considerable variations from one Member State to another and to know the organisation of the farming system is essential not only for productive purposes but also in case of implementation of a control strategy for managing the spread of certain diseases. Based on the characteristics of the husbandry and management systems, pig farms can be classified as commercial or industrial, backyards and free-ranging operations.

Commercial farms can be divided into multi-site farms, holdings specialised on one production step (farrowing, nurseries, or finishing) and on-site farms, which are premises that produce all production steps.

Based on the production phases, pig production can be managed by sub-production units that are connected to each other and that can be controlled by one or more operating units:

- Genetic material (semen, ova, embryos).
- Breeding – piglets, weights of 7-15 kg at an age of 21 to 45 days, respectively.
- Weaner – pig, weight 7-15 kg raised to 25-35 kg at an age of 56-84 days.
- Growing to finishing – refers to the stage where the animal gains weight from 25-35 kg to finishing weight. The target finishing weight (60 to 125 kg or more) depends on the prevailing market requirements and whether male piglets are castrated or left entire and the slaughter age depends on the growth rate of the pigs, which is largely determined by genotype, feed quantity and quality, and health status of the farm.
- Farrowing to finishing – includes all production phases, from breeding to slaughter weight. The feed may be self-produced (integrated unit) or purchased.
- Fully integrated system – pigs from birth to slaughter. These are commonly large-scale production systems where, after slaughter, the dressed carcasses may be sold to secondary processors prior to reaching retail outlets. In some limited cases the integrator may be involved in secondary processing of the pork, which may even be sold in company owned shops.

From a preventive point of view, multi-site farms present several advantages such as all-in-all-out practices that may be much more effectively implemented when a disease has to be eradicated (FAO/OIE/World Bank, 2010). However, since highly industrialised farms have a very high frequency of pig shipments, it is necessary for these farms to adopt stringent biosecurity measures to avoid the risk of spreading diseases through the movement of animals (see Chapter 10 for description of biosecurity measures). A breach of biosecurity in this type of farm can seriously affect the health status of many other related pig operations (Bellini *et al*, 2010).

Backyard farms are the most basic and traditional way of keeping pigs and the most common in developing countries and rural areas. In the majority of the eastern European countries, this small-scale pig production system prevails. Based on the data published by EUROSTAT in 2014, small-scale pig producers are mostly found in the 13 Member States that joined the EU since 2004 (EUROSTAT, 2014). However, these data refer to 2013, before the introduction of the ASFV in Eastern European countries, where since 2014 numerous outbreaks have been reported in the backyard sector, which resulted in a substantial reduction in the number of this type of farm. The higher density of pigs in small-scale confined production systems leads to a higher risk of pathogen circulation among herds. In backyard farms, pigs are generally purchased from several sources, such as other small pig producers, local markets or abroad, where the basic principles of biosecurity such as segregation and cleaning and disinfection are often poorly implemented. These are all dangerous practices that increase the risk of spreading pathogens (FAO/OIE/World Bank, 2010). Small-scale pig producers are normally involved in local trade and, in case of disease occurrence, this contributes to spreading the pathogen locally and, eventually, also to the wild boar population. Pig slaughter is usually carried out on the farm, although it may be restricted to proper slaughterhouses if there are local regulations on this issue (Jurado *et al*, 2018). The main biosecurity measures are not easy to implement in this type of farming system due to the minimal investment in infrastructure and the common practice of feeding pigs with kitchen waste (Bellini *et al*, 2016).

Outdoor pig farms are a traditional way of raising pigs in certain areas of southern Europe, such as the South Western Iberian Peninsula, Mediterranean Islands and Bulgaria where the climate is normally milder and there are conditions for keeping pigs outdoors. In general, local breeds (Black Iberian, East Balkan, Mangalica, Gascon, etc.) are used for this type of farming. In addition, in recent years, a new trend is developing in western European countries where outdoor pig producers raise their pigs in either certified free range or organic holdings. This new trend is facilitated by increasing availability of organic feed and positive economic projections for organic livestock (EC, 2019). Organic pig farms have grown significantly in Europe, also as a consequence of increased consumer awareness of food safety problems and environmental concerns. These farms have lower productivity but they can brand and sell pork for higher prices, and often have a larger portfolio of activities. However, its expansion is less widespread than what is observed in dairy and beef production systems. Among other reasons, it has been linked to a higher risk of disease spreading in outdoor housing. This is why in certain countries outdoor farming is strictly regulated by local veterinary authorities and are required to comply with biosecurity measures aimed at avoiding interaction between wild and domestic pigs. In the EU, keeping of pigs outside is prohibited in the territories affected by ASF.

7.5 EU classification of pig holdings in relation to African swine fever and biosecurity

Most of the ASF outbreaks in Europe have occurred in small pig holdings. Indeed, smallholders generally have lower levels of farm biosecurity than commercial farms and poor biosecurity is known to be a major limitation for disease prevention and control. In the current epidemiological context, the structure of the European swine industry requires differentiated biosecurity measures according to different risk levels of introduction and spread of ASF among the diversified farming systems.

A recent survey on biosecurity implementation was conducted by the World Organisation for Animal Health (OIE) with the aim of looking at the way biosecurity is applied across Europe. The survey covered the main fields related to the implementation of biosecurity in the countries to analyse current strengths and weaknesses and identify best practices (Bellini, 2018). The study revealed that biosecurity is mostly implemented on farmed animals, with poultry and pigs being the farming sectors with most frequent application of biosecurity practices. This is most likely linked to the recent epidemics of avian influenza and ASF in Europe. Additionally, it showed that biosecurity is normally targeted at commercial holdings, however, even other holdings that also have access to markets, for example for trade, should be included in the biosecurity programme. Although non-commercial farms can be a dead end in terms of disease spread, backyard units that sell animals at local or regional level can have a role in the spread of diseases. The respondents also highlighted the necessity of having mitigation measures to prevent diseases spreading from wild to domestic animals and this is particularly relevant for outdoor farming systems (Bellini, 2018).

In light of this, the Directorate-General for Health and Food Safety (DGSANTE) proceeded with the classification of pig farms into three categories as a first step for the development of a strategic approach for the management of ASF in the EU (SANTE/7113/2015-Rev 12), even before defining specific biosecurity measures:

- Non-commercial farms: farms where pigs are kept only for fattening for own consumption and neither pigs nor any of their products leave the holding.
- Commercial farms: farms which sell pigs, send pigs to a slaughterhouse or move pig products off the holding.
- Outdoor farms: farms which pigs are kept temporarily or permanently outdoor.

This classification, rather than taking into account the size of the farm or the type of establishment (breeding, fattening, etc.) considers the commercial attitude of the holdings. In this way it considers the risk of spreading the disease by trading pigs and the risk for the farm of being exposed to an external source of infection, such as the presence of infected wild pigs or ticks. In a nutshell, pig farms are categorised on the basis of the risk of introducing ASF and, based on that, targeted biosecurity measures have been established. See Chapter 10 for description of biosecurity measures.

7.6 Change in trends and consumer's demands

Meat consumption per capita in the EU has so far been on an upward trend, even though pig meat consumption per capita is declining in some EU Member States because consumers tend to favour poultry, which is cheaper and perceived as a healthier choice (Figure 7.1 and 7.2). More in general, EU meat consumption is expected to decline from 69.8 to 68.7 kg per capita by 2030 due to growing social and ethical concerns, environmental and climate worries, health claims, and an ageing European population. In fact, other trends in meat consumption are to be considered such as changing dietary patterns (flexitarians, vegetarians and vegans), especially among younger generations. Regarding food safety, 35% of EU citizens are concerned about the use of antibiotics, pesticides, environmental pollutants and food additives. Furthermore, there are rising concerns about climate change and animal welfare, and this is expected to lead to a slight decrease in meat consumption in the EU, where the number of vegetarians and vegans is particularly high in the younger generation (above 8% in Germany, France, Italy and Poland), and the number of flexitarians (consumers eating less meat) is increasing across all generations (EC, 2019).

Societal concerns lead also to an increasing demand for organic products, which is expected to support production growth in the medium term. Indeed, the 'world is becoming greener' and people show more concern about where their food comes from and how it is produced. The outdoor production systems have the potential to address such environmental and food safety concerns, while providing animal welfare and new opportunities for small, limited-resource farmers.

Acknowledgements

This publication is based on work from 'Understanding and combating African swine fever in Europe (ASF-STOP COST action 15116)' supported by COST (European Cooperation in Science and Technology).

References

Bellini, S., 2018. Application of biosecurity in different production systems at individual, country and regional levels. OIE Regional Commission of Europe, Tblisi (Georgia) September 2018. https://doi.org/10.20506/TT.2934

Bellini, S., Alborali, L., Zanardi, G., Bonazza, V. and Brocchi, E., 2010. Swine vesicular disease in northern Italy: diffusion through densely populated pig areas. OIE Revue Scientifique et Technique 29: 639-648.

Bellini, S., Rutili, D. and Guberti, V., 2016. Preventive measures aimed at minimizing the risk of African swine fever virus spread in pig farming systems. Acta Veterinaria Scandinavica 58: 82. https://doi.org/10.1186/s13028-016-0264-x.

Directorate General for Health and Food Safety (DGSANTE), 2015. African swine fever Strategy for Eastern Part of the European Union. SANTE/7113/2015-Rev 12, 2015. Available at: http://ec.europa.eu/food/sites/food/files/animals/docs/ad_control-measures_asf_wrk-doc-sante-2015-7113.pdf.

Directorate General for Health and Food Safety (DGSANTE), 2020. Available at: https://ec.europa.eu/food/animals/health/regulatory_committee_en.

European Commission (EC), 2019. EU agricultural outlook for market and income 2019-2030. Directorate General Agriculture and Rural Development (DGAGRI), Brussels. Available at: https://ec.europa.eu/info/news/eu-agricultural-outlook-2019-2030-african-swine-fever-continues-impact-global-meat-market-2019-dec-10_en.

European Commission (EC), 2020. For EU agricultural markets in 2020. Summer 2020. Edition No27. Directorate General Agriculture and Rural Development (DGAGRI), Brussels. Available at: https://ec.europa.eu/info/sites/info/files/food-farming-fisheries/farming/documents/short-term-outlook-summer-2020_en.pdf.

EUROSTAT, 2014. EUROSTAT – Statistics in Focus 15/2014. Pig farming in the European Union: considerable variations from one Member State to another. ISSN: 2314-9647.

EUROSTAT, 2016. Pork production up in the EU. Pork production up in the EU (19/09/2017). Available at: https://ec.europa.eu/eurostat/web/products-eurostat-news/-/DDN-20170919-1.

EUROSTAT, 2020. Web Site Pig Population – Annual Data. Available at: https://appsso.eurostat.ec.europa.eu/nui/show.do?dataset=apro_mt_lspig&lang=en.

Food and Agriculture Organization of the United Nations (FAO), 2016. Environmental Performance of Pig Supply Chains: Guidelines for assessment Livestock Environmental Assessment and Performance Partnership. FAO, Rome, Italy.

Food and Agriculture Organization of the United Nations (FAO), 2020. Food outlook – biannual report on global food markets: June 2020. Food Outlook, 1. FAO, Rome, Italy. https://doi.org/10.4060/ca9509e

Food and Agriculture Organization of the United Nations/World Organization for Animal Health/World Bank (FAO/OIE/World Bank), 2010. Good practices for biosecurity in the pig sector – Issues and options in developing and transition countries. FAO Animal Production and Health Paper N. 169. FAO, Rome, Italy.

Jurado, C., Martínez-Avilés, M., de la Torre, A., Štukelj, M., Cardoso de Carvalho Ferreira, H., Cerioli, M., Sánchez-Vizcaíno, J.M. and Bellini, S., 2018. Relevant measures to prevent the spread of African swine fever in the European Union domestic pig sector. Frontiers in Veterinary Science 5: 77. https://doi.org/10.3389/fvets.2018.00077

8. Management of wild boar populations in the European Union before and during the ASF crisis

F. Jori[1]*, G. Massei[2], A. Licoppe[3], F. Ruiz-Fons[4], A. Linden[5], P. Václavek[6], E. Chenais[7] and C. Rosell[8]

[1]ASTRE (Animal, Health, Territories, Risk and Ecosystems), CIRAD, INRAE, University of Montpellier, 34398 Montpellier, France; [2]National Wildlife Management Centre, Animal and Plant Health Agency, Sand Hutton, York YO41 1LZ, United Kingdom; [3]Department of Environmental and Agricultural Studies, Public Service of Wallonia, 5030 Gembloux, Belgium; [4]SaBio Group, Instituto de Investigación en Recursos Cinegéticos IREC (CSIC-UCLM-JCCM), 13005 Ciudad Real, Spain; [5]FARAH Research Center, Faculty of Veterinary Medicine, University of Liège, 4000 Liège, Belgium; [6]State Veterinary Institute Jihlava, Rantířovská 93/20, 586 01 Jihlava, Czech Republic; [7]National Veterinary Institute, 751 89 Uppsala, Sweden; [8]Minuartia Wildlife Consultancy / Department of Evolutionary Biology, Ecology and Environmental Sciences, University of Barcelona, 08028 Barcelona, Spain; ferran.jori@cirad.fr

Abstract

In recent decades, wild boar populations have been increasing worldwide due to several potential causes, including human-induced and natural environmental changes and biological and ecological factors. In Europe, this phenomenon has several economic, social and environmental implications such as the increase of agricultural and forest damage, road traffic accidents and potential ecological impact on animal and plant biodiversity. In addition, wild boar population growth and expansion can contribute to the maintenance and dissemination of infectious pathogens affecting animal and human health. In this context, the emergence of African swine fever (ASF) in Europe has become a serious challenge for animal disease control. The high susceptibility of wild boar to ASF infections and the capacity of the virus to remain infective in wild boar carcasses require a combination of wildlife management and veterinary strategies in order to eradicate this virus from EU forests. The goal of this chapter is to provide a thorough overview of those efforts. After illustrating the current situation of wild boar populations in Europe, the chapter describes the different methods applied by wildlife managers in the absence of ASF. Subsequently, the chapter reviews different approaches and tools applied in the context of ASF control, with a particular focus on the strategies implemented by countries that were successful in their eradication, such as Belgium and the Czech Republic. The last section of the chapter highlights areas that require future research to improve ASF management in natural wild boar populations, which remains a serious challenge for the large majority of countries in the EU.

Keywords: wild boar, control, surveillance, depopulation, fences

This publication is based upon work from COST Action CA15116, ASF-STOP, supported by COST (European Cooperation in Science and Technology).
www.cost.eu

EUROPEAN COOPERATION IN SCIENCE & TECHNOLOGY

Funded by the Horizon 2020 Framework Programme of the European Union

DOI 10.3920/978-90-8686-910-7_8,

8.1 Introduction

Highly prolific and extremely adaptable to different trophic resources and environments, wild boars (*Sus scrofa*) are present in large parts of the world. They probably constitute the most common and exploited ungulate species in Europe, where they have experienced a tremendous population increase in the last decades. The reasons for their demographic expansion and population growth are multifactorial including land use, climatic and environmental changes and human practices. Traditional hunting seems insufficient for reducing wild boar population sizes, and more efficient and innovative management strategies for population control are needed. The current high wild boar densities in Europe favour proliferation and dissemination of pathogens circulating in this species with potential spill over to domestic animals and humans. In this context, African swine fever virus (ASFV) genotype II found optimal conditions for spread and long-term maintenance among wild boar populations after the introduction to Europe in 2007. From 2014 to 2020, ASFV infected wild boar populations in eleven countries in the European Union (EU): Estonia, Lithuania, Latvia, Poland, Czech Republic, Bulgaria, Belgium, Romania, Hungary, Slovakia and Germany. So far, only two of these countries have eradicated, or are in the process of eradicating, the disease in free ranging wild boar populations. It is important to highlight that the introductions to new territories were not only the result of natural spread through wild boar movements, but often facilitated by human activities. With African swine fever (ASF) having being present in the European continent for more than a decade, a substantial amount of experience has been accumulated and there are many lessons to be learned about ASF management in natural wild boar populations that can be useful for other regions of the world exposed to the virus. The first section of this chapter describes the current demographic situation of wild boars in Europe, the drivers that have facilitated the observed rising trend in population sizes and their impact at the ecological, economic and sanitary levels. The second section presents different methods to manage wild boar populations in the absence of ASF. The next two sections review different methods that have been applied to manage and control the spread of ASF in different EU contexts by managing wild boar populations (Section 8.3) or by controlling the disease (Section 8.4). The last section highlights areas that require future research to improve disease management in free-ranging wild boar populations. A specific focus is on those countries that have achieved, or are in the process of achieving, ASF eradication with a combination of wildlife and disease management methods.

8.2 Current knowledge of wild boar populations in the EU

8.2.1 Wild boar demographic trends

At the beginning of the 20th century, many native wild boar populations in Europe had become extinct or occurred at very low densities. By the mid-1980s, wild boar numbers had increased dramatically and recolonised much of the species' former range.

Wild boars can thrive in a wide range of habitats, and the species' distribution is only limited by the availability of water in hotter climates and by harsh winters at higher latitudes. Wild boars are further expected to respond to global warming by increasing their geographical range northwards (Melis *et al.*, 2006). The species is already expanding in northern countries such as Norway and

Sweden (Thurfjell *et al.*, 2009), where availability of crops in summer and autumn allows wild boars to survive the winter months (Bieber and Ruf, 2005; Fernández-Llario and Carranza, 2000). In the UK, once extinct, wild boars are now widespread as a result of illegal releases or escapes from commercial breeding farms (Wilson, 2014).

Factors associated with this expansion include increased availability of anthropogenic food sources, urban sprawl encroaching on rural areas and a reduction in hunting pressure (Cahill *et al.*, 2012; Keuling *et al.*, 2016; Massei *et al.*, 2015). A review based on hunting statistics found a simultaneous sharp increase of wild boar numbers throughout Europe between the 1960s and the mid-1970s, followed by an apparent stabilisation of numbers in the following decade. Three decades later, a similar analysis based on hunting bags in 18 European countries showed that wild boar numbers continued to increase throughout Europe (Figure 8.1). In the same period, a decline in the number of hunters, observed across most of the countries, suggested that even if hunting is still the main cause of wild boar mortality, its potential to control populations is decreasing (Massei *et al.*, 2015). Therefore, despite a lower number of hunters, the total of wild boars harvested in these countries increased by 2.8 times in 20 years, passing from 864,000 to 2.5 million between 1992 and 2012. Excluding areas with unusual hunting bags due to disease control strategies, the number of wild boars hunted across Europe is still growing.

8.2.2 Density data

Strong variations in population densities are recorded within the European continent, even inside similar regions, e.g. in the Iberian Peninsula a range from less than three to over 30 wild boars per km^2 is registered. Food availability, such as crops or supplementary feeding, are the main factors affecting wild boar productivity and consequent local population densities.

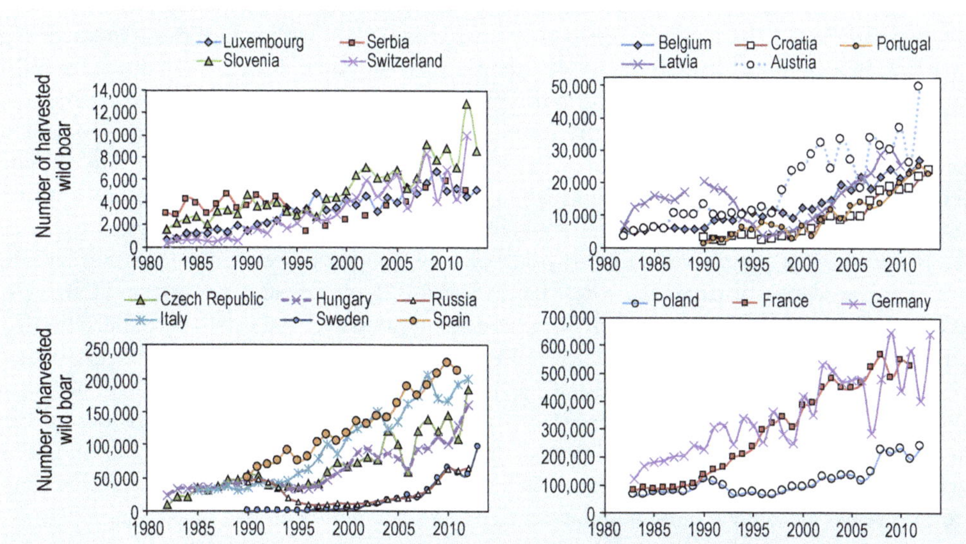

Figure 8.1. Wild boar hunting bags in European countries from 1980-2010 (Massei *et al.*, (2015) (courtesy of John Wiley and Sons, 02/06/2020).

Estimates of the size and density of wild boar populations are key factors to monitor and understand the epidemiology of wild boar diseases, such as ASF. The organisation and collection of hunting statistics and their analysis is essential not only for hunting management but also for developing wildlife policies. Hunting data are available and potentially comparable across Europe for use in predictive spatial models of wild boar density. They provide reasonable indicators of population trends. However, in practice, they are unreliable and patchy. Standardisation of procedures, methods and data collected across countries and regions provides opportunities for a common use of data across Europe and is a baseline prerequisite for a more sustainable management of the species and the effective control of diseases such as ASF. Three methods are recommended to estimate wild boar densities by the scientific consortium ENETWILD, specialised in monitoring wild species demographic dynamics in the EU (www.enetwild.com). The first one consists of the collection of hunting statistics based on hunting bags, hunting effort, size of the hunting grounds and, if possible, the number of sighted wild boars during driven hunts. This method has the advantage that most of the EU countries have available data that can be used and harmonised for regional large-scale abundance estimations. The second method consists of estimating densities with the widespread deployment of a camera trap grid, based on the Random Encounter Model approach, which can be used in relatively small areas. The third method consists of nocturnal density estimation by distance sampling along line transects using night vision equipment (applicable only in open areas with low vegetation cover).

8.2.3 Factors affecting population dynamics and growth

The significant increase and dispersal observed in wild boar populations is due to a combination of factors including extraordinary reproductive outputs, low mortality rates and environmental and sociological changes that have been occurring in the last decades and that have a cumulative effect in wild boar reproduction and survival. With the highest reproductive rate among all ungulates, annual population growth in this species may exceed 200% (Bieber and Ruf 2005; Keuling *et al.*, 2013). Wild boars are primarily seasonal breeders, with typical peaks in farrowing during winter/early spring and to a lesser extent, in late summer (Rosell *et al.*, 2012). In most populations, females produce one litter per year and typical litter sizes range between four and seven piglets (Bieber and Ruf, 2005). In Europe, a high proportion of females can reproduce in their first year and an earlier onset of puberty is observed when sows reach 27-33 kg of body weight (Malmsten *et al.*, 2017; Servanty *et al.*, 2011). Both the number of sows reproducing and litter sizes are associated with the availability of high energy food (Servanty *et al.*, 2009). The main causes of natural mortality in wild boars are food scarcity, extreme weather and, in some areas, predation by large carnivores (Bieber and Ruf, 2005; Fernández-Llario and Carranza, 2000). Adult survival rates in good environmental conditions are typically over 70%, although in poor conditions they can drop to 25-58% (Bieber and Ruf, 2005). The highest mortality rates are reported for piglets under 4 months of age. In addition to natural causes, wild boar mortality is mainly associated with anthropogenic factors, with hunting and road traffic collisions having the greatest impact on survival (Keuling *et al.*, 2013). In a population managed by hunting, wild boars rarely survive their second year of life. The generation times of wild boar are ranging from 2.3 years in heavily hunted populations to 3.6 years in lightly hunted ones (Servanty *et al.*, 2011). In most European countries, wild boar population recruitment is higher than mortality, leading to consistent annual population increases (Keuling *et al.*, 2013; Massei *et al.*, 2015). As survival depends on the availability of high-energy food (Bieber and Ruf, 2005), anthropogenic food sources such as crops and supplementary food can offset natural mortality (Veeroja and Männil, 2014). For instance, in the Czech Republic, cereals constituted more than 50% of the stomach-content biomass of wild boar. The sources of these cereals included both agricultural crops and, to a large extent, supplementary feeding provided by hunters (Ježek *et al.*, 2016). From that perspective, habitat alterations such as agricultural expansion are beneficial to wild boar population growth. However, expanding populations of wild boars are also adapting to urban and peri-urban environments because they provide access to food sources all year round, which ultimately increases annual population growth (Cahill *et al.*, 2012). Global warming is also leading to an increase in mast seeding frequency (acorns) resulting in a higher proportion of breeding females and a higher growth rate of the population inhabiting deciduous forests (Touzot *et al.*, 2020).

8.3 Overabundance of wild boar population and its consequences

8.3.1 Impacts on biodiversity

Wild boars are opportunistic omnivores with a diet mainly composed of plants but also including animal matter. Feeding behaviour includes foraging to collect fruits directly from the plants or fallen on the ground, occasionally eating grasses and leaves and rooting to feed on roots, fungi or invertebrates. Rooting activity causes important disturbances to soil and plant communities, inducing changes on habitat quality and ecosystem dynamics (Ballari and Barrios-Garcia, 2013). Impacts on forest regeneration have been reported, particularly affecting non-dominant tree species, which may have a negative impact on forest biodiversity (Bongi *et al.*, 2017) together with the reduction and modification in spatial patterns of vegetation recruitment. Damage to reforestation areas includes feeding on roots of nursery seedlings, which attract wild boars due to their rich nutrient content. Rooting also affects alpine and subalpine grasslands, modifying the diversity and heterogeneity of this habitat (Bueno *et al.*, 2010). However, although most studies indicate that wild boars are associated with decreased plant biomass, conflicting evidence exists concerning the effect of rooting on plant species diversity and composition. For instance, in Sweden the number of plant species increased following wild boar rooting (Welander, 2000). Similarly, in the Spanish Pyrenees alpine grasslands, wild boar rooting created large gaps that increased plant community heterogeneity and maintained high levels of plant diversity (Bueno *et al.*, 2010).

In addition to their impact on vegetation diversity, wild boars can also have an impact on invertebrate and vertebrate fauna. The destruction of nests and predation on eggs of endangered ground nesting birds or reptiles is known as one of the most deleterious effects of wild boar (Graitson *et al.*, 2019). Overall, large populations of wild boar have the potential to produce an environmental impact on animal and plant biodiversity (Barrios-Garcia and Ballari, 2012).

8.3.2 Damage to crops and pastures

Damage to croplands is one of the most important sources of conflict between agricultural farmers and wild boars, involving significant economic losses to compensate and prevent agricultural damage. The wild boar diet comprises a wide range of crops including maize, sunflower, rice, potatoes, and many species of fruit, such as grapes, with strong seasonal and regional variations.

The amount of crops consumed depends on availability, and represents a range of 37-88% of the wild boar diet in Europe (Barrios-Garcia and Ballari, 2012). Crops provide high-energy food, which has been associated with an increase of mean litter size compared with populations feeding in forests, contributing to the population growth rate (Rosell *et al.*, 2012). Wild boars can also affect pastures by ground rooting activities, which can extend to significant depths and widths and hamper the movements of agricultural machinery or reduce the value of the areas available for cattle grazing (Bueno *et al.*, 2010).

8.3.3 Traffic accidents

The rising numbers of wild boars throughout Europe have resulted in growing numbers of road collisions, although there is a marked regional variation. In some Mediterranean regions, wild boars are responsible for up to 85% of all traffic accidents involving wildlife. In northern and central Europe, wild boars make up a smaller portion of the total accidents, although a sharp increase in vehicle collisions has been observed in some countries, such as Sweden, where these accidents increased by 250% between 2003 and 2011 (Gren *et al.*, 2016). Wild boar-vehicle collisions show marked temporal patterns with higher frequencies from October to January, coinciding with the main rut period of wild boar and the hunting season (Langbein *et al.*, 2011). This is also a social and economic concern, the average cost of wild boar collisions being estimated between 2,700 and 9,000 €. Furthermore, in Spain wild boar accidents cause three times more injuries per collision than those induced by other solitary species (such as roe deer), probably due to the gregarious behaviour of wild boars (Sáenz-de-Santa-María and Tellería, 2015).

8.3.4 Urban incursions

In recent years, the number of conflicts due to incursions of wild boars into urban settlements has been increasing in many large cities in Europe as well as in other parts of the world (Pei *et al.*, 2010). By 2010, at least 44 cities in 15 countries had reported problems related to the presence of wild boar or feral pigs (Cahill *et al.*, 2012). Attracted by waste containers and litter bins, food offered by people or pet food from cat colonies, wild boars present in natural areas adjacent to cities progressively colonise urban environments, often using corridors such as streams or other green areas. Apart from damage to grass, garden irrigation systems and other urban infrastructure, some of the most significant problems arising from these incursions are the risk of attacks to humans or the potential transmission of zoonotic pathogens to people.

8.3.5 Disease implications

Many ecologists consider diseases to be a natural process to regulate wildlife populations when these grow beyond their carrying capacity. Overabundance reduces available trophic resources, which induces weakness and death of the weakest individuals in the process of natural selection. In addition, lack of trophic resources induces stress, which has a negative impact on the immune system, facilitating multiplication of potential pathogens. There are many examples illustrating the association between overabundance and occurrence of disease in ungulate populations (Gortázar *et al.*, 2006; O'Brien *et al.*, 2006). Wild boars are exposed to a large diversity of pathogens and some of them can affect other species of mammals, including humans (Jori *et al.*, 2018, 2020; Ruiz-Fons, 2017). The current situation of overabundance of wild boars in Europe is particularly favourable for the emergence and spread of infectious diseases in the large continuous ecosystems in Europe (Pittiglio *et al.*, 2018). During the last 30 years, the number of disease notifications in wild boars in Europe has significantly increased. In this context, the challenge represented by the current ASF crisis in Europe is not surprising.

8.4 Reducing wild boar population numbers in the absence of African swine fever

In the context of the current overabundance, there are several methods to manage wild boar populations with the common goal of mitigating their economic, ecological or agricultural impact. However, in principle, these measures are not designed to control the spread of infectious diseases in a wild boar population.

8.4.1 Hunting

Recreational hunting is the most widely method used to control wild boar numbers in Europe. This includes hunting on foot or from high seats and driven hunts using dogs, the latter mostly employed in areas with dense vegetation (Geisser and Reyer, 2005; Massei *et al.*, 2011). Some methods allow targeting specific age classes and sexes. As wild boar behaviour is essentially nocturnal, individual hunting at night is permitted in some countries. Depending on the region, this is done by using a light source or night vision devices. As juvenile survival has the largest effect on recruitment, applying increased hunting pressure on juveniles appears to be the most effective option for reducing wild boar numbers (Geisser and Reyer, 2005; Servanty *et al.*, 2009). A harvest model developed for a French wild boar population showed that the most efficient way to limit the growth rate was to target medium-sized females (Gamelon *et al.*, 2012). However, recreational hunters tend to select large males rather than females or juveniles when hunting from high seats or waiting for wild boars to visit bait stations (Keuling *et al.*, 2013). Recreational hunters play an important role in decreasing wild boar numbers, and their in-depth knowledge of local areas may help in early detection of carcasses during disease outbreaks (EFSA *et al.*, 2018a). However, hunters are also responsible for bad practices such as illegal releases and translocations of wild boars aimed at re-stocking populations, and the use of supplementary feeding to maintain high densities of wild boar (Oja *et al.*, 2015, 2017).

Hunting may affect the spatial behaviour of wild boars by increasing dispersal and long-distance movements (Casas-Díaz *et al.*, 2013). Some authors found that repeated hunting and direct chasing with dogs increased movement in wild boar social groups (Scillitani *et al.*, 2010). These behavioural changes may in turn increase contact between individuals and have negative consequences for disease transmission (Keuling *et al.*, 2008). Keuling *et al.* (2013) suggested that for hunting to produce a marked effect on wild boar population size, at least 65% of the starting population would need to be removed each year. When densities of wild boars are high, shooting can be an efficient way of removing large numbers of animals in a short time. Removing wild boars at a local scale may reduce their impacts in the short-term, but may attract more individuals from the surrounding areas and may thus not be effective at reducing the long-term effects (Tolon *et al.*, 2009). Therefore, coordinated wild boar population control should be carried out, ideally, at regional scale.

8.4.2 Trapping

Wild boars are relatively easy to trap, particularly when they occur at high densities. Trapping has been used to control populations of wild boars and feral pigs, often in combination with shooting. Trapping can remove large numbers of animals in a relatively short time (Massei *et al.*, 2011). Many types of traps exist on the market, ranging from single animal traps to corrals that can hold large groups (Figure 8.2). The latest generation of traps are equipped with new technologies that allow trap operators to become informed by cell phone message or e-mail when an animal has entered the trap. Wild boars can be captured using drop-nets, suspended over bait piles and triggered remotely. Capturing whole sounders simultaneously can reduce the risk of disease transmission by avoiding social disruption. Trapping success is affected by many factors including wild boar density, season, trap type and location, density of traps, trap effort and bait type (Parkes *et al.*, 2010). Trapping requires intense labour to achieve substantial population reduction. This implies checking traps at least once per day to ensure captured animals are humanely culled and that non-target species are released. In residential areas, or where shooting may be publicly opposed or illegal, trapping is useful for removing live animals. However, in this context, people against lethal animal measures can easily damage or boycott capture operations.

In recent decades, animal welfare concerns have shifted public attitudes towards non-lethal methods to reduce wild boar numbers (Massei and Cowan, 2014). For instance, 44% of interviewed Berlin residents believed wild boar numbers should be reduced, although 67% of these respondents were against any lethal methods of removal (Kotulski and König 2008). Similar

Figure 8.2. Self-constructed round-shaped trap with an eight-metres diameter used in Belgium to capture large wild boar sounders. The time to build a trap was around three hours and the material needed for its construction costed around 1000 Euros. The network of around 150 traps allowed culling more than 1,300 wild boar (photo: Public Service of Wallonia, Belgium).

views have been expressed in Barcelona, where plans for lethal removal of wild boars had to be withdrawn due to public opposition (Cahill *et al.*, 2012). Trapping may be an appropriate method to remove problematic animals or as additional technique to capture juveniles, but it is considered relatively inefficient for population control on a large scale.

8.4.3 Translocations

Translocation of wildlife is often advocated by the public to resolve human-wildlife conflicts. Translocation is labour intensive, as it involves trapping and transport of animals. In many cities in Europe, translocation is used to remove problematic wild boar families (sounders) that become habituated to urban environments. However, there is little scientific evidence to support the humaneness of translocations, which often have adverse effects on the animals moved, such as malnutrition, dehydration, increased predation and subsequent immunosuppression (Massei *et al.*, 2010). There is limited research to confirm the success of translocations to new environments. Translocations also have the potential to spread diseases within and between populations (Kock *et al.*, 2010), and the practice is illegal in many European countries. Therefore, translocation is not recommended as a realistic and valid solution to solve wild boar overpopulation.

8.4.4 Fertility control

Fertility control has been increasingly advocated as a humane alternative to lethal control of wild boar (Massei and Cowan 2014; Pepin *et al.*, 2017). Immunocontraceptives formulated as single-shot intramuscular injections, stimulate the production of antibodies against the Gonadotropin-Releasing-Hormone (GnRH), responsible for ovulation and spermatogenesis. As a result, treated females do not reproduce. In captive trials, a single dose of GnRH-based immunocontraceptive induced infertility in female wild boar for at least four to six years with no adverse side effects on their behaviour (Quy *et al.*, 2014). Contraceptives could be considered for contexts where lethal control is not an option, such as in areas where shooting is not practical or publicly acceptable (Massei and Cowan, 2014). The development of orally delivered contraceptives has shown some effect in maintaining low density of wild boar after the implementation of culling programs (Pepin *et al.*, 2017) and this application is currently being investigated. Nevertheless, contraceptives are not currently registered yet to be used in feral pig or wild boar in Europe. In addition, considering that their effects are only noticeable in the long term, they are not considered a method of choice in case of an urgent need to reduce wild boar population size.

8.4.5 Use of toxicants

Poison baits are commonly used as an alternative in combination with other methods in countries where wild boars or feral pigs are invasive alien species and considered as pests (Lavelle *et al.*, 2018; Snow *et al.*, 2016). These methods are not allowed in the EU, but can be derogated in the case of emergencies (Regulation 528/2012, Article 55). The derogations must consider humaneness, environmental impact, the impact on non-target species and biodiversity (Gentle *et al.*, 2014). In Australia and USA, the best result has been obtained using sodium nitrite, which is orally absorbed, humane (quick and painless), with low environmental residues and no secondary poisoning risks (Snow *et al.*, 2016, 2017). The risks of consumption by non-target species can be reduced using selective delivery systems that limit the access of sympatric species to the bait.

Experience in Australia and the USA shows that this is a very efficient and cost-effective method to reduce feral pig populations.

8.4.6 Comparison of methods

Several studies on wild boar control measures have highlighted the need to combine methods to ensure success of population reduction (for a review see Massei *et al.*, 2011). Each technique presents different advantages and disadvantages (Table 8.1) that should be considered when selecting a population control method. Different study sites, contexts and goals require the combination of different techniques; in addition, methods that are useful to reduce numbers in the absence of ASF (e.g. drive hunts) are not necessarily recommended when the disease is present.

8.5 Methods influencing wild boar movement and behaviour

In this section, we present different methods to physically separate wild boar populations or control its movements through food provision or erection of physical barriers.

8.5.1 Artificial feeding

Diversionary and supplementary feeding are used respectively to decrease wild boar damage to crops, by encouraging animals to feed elsewhere, or to concentrate wild boar in a particular area and increase the number of animals available for hunting (Geisser and Reyer 2005; Ježek *et al.*, 2016; Massei *et al.*, 2011). In France, it was found that using maize in diversionary feeding resulted in a 60% reduction in wild boar damage to vineyards (Calenge *et al.*, 2004). Other studies found that diversionary feeding resulted in either limited or no reduction of damage to crops, and that it may cause adverse effects on biodiversity such as a significant increase on predation risk of ground bird nests in the proximity of sites with supplementary feeding (Oja *et al.*, 2015). One of the disadvantages of supplementary feeding is that the food needs to be provided on a regular basis to act as a suitable attractant, which can incur high costs for labour and resources. Moreover, supplementary feeding may increase survival rates and enhance reproductive success of wild boar (Geisser and Reyer, 2005), thus contributing to population growth. For instance, in the Czech Republic, Ježek *et al.* (2016) suggested that the 84,665 feeding sites (12/1000 ha) present on the national territory could significantly contribute to increased survival rates in this species. In addition, artificial feeding can lead to increased pathogen transmission at feeding points (Ježek *et al.*, 2016; Navarro-Gonzalez *et al.*, 2013). In the context of the current ASF epidemic, the European Food Safety Authority, EFSA has recommended a complete ban on feeding of wild boar to reduce habitat carrying capacity (EFSA *et al.*, 2018b; EFSA *et al.*, 2020)

8.5.2 Fencing and other measures to reduce wild boar movement

Fences are widely used to prevent wild boar access to croplands and infrastructures, such as highways, high-speed railways and airfields (Figure 8.3). Different types of fences can be effective against wild boar if they are appropriately designed and regularly maintained (Rosell *et al.*, 2019). Fencing type should be selected according to management goals, local conditions and duration of the required effects.

Table 8.1. Advantages and disadvantages of main methods for reducing wild boar populations.

Method	Advantages	Disadvantages
Hunting	• Reduction of large numbers in a short time • No cost for governments • Possible to involve members of local rural communities in wild boar management • Possible to target large females	• Growing social opposition against hunting • Bad practices associated with hunting (e.g. supplementary feeding) • Driven hunts with dogs may increase wild boar dispersal • Low acceptance from hunters to eliminate their game resource (recreational hunting)
Trapping	• Possible to capture complete sounders • Allow high level of biosecurity • Culling can be done under humane conditions	• Low acceptance from hunters to eliminate their game resource • Non-selective method: juveniles captured first and more easily than adults • Non-specific method: Other mammals can be captured in the trap • Requires a certain level of expertise • Traps may be tampered with by the public • Some animal welfare cost
Translocations	• Publicly acceptable • Fast-acting at local level • Usable in residential areas	• Labour-intensive • May translocate pathogens/diseases • Animal welfare costs of trapping, transport, and post release survival • May encourage illegal introduction of wild boar
Fertility control by injectable contraceptives	• Injections with a long-term effect are feasible • No social disruption • Usable in residential areas • Species-specific • May decrease disease transmission	• Slow acting at population level Applicable only at small scale and experimentally • Expensive due to trap-inject-release effort • Same animal welfare cost as trapping
Fertility control by oral contraceptives	• No social disruption • Usable in residential areas • Species-specific bait delivery system available • May decrease disease transmission • Applicable at large scale • Relatively inexpensive • Publicly acceptable	• Not commercially available • Slow acting at population level • May concentrate animals around bait points
Toxicants	• Reduce high numbers of animals within a short time	• Not allowed in Europe • Products are not specific to wild boar • Low social acceptance • Possible animal welfare costs

Figure 8.3. Some examples of wild boar proof fences: (A) knotted fences suitable for mitigating damage to pastures are deployed in Catalonia, Spain (photo: C. Rosell); (B) reinforcement fence installed on an existing perimeter fence (photo: F. Navàs).

Mesh wire perimeter fences have been extensively monitored on transport infrastructure. The most effective fences against wild boar are knotted rectangular mesh fences buried to a depth of 20-25 centimetres (cm) into the soil (Figure 8.3A). Progressive density is recommended with a distance of 15 cm between vertical wires and from 5-10 cm at the bottom to 15-20 cm at the top between horizontal wires. Wild boar fences must reach 140-160 cm height. Quarterly maintenance is required.

Reinforcement of existing fences is required when these are not buried or are constructed with chain-link mesh that can be bent by individuals opening holes and passing through them. A proven, effective reinforcement consists of panels of rigid welded mesh (five cm between vertical and 30 cm between horizontal wires), installed outside existing fences and anchored to them up to a height of 60-90 cm (Figure 8.3B). First, the horizontal wires at the bottom need to be cut, leaving vertical spikes that must be stuck into the ground to a depth of 20-25cm.

Electric fencing consists of a power supply system and conductive wires that cause an electric discharge to wild boars upon physical contact with them, dissuading them from crossing the electrified line. A minimum of two wires installed between 25 and 50 cm above ground level are required to deter wild boar access. Appropriate energiser, providing a pulse of 4-8 Joules and with 12 Volts battery, is required to deter wild boar movements. The system needs frequent maintenance to avoid vegetation touching the wires.

Maintenance is a critical issue to guarantee long-term effectiveness of all fence types. Fences must be periodically inspected and damage repaired. While fences along roads and railways are maintained by infrastructure operators, extensive fences are difficult to maintain and many vulnerable points appear, for example river crossings. In addition, fences impact biodiversity by preventing wildlife movements and migrations of other species and reducing ecological connectivity (Woodroffe *et al.*, 2014).

Odour repellents can equally be applied to reduce wild boar movements (Bíl *et al.*, 2018). Several substances are commercialised with different levels of efficiency: while some products show a temporary effect for several weeks, others have not proven to be effective at all. Often, wild boars become habituated and their effectiveness is lost after several applications. Hence, their use is not recommended if a long-term effect is needed.

Sound repellents activated by movement sensors are being applied to prevent animal-vehicle collisions and to protect croplands. The effectiveness varies depending on each particular device. Those not activated by sensors that emit periodic 'scaring sounds' have proven to be ineffective.

8.5.3 Wildlife corridors in transport infrastructure

Habitat fragmentation has been identified as one of the main factors causing biodiversity loss. The barrier effect of linear transport infrastructure is reduced by the construction of wildlife corridors that are combined with guiding fences and permit reduction of wildlife mortality due to animal-vehicle collisions while increasing ecological connectivity. Wild boars use many types of corridors: wildlife underpasses and overpasses as well as multi-use structures combining wildlife use with drainage, forestry or cattle paths (Iuell, 2003). A preference for large overpasses and underpasses has been recorded, but in some areas wild boars also use narrow structures. Wildlife corridor management may be a critical issue providing (or not) possibilities for wild boars to move across the landscape. In a fragmented landscape, wildlife corridors are important to ensure long-term population persistence. Some techniques could be used to avoid corridors being used by wild boars for short periods such as ASF outbreaks, when wild boar movements should be restricted.

8.6 Management of wild boar populations applied in the context of African swine fever control

This section describes a series of measures applied to manage ASF in different EU countries, in some cases in the context of a focal introduction, and in others in a context of wider dissemination. Strategies described are a combination of wild boar management and disease control methods. Despite the fact that successful eradication of ASF virus after a focal introduction remains the exception rather than the rule, these measures have proven useful to contain the spread of the diseases to different degrees. It is worth emphasising that there is no single recipe for successful control or eradication of ASF. Based on the EU experience, the success of restoring disease freedom after an ASF introduction in a given territory has been the result of a combination of procedures that had to be changed and adapted continuously according to the epidemiological progress of the disease and to the geographical and socio-ecological context of the territory.

8.6.1 Zoning

According to international animal health regulations, zoning or regionalisation for disease control purposes allows the identification of specific geographical areas within a country or region as having a defined status with respect to a particular disease linked to specific disease control actions (OIE, 2019). These areas (sometimes also called zones) are often concentric around a confirmed or suspected focus of infection, with the most intensive disease control activities taking place in the inner area. An infected area is the area where the disease is present and a free area is an area where the disease is absent. Buffer areas are created between an infected and a free area. In some cases, they can be surrounded by another concentric area placed under a higher level of surveillance called control or surveillance area (Figure 8.4).

Based on the EU Council Directive 2002/60/EC of 27 June 2002, in the case of confirmation of ASF in wild boars in an EU country, the identification and establishment of an infected area (sometimes called high-risk area) should be implemented as soon as possible. In addition, the Strategic approach to the management of ASF for the EU (working document SANTE/7113/2015 – Rev 12) has been developed with the aim of establishing harmonised measures in response to the epidemiological situation with regard to ASF in the EU. This strategy is based on the scientific output from the European Food Safety Agency (EFSA *et al.*, 2018b) and states that this implementation must be adapted to local circumstances depending on the country or region in question. The measures implemented in each of the management areas, as well as the names and size of the areas and the timing of implementation of measures, may differ between different countries and can change over time based on the epidemiological situation (Dixon *et al.*, 2020) and the effect of the measures implemented (Table 8.2).

8.6.2 Methods to restrict wild boar movement

Using fences to reduce the risk of ASF spread through wild boar movements might be useful in case of a localised point source incursion. The aim is not to completely halt wild boar movements, which might be unrealistic if the fence perimeter is long and the terrain is rough, but rather to reduce wild boar movements as much as possible. In addition to fences, habitat management and ban of forest activities can also contribute to reduce wild boar mobility outside an infected area.

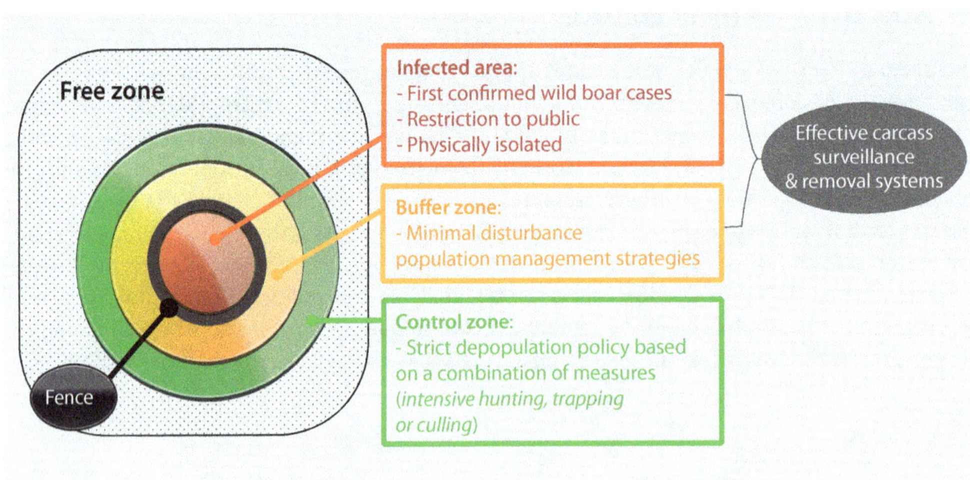

Figure 8.4. Schematic representation of disease management areas suggested to define concrete geographic areas where specific disease control strategies will be implemented in order to address focal ASF incursions, based on OIE recommendations (Source: Dixon *et al.*, 2020 based on OIE, 2019). The size of the areas and the combination of measures to be implemented in each area require regular monitoring and updates, based on the progress of the epidemiological context. The fence indicated is a control strategy that could be used to both prevent movement of wild boars and delineate the restricted area.

Table 8.2. Summarised information on different regionalisation and disease management strategies proposed by certain European countries for the control of ASF after detection of an outbreak in wild boar population.

Country	Terminology for infected and non-infected areas	Restrictions and applied measures
Czech Republic	• Highest risk area (core area) includes all the infected cases (fenced area). • High risk area (buffer area) around the highest risk sub-area, calculated considering the maximum annual increase of the home ranges of the wild boars living in the fenced area. • Low risk area infected area with low risk, corresponded to demarcation of Part II (according to Implementing Decision (EU) 2017/1162). • Intensive hunting area defined by the layout of highways (8,500 km^2)	• Fencing around infected/highest risk area • All infected area: – increased passive surveillance (rewards are given to those persons finding wild boar carcasses); – hunting ban (any species, any hunting method); – ban on wild boar feeding; – ban on entrance for the general public into the ASF highest risk areas; – disposal and sampling of wild boar carcasses in selected rendering plants; – individual hunting (snipers) at the final stage of the epidemic phase; – sampling and testing for both ASF and CSF in any wild boar carcass found; – intensive hunting area: intensive hunting.
Bulgaria	• Infected area of 20 km radius around the case. • Buffer area 20 km around the infected area. • Surveillance areas Whole region/province.	• Ban on hunting, possibility of sanitary shooting of wild boar by appointed hunters trained on biosecurity and trapping. • Searching for wild boar carcasses by appointed hunters, trained on biosecurity. • Ban on movement of domestic pigs, semen, ova, embryos, meat or products originating from the infected area. • Sampling of domestic pigs (for serological and virological testing) within the infected area. • Enhanced biosecurity measures.
Belgium	• Infected area (IA) based on a radius of 15 km around the place where the first confirmed positive carcass was discovered. • A month later, the IA was subdivided into three concentric areas: – core area corresponding to an area where all the carcasses of wild boars positive for the virus were found; – buffer area corresponding to a 6 km strip around the Core area completely fenced; – an enhanced observation area in the distal part, its size is variable and its perimeter can be established following physical or administrative limits.	• In addition, the territory surrounding these three areas was the subject of vigilance measures, namely: – hunting, movement and logging restrictions; – passive surveillance; – active search for wild boar carcasses; – compulsory reporting of any wild boar found dead. • The Administration took the necessary measures to ensure that virus detection was carried out on any wild boar found dead. All carcasses were compulsorily destroyed under official control.

8.6.2.1 Transboundary fencing

Since the introduction and spread of ASF in the EU, long distance transboundary fencing has been used by several countries to protect their national territories from virus incursion from their neighbours, despite their questionable efficiency and negative environmental impacts (Section 8.5.2). Bulgaria, for instance, erected a 133 km fence along the border with Romania. Similarly, shortly after the introduction of the virus in Belgium in September 2018, France erected a total 170 km of fences along the Belgian border. At the end of 2019, Germany erected a 120 km electric fence, after several infected wild boar cases were found in Poland less than 50 km from the German border.

Fences are politically tempting in the context of an emerging disease (Mysterud and Rolandsen, 2019) because they are a highly visible measure and, in the short-term, they can efficiently reduce disease transmission by direct contact. Unfortunately, when a long transboundary fence is built as a response to an emergency, plans for measuring its efficiency, calculating its maintenance costs and assessing its biological impact in terms of wildlife conservation are rarely considered (Jakes *et al.*, 2018). Therefore, it is recommended to perform a cost-benefit analysis and seek advice on the potential environmental consequences before taking the decision to implement such a high impact and resource consuming measure.

8.6.2.2 Focal fencing

Focal fencing should be applied as quickly as possible after an outbreak or focal introduction has been detected and the infected area has been defined and can be surrounded by a physical barrier. The objective of such fences is to prevent wild boar population movements in and out of the infected area. The type of fence can vary as long as it provides movement restriction (Section 8.5.2). Since this focal fence is unlikely to encompass the entire home ranges of wild boar sounders living in the infected area, attempts of some individuals to cross the fence can be expected. Therefore, a high level of monitoring and surveillance to prevent, detect and eliminate potential infected animals crossing the fence is recommended. Fencing around a focal introduction can save some time, especially when densities are high. Indeed, even if some positive cases are detected later outside the fence, partitioning the infected area with physical barriers can be useful for improving the efficiency of culling activities in the post-epidemic phase, as shown in Belgium (Figure 8.3). In other cases, focal fencing combined with intensive hunting can be useful to reduce or eliminate wild boar populations from a given area such as implemented in Belgium and the Czech Republic (Figures 8.5 and 8.6).

8.6.2.3 Habitat management

Section 8.5.1 of this chapter provided a short overview of the effects of wild boar habitat management on the dynamics of wild boar population. It should be noted that those measures do not have an immediate effect on the population and are only noticeable after one or two generations. Therefore, they have in general very little utility in the context of an ASF emergency where urgent effects in the short term are needed (Jori *et al.*, 2020). The only exception is the provision of diversionary feeding and baiting in a given area to prevent wild boar movements. This approach was used in the Czech Republic to aggregate potentially infected animals in the infected

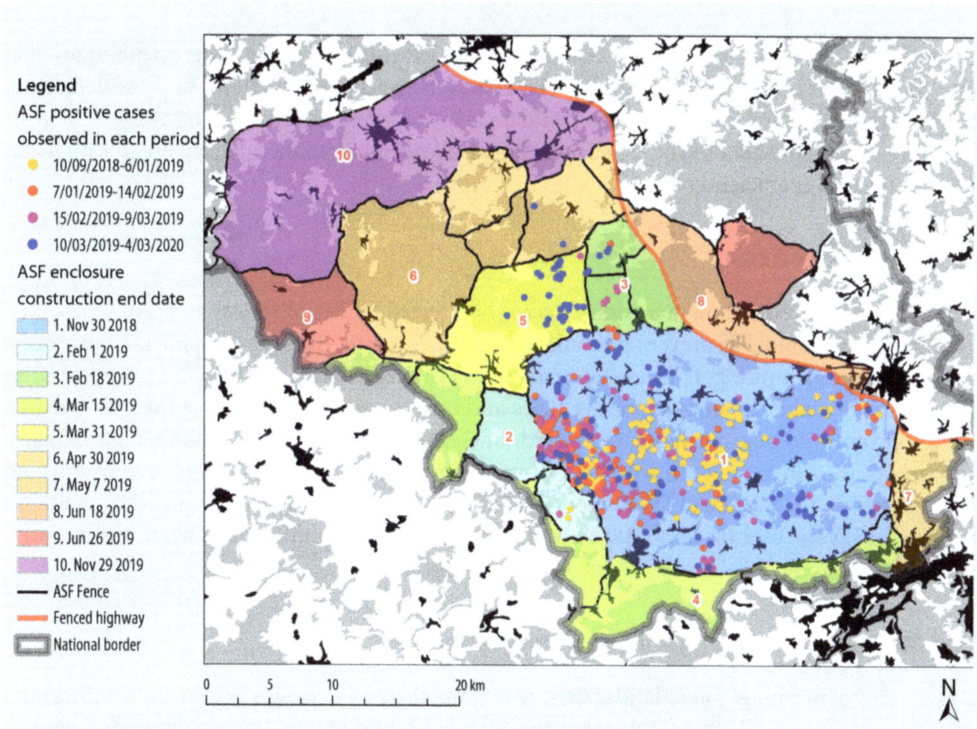

Figure 8.5. Cartographic representation of wild boars positive for African swine fever (dots) in Belgium and the network of fences (black lines) intended to slow down the spread of the disease. The colour codes of the dots correspond to periods ending with the first observation of a fence crossing. The colour codes of the enclosures (fenced areas) correspond to the completion dates of the fence. Fences in France and Luxembourg (not shown) are connected to the Belgian network (map provided by Public Service of Wallonia, Belgium).

Figure 8.6. Maps showing details of the infected area in the Czech Republic: (A) location of the different areas (high-risk, low-risk and intensive hunting area) in the east of the country; (B) details of the fenced high-risk area with locations of positive (red dots) and negative (green dots) animals culled or found (maps provided by Petr Vaclavek).

area to facilitate culling. In that case, crop fields were left unharvested in order to maintain wild boar populations in the area and prevent dispersion. Nevertheless, it should be considered that supplementary feeding could also attract wild boar living in the surrounding non-infected areas. Therefore, this method should only be applied if food availability in the adjacent areas is high and the infected area is efficiently fenced.

8.6.2.4 Forest access ban

Following ASF introduction, provisional access to infected forests or forests adjacent to infected areas should be banned to prevent potential spread of the virus. This ban is applied to any forest activity, including hunting, trekking or extractive activities in order to limit the transportation of contaminated objects such as tools, vehicles and other equipment outside of the infected area and to prevent wild boar movements. Indeed, in Germany, collective driven hunts, particularly with dogs, led to an increase in wild boar home ranges and the potential dispersion of infected animals outside the focal area (Keuling *et al.*, 2008). Therefore, it is recommended to maintain this ban for several months and lift it only at the end of the epidemic phase, when the majority of the animals in the focal area have succumbed to the virus.

8.6.3 Methods for reducing wild boar populations

In large and continuous forest landscapes, it is unrealistic to consider achieving a substantial reduction in the numbers of the targeted wild boar population. However, under certain circumstances and by applying a combination of zoning and fencing, lethal population management methods have been successful in reducing wild boar populations effectively. Recent studies suggest that in an ASF outbreak the natural reduction of the population is fairly drastic and the reduction is enhanced by but not dependent upon control measures (Morelle *et al.*, 2020).

8.6.3.1 Hunting to control populations

Depending on the evolution of the epidemiological situation, different and even opposed hunting strategies can be applied in the same territory. In infected areas that are not physically contained by fences or geographical barriers (highways, rivers), hunting should be avoided by all means, in order to limit the spread of the disease (Section 8.6.2.4). However, if an infected population is circumscribed into a small area, hunting to reduce population size can be attempted at the end of the epidemic phase, when the remaining wild boar populations are limited. The same approach can also be applied in the buffer area around the infected area to reduce population size as much as possible. Different measures to reduce wild boar numbers can be applied, alone or in combination, with the help of a coordinated multidisciplinary team involving specifically trained local hunters, wildlife managers, animal health authorities and other government officials (Guberti *et al.*, 2019).

8.6.3.2 Intensive hunting policy

In areas surrounding an infected area, it can be beneficial to attempt a drastic reduction of the wild boar population by any possible method. The most obvious and accepted method in the EU is to intensify hunting pressure. This can be achieved for instance by increasing the number of hunters engaged in radical population reduction policies. Experience from infected countries suggests that it might be difficult to obtain compliance of hunting clubs and associations to implement unselective shooting and intensive hunting measures. Therefore, it might be necessary to hire professional or elite shooters or staff from other government institutions with professional shooting skills (army, police). The use of specialised equipment, such as silencers and night vision devices has proven successful to improve hunting efficiency with minimal wild boar disturbance and dispersal (Box 8.1). To increase efficiency, culling efforts should be concentrated and sustained in time and space. The use of fences to restrict animal movements can facilitate the reduction of wild boar numbers more efficiently. Such an intensified culling operation must be carefully planned with the perspective to remove and destroy a large number of carcasses in a limited amount of time. Therefore, anticipating sufficient human and material resources is recommended.

8.6.3.3 Trapping

Because traditional hunting techniques are not sufficient to drastically reduce wild boar populations (Section 8.4.2), it can be useful to combined hunting with other tools such as trapping. The use of large traps designed to capture several individuals at every catch is particularly interesting in this regard. In Belgium, circular corrals with a diameter of eight metres (Figure 8.2), were used with success. A network of traps can be placed every 300 hectares on former feeding points in large forested areas far from crop fields or close to points of interest for wild boar, such as mud puddles. Compared to other lethal methods, such as hunting, trapping has the advantage of allowing easier implementation of biosecurity measures as well as a higher level of animal welfare, and is thus probably better accepted by the general public. However, in EU countries some hunters could be against the concept of culling (i.e. shooting) trapped animals and might boycott trapping efforts. In addition, as mentioned in Section 8.4.2, effectiveness of the method is variable depending on the availability of alternative trophic resources. Therefore, its use might be restricted to suitable seasons and areas. Finally, it is important to highlight that the implementation of this method requires considerable time, effort and human resources, since baiting and trapping efforts need to be monitored and maintained regularly.

8.6.4 Combination of different methods

All the methods reviewed in this section can be applied alone or in combination. There is no single formula and their application and combination in space and time will depend on the goals imposed by the specific epidemiological and geographical context. In the Belgian control area (Box 8.2), for instance, combining intensive hunting and trapping with fencing provided efficient results. Conversely, in the buffer area, only trapping was performed before the arrival of the epidemic wave.

Box 8.1. Case of the Czech Republic.

The focal introduction of ASF in the Czech Republic is the only case in which ASF spread in wild boar has been successfully eradicated to date in the EU. The focal character of the incursion facilitated early detection and timely implementation of strict measures in a relatively localised affected area, resulting in very limited spread of the disease. Measures were continuously adjusted, based on the epidemiological situation and disease progress. Passive surveillance through collection and testing of wild boar carcasses was considered key to the final success, allowing immediate definition of the perimeter of the infected area after the disease had been confirmed. The wild boar population within the infected area was initially kept as undisturbed as possible. In addition, 115 hectares of crops were left unharvested in order to keep wild boars inside the core area. Active surveillance and depopulation of wild boars in the infected area started only after the epidemic had decreased. At the end of this process, active searching for carcasses was considered very important in order to ensure the removal of any potential long-term source of infection in the area. In addition to these population management procedures, measures such as fencing and intensive hunting were applied.

Zoning: Demarcation of the highest-risk area, called 'core area' in Czech Republic, was combined with intensive passive surveillance and carcass search. The core area was defined by a polygon that encompassed all found dead ASF-positive wild boar. Upon identification of the core area, a buffer area was defined around it. The determination of the buffer area was performed with the help of hunting experts and it was based on wild boar annual home range and natural corridors in the landscape (roads, villages, etc.).

Fencing: Ten kilometres of electric fence (voltage 6,500-11,000 V), combined with odour fences (Bíl *et al.*, 2018), were placed on the outer periphery of the highest-risk area (57 km^2) in order to limit movement of wild boar outside these boundaries as much as possible (Figure 8.4). It was difficult to assess the contribution of fences to the eradication of the disease but given the successful outcome, it was assumed they had a positive effect. There were 11 positive animals detected outside the fenced area, suggesting potential fence leakage in some places. Nevertheless, these were isolated cases localised in the neighbourhood of a village that could not be entirely fenced off. It concerned one small sounder of wild boars living in that area. Local hunters had been observing some night-time movement of wild boars through the village. The activity of police snipers that were part of the eradication effort might also have provoked unusual behavioural patterns in wild boars in the fenced area. The construction of a massive wire fence was discussed but finally discarded because of its high costs and the long time estimated for its deployment. Since the area concerned hosted a high density of wild boars, the construction of a massive wire fence over a long perimeter including private land would have been too slow and problematic.

Intensive hunting: The hunting ban in the fenced high-risk area was lifted at the end of the epidemic peak (11 weeks after the first detected cases), with the aim of depopulating the area as quickly, silently, and efficiently as possible. The government authorities decided to use police snipers trained in hunting biosecurity that split into eight two-man teams shooting wild boar at three-day intervals. All the shot animals were safely moved to the nearest road, and then transported to be sampled at the rendering plant. Targeted wild boars were located with helicopters and drones, supported by ground teams equipped with night vision devices. Snipers used weapons with silencers and intended to be as precise as possible in aiming at the head in order to minimise bleeding and environmental contamination with the virus. The ground veterinary teams immediately received GPS locations of shot animals for prompt carcass collection and disposal. During the10-week operation, police snipers shot 157 wild boar and the infected area was almost completely depopulated. This experience confirmed the efficiency of implementing intensive hunting at the final stages of the epidemic phase in order to contribute to disease eradication.

Box 8.2. Highlights of the Belgian experience in the control of ASF.

In Belgium, installation of fences was part of the ASF control strategy from the first case notification in September 2018. The fences used were 120 cm high wire mesh. In total, about 300 km were erected on an area of 1,100 km^2, to which could be added the fence of the Brussels-Luxemburg motorway of around 40 km. From the outset, the fences were seen as a means to hinder the dispersal of wild boars and not as impassable obstacles. The fences contained multiple weak points, such as gates and rivers, which were not secured.
The strategy for the construction of the fencing network was considered in a dynamic way. Initially, the fences were installed to break the epidemic wave, which was progressing rapidly given the continuity of the forest massif and the abundance of wild boars. The first layer of fences was built in November 2018. The location of the fences was based on the results of surveillance to avoid fencing within the infected area. The hoped-for interruption of wild boar movements was achieved, but fence crossings did occur especially in rural areas where the number of gates was higher. This resulted in an expansion of the infected area on three occasions (January, February and March 2019). Each enlargement automatically resulted in the installation of new fences to contain these new incursions, without necessarily waiting for surveillance results, in order to save time (Figure 8.5).
Once the epidemic phase was over (April 2019), new fences were erected approximately 5-10 km from the edge of the infected area. At the same time, the construction of other fences ensured the transboundary connection with French and Luxembourg fence networks. In this way, the former infected area was completely circumscribed by simple fence lines combined with 20 huge management enclosures. This complete containment of the remaining wild boar population in the infected area facilitated the efficient implementation of passive surveillance and depopulation efforts. During the epidemic, in the infected areas, organised searches for carcasses and trapping were the only operations conducted (until May 2019). During the post-epidemic (after May 2019), night shooting with generalised use of baiting points and camera alerts was also implemented to cull the remaining and scarce wild boar specimens. In the free areas, in addition to passive surveillance, a succession of depopulation methods was used according to the season: driven hunts in autumn and winter, trapping in spring and summer, and night shooting all year long. The combination of these techniques, consistently applied within the fenced areas, allowed almost complete reduction of the population of wild boars. More than 6,000 wild boars were removed from September 2018 to March 2020 in the ASF management area (Part II + I = 1,100 km^2): around 3,100 were shot in driven hunts, 1,200 by trapping, 500 by night shooting and 1,200 were found dead.
The development of a dynamic fence network, in combination with depopulation measures and organised search for carcasses, considerably helped to reduce the spread of the disease.

8.7 Disease management methods applied to wild boar populations

8.7.1 Passive surveillance

In the context of ASF, passive surveillance is the process of monitoring the presence of wild boar carcasses in the forest, which is generally implemented with the help of a team of experienced forest workers or hunters (Jori *et al.*, 2020). This approach is considered to be an essential step in ASF prevention, control and eradication measures (Chenais *et al.*, 2019; Gervasi *et al.*, 2019; Guberti *et al.*, 2019). In free areas, passive surveillance is key for early detection and all of the primary ASF cases in wild boar in the newly infected countries during the current epidemic were detected by this method (EFSA *et al.*, 2018b). A passive surveillance system is effective in field conditions if there is a clear suspected case definition, strong collaboration with field operatives, a procedure for managing samples and carcasses and a follow-up of the results. The official suspected case definition of ASF in the domestic pig sector is based on evidence of the disease indicated by tangible facts such as symptoms, lesions or response to diagnostic tests: '[…] *any pig or pig carcass exhibiting clinical symptoms or showing post-mortem lesions or reactions to laboratory tests carried out in accordance with the diagnostic manual which indicate the possible presence of African swine fever*' (EU Council Directive 2002/60/EC of 27 June 2002).

However, when dealing with free ranging wildlife, it seems more convenient to employ a case definition based on the suspicion of a potential case, with a lower requirement for evidence of the pathogen: '[…] *any found carcass out of the context of hunting, including road killed animals and any diseased wild boar shot for sanitary reasons*'. The broader the suspected case definition, the more sensitive the surveillance system will be. The inclusion of road killed animals is a matter of debate (Schulz *et al.*, 2019). This decision depends on the competent authorities and on the human and material resources available to manage these carcasses. In contrast, wild boars shot for sanitary reasons (sick animals or individuals presenting abnormal behaviour) have to be included as a priority in passive surveillance. A structured procedure for managing carcasses and transporting samples from the field to the laboratory is crucial (Chapter 5). In some field investigations, some hunters have been trained in biosecurity procedures before taking samples and removing carcasses for disposal. Within an ASF crisis context, it is advisable that sampling and carcass handling are carried out by competent authorities or trained personnel. Practically, a hotline or similar online tool must be set up and widely disseminated to facilitate the notification of discovered carcasses. The competent authorities or services (veterinary and forestry services) are then responsible for the next steps, which include tracing, transporting the carcasses to the laboratory, sampling and destroying the carcass under the required biosecurity procedures to prevent secondary spread. This procedure must be operational before any outbreak, and sustainable in the long term. Follow-up of passive surveillance results is needed to assess the effectiveness of the system. Indeed, in absence of ASF, the discovery of a carcass is often an incidental event and it is difficult to determine if the rate of discovery corresponds to the rate of natural mortality. Assuming natural mortality of 10% per year (Keuling *et al.*, 2013) and the hypothesis that only 10% of carcasses are detected, it has been suggested that effective passive surveillance should be able to detect 1% of the whole estimated wild boar population in a given area, in the absence of ASF. However, these suggested figures are only a bulk estimation based on expert opinion (Guberti *et al.*, 2019). More evidence-based assessment is currently needed to ascertain what level of passive surveillance is required to ensure the early detection of index cases in wild boar and to what extent that target would be feasible and realistic to be implemented in practice.

In infected areas, passive surveillance is also crucial for ASF control. Active search for carcasses, removal and testing allows delimitation of the infected area, helps to follow the epidemic phases and reduces the viral load in the environment. Indeed, the high tenacity of the virus in the carcass and in the environment is considered the main reason for maintenance and spread of the virus even in areas with very low wild boar densities (Guberti *et al.*, 2019). ASF virus (ASFV) has been shown to persist in blood (540 days at 4 °C), spleen (204 days), bone marrow (180 days), skin/fat (300 days) and frozen meat (1000 days). The virus further survives the process of putrefaction, with longer persistence in carcasses in cold winter temperatures than in summer. Moreover, soil and water (mud puddles in which carcasses can be found) from underneath a contaminated carcass may also play a role in the persistence of the virus (reviewed in Chenais *et al.*, 2019). These carcasses are exposed to scavengers including wild boars, with studies showing evidence of direct contact (wild boar sniffing and poking carcasses) and even confirmed cannibalism (Cukor *et al.*, 2020a; Probst *et al.*, 2017). Any type of contact is likely to transmit the virus. Hence, carcass removal should be carried out as soon as possible. In infected areas, passive surveillance should be systematically implemented to increase the probability of finding carcasses. Such enhanced passive surveillance is time-consuming and requires significant human resources, especially if the infected areas are large. Carcasses may be located in inaccessible environments or hidden both in summer (in high vegetation) and in winter (under snow). However, recent studies demonstrated that there are specific environments where dead wild boars are most likely to be found. Sick wild boars have feverish and depressed behaviour that leads them to search for cooler and moist areas. Therefore, selected deathbeds will be in forests rather than in open landscapes, near water sources and in quiet places away from roads (Cukor *et al.*, 2020b; Morelle *et al.*, 2019).

During the ASF epidemic in Belgium, active search for carcasses was designed (cartographic monitoring) and implemented by the regional authorities with occasional participation of the army. Hunters and landowners were requested to notify any dead wild boar. Practically, a map of priorities was adapted every two weeks according to previous results (localisation of the last positive fresh cases, passive surveillance results from previous weeks) and sent to the field coordinators. Some criteria were also considered (such as topographic wetness index, heat load index and distances from rivers) to target specific environments according to the ASF carcass model previously described (Morelle *et al.*, 2019). This model was built to respond to the emergency and was based on the first 200 positive cases in Belgium and on the datasets from the Czech Republic and Poland. As soon as the carcass was detected, the precise geographic location was reported to the team in charge of carcass removal and destruction.

8.7.2 Carcass management and biosecurity

At all stages of passive surveillance, biosecurity procedures must be respected to avoid spreading ASFV. It is also essential to prohibit access to pig holdings by any person who has been in contact with a wild boar (living or dead) in the prior 48 hours.

Even in free areas, any dead wild boar must be considered as potentially ASF positive especially in areas bordering infected areas. Ideally, people (hunters, forest rangers or walkers) who discover a dead wild boar should notify competent services without approaching or touching the carcass. Samples for ASFV analysis can be collected at the discovery site, at a collection centre or at the rendering plant. After sampling the carcass can be safely destroyed *in situ* (deep burial or burning if authorised by competent authorities) or individually packed and sent to a rendering plant following biosecurity procedures. It is important that forestry officers have received prior training in biosecurity for carcass management.

In infected areas, biosecurity measures must be strengthened. For carcass searches, workers must be equipped with boots and clothes specifically dedicated to this activity and which will be cleaned and disinfected daily. Carcass removal, packaging and transport require qualified personnel and specific equipment (boots, overalls, gloves, masks and packing material). Transport vehicles must be exclusively dedicated to this task in the infected area. These teams need to be trained to pack and transport carcasses according to strict biosecurity rules. Approved biocides must be used for disinfection of materials, vehicles (after cleaning) and the soil under the carcass. All people involved must undergo preliminary biosecurity training adapted to their tasks.

8.7.3 Stakeholder communication

Achieving control of ASF in the wild requires engagement with a large panel of stakeholders (Jori *et al.*, 2020). It is essential to identify and involve these stakeholders already during the contingency planning. To engage efficiently it is further important to understand what motivates each stakeholder group to participate in detecting and controlling ASF. Sincere interest, participatory dialogue and open communication are needed to obtain such understanding. In this regard, regular, relevant and transparent communication can actually be an incentive for certain stakeholders. Good communication and feedback to hunters have for example been reported as more motivational than other (financial) incentives. To boost motivation for early detection and control it can be useful to raise stakeholders' awareness about the different advantages of controlling the disease, such as the economic impact for the pork industry or the access and freedom of use of the forest and wild boar hunting activity (Jori *et al.*, 2020). Among the various stakeholders, hunters stand out as especially important for early detection, and at later stages surveillance and control of ASF. However, relying only on hunters for early detection can be challenging. Indeed, given the drastic measures that will be implemented in the event of a positive case being discovered, namely a ban on access to the forest, many hunters might be reluctant to collaborate or even tempted to hide potential cases. A good level of communication with hunters and hunter organisations is instrumental and thus deserves special attention (Keuling *et al.*, 2016).

Considering the epidemiological patterns of national and international spread of ASF in wild boars (as described in Chapter 9), raising awareness of ASF transmission risks among different sectors is paramount for prevention, early detection and for achieving successful control (Chenais *et al.*, 2019). Those sectors should include the general public, but also sectors involved in domestic pig farming, wildlife management or different forest activities (hunters, runners, dog walkers, ornithologists, forests managers, berry pickers, etc.). Communication messages should be as simple as possible, and include the global dimension of ASF, its epidemiology and ecology, the economic impact and the absence of treatment and vaccine. Scientific messages must be communicated in a language that is reachable and adapted to non-specialists (Dietz, 2013). Communication further needs to be adapted to the different stakeholders and tailored for long-term application. If ASF becomes established in a territory and the crisis lasts for several years, there is a risk that stakeholders will become habituated to the endemic disease situation.

In this case, compliance with prevention and control measures is likely to decrease, requiring repetition and adaptation of the messages. It is also important to keep awareness and application of preventive measures high over a prolonged time in areas at risk.

8.8 Need for future research

8.8.1 Future trends for improving carcass detection and reporting

In the context of an outbreak, the tendency of diseased animals to hide can considerably reduce the success of finding wild boar carcasses, and thus influence both the efficiency of passive surveillance and the motivation of the persons involved in carcass searches. Methods and ideas for improving the probability of wild boar carcass and disease detection will differ between countries. Methods can include having voluntary hunters (paid or unpaid) to search for carcasses even during periods with low hunting activities, provision/availability of easy-to-use sampling material and infrastructures for carcass disposal. In many countries, hunters are asked to volunteer to search for carcasses. If this community does not see any advantage in reporting or taking samples from dead wild boar, passive surveillance will not be successful. There is a need to develop research aimed at improving the efficiency of carcass detection, including the perception of hunters towards this activity, the use of mobile phone applications, and active search with trained dogs and the use of computer modelling to identify areas with a higher probability of carcass presence.

8.8.2 Oral administration of pharmaceuticals to wild boar

Oral administration of pharmaceuticals (ASF vaccine, contraceptives or other products to reduce wild boar populations) will require efficient administration methods to deliver the products in the wild using baits and selective delivery systems. For a bait to be efficient, it should be attractive, palatable and accessible to wild boar, but also unavailable or safe for other sympatric species living in the same ecosystem. It should also be resistant to local climatic and environmental conditions. Because wild boars have access to other natural sources of food, the best season for animals to become interested in baits needs to be identified. This season is usually coincident with the period of food scarcity and may vary depending on the forest structure, the climate and geographic conditions, as well as the presence of other sympatric competitor wildlife species. In Mediterranean ecosystems it is usually summer, while in continental Europe it is mostly winter. In the EU, extensive experience in wild boar baiting and oral vaccination was achieved during the wild boar classical swine fever vaccination programmes in Germany and France (Rossi *et al.*, 2015). Ideally, the bait needs to be provided with selective delivery systems that prevent their consumption by other non-target species (Campbell *et al.*, 2011; Massei *et al.*, 2010). Some other systems have been tested to deliver vaccines against bovine tuberculosis in wild boar in Spain (Beltrán-Beck *et al.*, 2012; Díez-Delgado *et al.*, 2018). Some systems developed in Australia and the US to administer toxicants to feral pigs could potentially be used to administer other pharmaceutical products (Lavelle *et al.*, 2018; Snow *et al.*, 2016). In any case, those systems need to be tested in different ecosystems and climatic conditions before being widely implemented. For instance, a bait developed and tested in humid cold conditions might be unsuitable in a hot and dry environment. Similarly, some selective delivery systems such as those tested with success in Australia or the UK, might not work so selectively in some European habitats where wild boars coexist with other large mammals such as bears or wolves.

8.8.3 New technologies to record and restrict wild boar movements

Fencing to reduce wild boar movements is a tool that contributes to the control and eradication of an ASF outbreak. Wire mesh and electric fences have proven effective if appropriate technical specifications are followed. Other innovative methods to deter wild boar are being tested based on different stimuli that could be olfactory, light or sound. Many of them have been revealed as ineffective while others provide acceptable results that should be further developed. Devices provided with sensors and activated only when wild boar are recorded are being investigated. The use of drones has also a high potential to assess wild boar location and help in wild boar management. Those techniques are not only applicable in the context of an ASF outbreak but also to reduce wild boar damage to croplands and conflicts in urban areas.

8.9 Final remarks

The battle against ASF in the wild boar population in the EU has been a source of frustration, but also an opportunity to learn more about this complex disease, its epidemiology and dynamics in a new environment. For many animal health authorities, wildlife management services and hunting associations the control of ASF in free ranging populations remains a major challenge. The ability of the virus to persist in the environment complicates the control and eradication of the virus from forest ecosystems in the currently affected European countries (Chenais *et al.*, 2018). In addition to long-term persistence, awareness about the disease needs to remain high throughout Europe due to the common occurrence of long-distance jumps with unexpected disease emergence of the virus into new territories. The experience of the Czech Republic and Belgium suggests that preparedness, early detection and prompt action are instrumental in containing the spread of the disease and can serve as examples of success stories. However, the risk of spread throughout the continent and the associated challenge to prevent this risk remain very high.

EU expertise accumulated in the field of wild boar management during decades has been an invaluable help for deploying strategies to contain and monitor the disease within wild boar populations. The need for an urgent response in the field has obliged animal health authorities, hunting associations and wildlife managers to work together with a common goal, and to share expertise, methods and tools arising from these different disciplines. In this process, the need to engage and maintain a high level of communication has been repeatedly emphasised. Another fundamental aspect in this multidisciplinary collaboration has been the importance of passive surveillance in order to detect the first cases, monitor the progress of the disease or even evaluate or confirm the success of control methods within a given wild boar population. This area still has room for improvement and can benefit from applied research in different fields such as the use of searching dogs, the development of innovative techniques to restrict wild boar movements, or the use of mobile phone technology or computer modelling to improve carcass detectability.

Finally, it is worth mentioning that despite wild boars being fascinating animals and a fundamental part of Eurasian forest ecosystems, the increasing population trend in Eurasia and its consequences are raising growing concerns. Improving our capacity to reduce wild boar population numbers, mitigate its effects, including the control of ASF and other wild boar pathogens will certainly be one of the most serious challenges for wildlife management and animal health disciplines in the next decades.

Acknowledgements

This publication is based on work from 'Understanding and combating African swine fever in Europe (ASF-STOP COST action 15116)' supported by COST (European Cooperation in Science and Technology).

References

Ballari, S., and Barrios-Garcia, M.N., 2013. A review of wild boar Sus scrofa diet and factors affecting food selection in native and introduced ranges. Mammal Review 44. https://doi.org/10.1111/mam.12015

Barrios-Garcia, M.N. and Ballari, S.A., 2012. Impact of wild boar (*Sus scrofa*) in its introduced and native range: a review. Biological Invasions 14: 2283-2300. https://doi.org/10.1007/s10530-012-0229-6

Beltrán-Beck, B., Ballesteros, C., Vicente, J., de la Fuente, J. and Gortázar, C., 2012. Progress in oral vaccination against tuberculosis in its main wildlife reservoir in Iberia, the Eurasian wild boar. Veterinary Medicine International 2012: 978501. https://doi.org/10.1155/2012/978501

Bieber, C. and Ruf, T., 2005. Population dynamics in wild boar *Sus scrofa*: ecology, elasticity of growth rate and implications for the management of pulsed resource consumers. Journal of Applied Ecology 42: 1203-1213. https://doi.org/10.1111/j.1365-2664.2005.01094.x

Bíl, M., Andrášik, R., Bartonička, T., Křivánková, Z. and Sedoník, J., 2018. An evaluation of odor repellent effectiveness in prevention of wildlife-vehicle collisions. Journal of Environmental Management 205: 209-214. https://doi.org/10.1016/j.jenvman.2017.09.081

Bongi, P., Tomaselli, M., Petraglia, A., Tintori, D. and Carbognani, M., 2017. Wild boar impact on forest regeneration in the northern Apennines (Italy). Forest Ecology and Management 391: 230-238.

Bueno, G., Barrio, I.C., García-González, R., Alados, C. and Gómez-García, D., 2010. Does wild boar rooting affect livestock grazing areas in alpine grasslands? European Journal of Wildlife Research 56: 765-770. https://doi.org/10.1007/s10344-010-0372-2

Cahill, S., Llimona, F., Cabañeros, L. and Calomardo, F., 2012. Characteristics of wild boar (*Sus scrofa*) habituation to urban areas in the Collserola Natural Park (Barcelona) and comparison with other locations. Animal Biodiversity and Conservation 35: 221-233.

Calenge, C., Maillard, D., Fournier, P. and Fouque, C., 2004. Efficiency of spreading maize in the garrigues to reduce wild boar (*Sus scrofa*) damage to Mediterranean vineyards. European Journal of Wildlife Research 50: 112-120.

Campbell, T.A., Long, D.B. and Massei, G., 2011. Efficacy of the boar-operated-system to deliver baits to feral swine. Preventive Veterinary Medicine 98: 243-249. https://doi.org/10.1016/j.prevetmed.2010.11.018

Casas-Díaz, E., Closa-Sebastià, F., Peris, A., Miño, A., Torrentó, J., Casanovas, R., Marco, I., Lavín, S., Fernández-Llario, P. and Serrano, E., 2013. Recorded dispersal of wild boar (*Sus scrofa*) in Northeast Spain: Implications for disease-monitoring programs. Wildlife Biology in Practice 9: 19-26. https://doi.org/10.2461/wbp.2013.ibeun.3

Chenais, E., Depner, K., Guberti, V., Dietze, K., Viltrop, A. and Ståhl, K., 2019. Epidemiological considerations on African swine fever in Europe 2014-2018. Porcine Health Management 5: 6. https://doi.org/10.1186/s40813-018-0109-2

Chenais, E., Ståhl, K., Guberti, V. and Depner, K., 2018. Identification of wild boar-habitat epidemiologic cycle in African swine fever epizootic. Emerging Infectious Diseases 24: 810-812. https://doi.org/10.3201/eid2404.172127

Cukor, J., Linda, R., Václavek, P., Mahlerová, K., Šatrán, P. and Havránek, F., 2020a. Confirmed cannibalism in wild boar and its possible role in African swine fever transmission. Transboundary and Emerging Diseases 67: 1068-1073. https://doi.org/10.1111/tbed.13468

Cukor, J., Linda, R., Václavek, P., Šatrán, P., Mahlerová, K., Vacek, Z., Kunca, T. and Havránek, F., 2020b. Wild boar deathbed choice in relation to ASF: Are there any differences between positive and negative carcasses? Preventive Veterinary Medicine 177: 104943. https://doi.org/10.1016/j.prevetmed.2020.104943

Dietz, T., 2013. Bringing values and deliberation to science communication. Proceedings of the National Academy of Sciences of the USA 110: 14081-14087. https://doi.org/10.1073/pnas.1212740110

Díez-Delgado, I., Sevilla, I.A., Romero, B., Tanner, E., Barasona, J.A., White, A.R., Lurz, P.W.W., Boots, M., de la Fuente, J., Dominguez, L., Vicente, J., Garrido, J.M., Juste, R.A., Aranaz, A. and Gortazar, C., 2018. Impact of piglet oral vaccination against tuberculosis in endemic free-ranging wild boar populations. Preventive Veterinary Medicine 155: 11-20. https://doi.org/10.1016/j.prevetmed.2018.04.002

Dixon, L.K., Stahl, K., Jori, F., Vial, L., and Pfeiffer, D.U., 2020. African swine fever epidemiology and control. Annual Review of Animal Biosciences 8: 221-246. https://doi.org/10.1146/annurev-animal-021419-083741

European Commission (EC), 2002. Council Directive of 27 June 2002 laying down specific provisions for the control of African swine fever and amending Directive 92/119/EEC as regards Teschen disease and African swine fever. Official Journal of the European Union L 192: 27-46. Available at: https://eur-lex.europa.eu/legal-content/EN/TXT/?uri=OJ:L:2002:192:TOC

European Food Safety Authority (EFSA), 2018a. Epidemiological analyses of African swine fever in the European Union (November 2017 until November 2018). EFSA Journal 16: e05494. https://doi.org/10.2903/j.efsa.2018.5494

European Food Safety Authority (EFSA), 2018b. African swine fever in wild boar. EFSA Journal 16: e05344. https://doi.org/10.2903/j.efsa.2018.5344

European Food Safety Authority (EFSA), 2020. Epidemiological analyses of African swine fever in the European Union (November 2018 to October 2019). EFSA Journal 18: e05996. https://doi.org/10.2903/j.efsa.2020.5996

European Union (EU), 2012. Regulation (EU) no 528/2012 of the European Parliament and of the Council of 22 May 2012 concerning the making available on the market and use of biocidal products. Available at: https://eur-lex.europa.eu/legal-content/EN/TXT/PDF/?uri=CELEX:32012R0528&from=EN

European Union (EU), 2015. African swine fever Strategy for Eastern Part of the European Union. SANTE/7113/2015-Rev 12. Directorate General for Health and Food Safety. Available at: http://ec.europa.eu/food/sites/food/files/animals/docs/ad_control-measures_asf_wrk-doc-sante-2015-7113.pdf

European Union (EU), 2017. Commission Implementing Decision (EU) 2017/1162 of 28 June 2017 concerning certain interim protective measures relating to African swine fever in the Czech Republic. Available at: https://eur-lex.europa.eu/legal-content/EN/TXT/PDF/?uri=CELEX:32017D1162&from=EN

Fernández-Llario, P. and Carranza J., 2000. Reproductive performance of the wild boar in a Mediterranean ecosystem under drought conditions. Ethology, Ecology and Evolution 12: 335-343. https://doi.org/10.1080/08927014.2000.9522791

Gamelon, M., Gaillard, J.-M., Servanty, S., Gimenez, O., Toïgo, C., Baubet, E., Klein, F. and Lebreton, J.-D., 2012. Making use of harvest information to examine alternative management scenarios: a body weight-structured model for wild boar. Journal of Applied Ecology 49: 833-841. https://doi.org/10.1111/j.1365-2664.2012.02160.x

Geisser, H. and Reyer, H.-U., 2005. The influence of food and temperature on population density of wild boar *Sus scrofa* in the Thurgau (Switzerland). Journal of Zoology 267: 89-96. https://doi.org/10.1017/S095283690500734X

Gentle, M., Speed, J. and People, A., 2014. Impacts on nontarget avian species from aerial meat baiting for feral pigs. Ecological Management & Restoration 15: 222-230. https://doi.org/10.1111/emr.12132

Gervasi, V., Marcon, A., Bellini, S., and Guberti, V., 2019. Evaluation of the efficiency of active and passive surveillance in the detection of African swine fever in wild boar. Veterinary Sciences 7: 5. https://doi.org/10.3390/vetsci7010005

Gortázar, C., Acevedo, P., Ruiz-Fons, F. and Vicente, J., 2006. Disease risks and overabundance of game species. European Journal of Wildlife Research 52: 81-87. https://doi.org/10.1007/s10344-005-0022-2

Graitson, E., Barbraud, C. and Bonnet, X., 2019. Catastrophic impact of wild boars: insufficient hunting pressure pushes snakes to the brink. Animal Conservation 22: 165-176. https://doi.org/10.1111/acv.12447

Gren, I.M., Häggmark-Svensson, T., Andersson, H., Jansson, G. and Jägerbrand, A., 2016. Using traffic data to estimate wildlife populations. Journal of Bioeconomics 18: 17-31.

Guberti, V., Khomenko, S., Masiulis, M. and Kerba, S., 2019. African swine fever in wild boar ecology and biosecurity. Animal Production and Health Manual Vol. 22. FAO, OIE and EC, Rome, Italy.

Iuell, B. (ed.), 2003. Wildlife and traffic. A European handbook for identifying conflicts and designing solutions. KNNV Uitgeverij, Zeist, the Netherlands. Available at: https://handbookwildlifetraffic.info/

Jakes, A.F., Jones, P.F., Paige, L.C., Seidler, R.G. and Huijser, M.P., 2018. A fence runs through it: A call for greater attention to the influence of fences on wildlife and ecosystems. Biological Conservation 227: 310-318. https://doi.org/10.1016/j.biocon.2018.09.026

Ježek, M., Holá, M., Kušta, T. and Červený, J., 2016. Creeping into a wild boar stomach to find traces of supplementary feeding. Wildlife Research 43: 590-598.

Jori, F., Chenais, E., Boinas, F., Busauskas, P., Dholllander, S., Fleischmann, L., Olsevskis, E., Rijks, J.M., Schulz, K., Thulke, H.H., Viltrop, A. and Stahl, K., 2020. Application of the World Café method to discuss the efficiency of African swine fever control strategies in European wild boar (*Sus scrofa*) populations. Preventive Veterinary Medicine 185: 105178. https://doi.org/10.1016/j.prevetmed.2020.105178

Jori, F., Payne, A., Ståhl, K., Nava, A., and Rossi, S., 2018. Wild and feral pigs: disease transmission at the interface between wild and domestic pig species in the Old and the New World. In: Melletti, M. and Meijaard, E. (eds.) Ecology, evolution and management of wild pigs and peccaries. Implications for conservation, Cambridge University Press, Cambridge, UK, pp. 388-403.

Keuling, O., Baubet, E., Duscher, A., Ebert, C., Fischer, C., Monaco, A., Podgórski, T., Prevot, C., Ronnenberg, K., Sodeikat, G., Stier, N. and Thurfjell, H., 2013. Mortality rates of wild boar *Sus scrofa* L. in central Europe. European Journal of Wildlife Research 59: 805-814. https://doi.org/10.1007/s10344-013-0733-8

Keuling, O., Stier, N., and Roth, M., 2008. How does hunting influence activity and spatial usage in wild boar? European Journal of Wildlife Research 54: 729-737. https://doi.org/10.1007/s10344-008-0204-9

Keuling, O., Strauβ, E. and Siebert, U., 2016. Regulating wild boar populations is 'somebody else's problem'! – Human dimension in wild boar management. Science of the Total Environment 554-555: 311-319. https://doi.org/10.1016/j.scitotenv.2016.02.159

Kock, R.A., Woodford, M.H. and Rossiter, P.B., 2010. Disease risks associated with the translocation of wildlife. Revue Scientifique et Technique (International Office of Epizootics) 29: 329-350. https://doi.org/10.20506/rst.29.2.1980

Kotulski, Y. and König, A., 2008. Conflicts, crises and challenges: wild boar in the Berlin City – a social empirical and statistical survey. Natura Croatica 17: 233-246.

Langbein, J., Putman, R. and Pokorny, B., 2011. Traffic collisions involving deer and other ungulates in Europe and available measures for mitigation. In: Apollonio, M., Andersen, R. and Putman, R. (eds.) Ungulate management in Europe: problems and practices. Cambridge University Press, Cambridge, UK, pp. 215-259.

Lavelle, M.J., Snow, N.P., Halseth, J.M., VanNatta, E.H., Sanders, H.N. and VerCauteren, K.C., 2018. Evaluation of movement behaviors to inform toxic baiting strategies for invasive wild pigs (*Sus scrofa*). Pest Management Science 74: 2504-2510. https://doi.org/10.1002/ps.4929

Malmsten, A., Jansson, G., Lundeheim, N. and Dalin, A.-M., 2017. The reproductive pattern and potential of free ranging female wild boars (*Sus scrofa*) in Sweden. Acta Veterinaria Scandinavica 59: 52. https://doi.org/10.1186/s13028-017-0321-0

Massei, G. and Cowan, D., 2014. Fertility control to mitigate human-wildlife conflicts: a review. CSIRO Wildlife Research 41: 1-21. https://doi.org/10.1071/WR13141

Massei, G., Coats, J., Quy, R., Storer, K. and Cowan, D.P., 2010. The boar-operated-system: a novel method to deliver baits to wild pigs. Journal of Wildlife Management 74: 333-336.

Massei, G., Kindberg, J., Licoppe, A., Gačić, D., Šprem, N., Kamler, J., Baubet, E., Hohmann, U., Monaco, A., Ozoliņš, J., Cellina, S., Podgórski, T., Fonseca, C., Markov, N., Pokorny, B., Rosell, C. and Náhlik, A., 2015. Wild boar populations up, numbers of hunters down? A review of trends and implications for Europe. Pest Management Science 71: 492-500. https://doi.org/10.1002/ps.3965

Massei, G., Roy, S. and Bunting, R., 2011. Too many hogs? A review of methods to mitigate impact by wild boar and feral hogs. Human-Wildlife Interactions 5: 80-99.

Melis, C., Szafrańska, P.A., Jędrzejewska, B. and Bartoń, K., 2006. Biogeographical variation in the population density of wild boar (*Sus scrofa*) in western Eurasia. Journal of Biogeography 33: 803-811. https://doi.org/10.1111/j.1365-2699.2006.01434.x

Morelle, K., Bubnicki, J., Churski, M., Gryz, J., Podgórski, T. and Kuijper, D.P.J., 2020. Disease-induced mortality outweighs hunting in causing wild boar population crash after African swine fever outbreak. Frontiers in Veterinary Science 7: 378. https://doi.org/10.3389/fvets.2020.00378

Morelle, K., Jezek, M., Licoppe, A. and Podgorski, T., 2019. Deathbed choice by ASF-infected wild boar can help find carcasses. Transboundary and Emerging Diseases 66: 1821-1826. https://doi.org/10.1111/tbed.13267

Mysterud, A. and Rolandsen, C.M., 2019. Fencing for wildlife disease control. Journal of Applied Ecology 56: 519-525. https://doi.org/10.1111/1365-2664.13301

Navarro-Gonzalez, N., Fernández-Llario, P., Pérez-Martín, J.E., Mentaberre, G., López-Martín, J.M., Lavín, S., and Serrano, E., 2013. Supplemental feeding drives endoparasite infection in wild boar in Western Spain. Veterinary Parasitology 196: 114-123. https://doi.org/10.1016/j.vetpar.2013.02.019

O'Brien, D.J., Schmitt, S.M., Fitzgerald, S.D., Berry, D.E. and Hickling, G.J., 2006. Managing the wildlife reservoir of *Mycobacterium bovis*: The Michigan, USA, experience. Veterinary Microbiology 112: 313-323. https://doi.org/10.1016/j.vetmic.2005.11.014

Oja, R., Velström, K., Moks, E., Jokelainen, P. and Lassen, B., 2017. How does supplementary feeding affect endoparasite infection in wild boar? Parasitology Research 116: 2131-2137. https://doi.org/10.1007/s00436-017-5512-0

Oja, R., Zilmer, K. and Valdmann, H., 2015. Spatiotemporal effects of supplementary feeding of wild boar (*Sus scrofa*) on artificial ground nest depredation. PLoS One 10: 0135254.

Parkes, J., Ramsey, D., Macdonald, N., Walker, K., McKnight, S., Cohen, B. and Morrison, S., 2010. Rapid eradication of feral pigs (*Sus scrofa*) from Santa Cruz Island, California. Biological Conservation 143: 634-641. https://doi.org/10.1016/j.biocon.2009.11.028

Pei, K., Lai, Y., Corlett, R. and Suen, K-Y., 2010. The larger mammal fauna of Hong Kong: species survival in a highly degraded landscape. Zoological Studies 49: 253-264.

Pepin, K.M., Davis, A.J., Cunningham, F.L., VerCauteren, K.C. and Eckery, D.C., 2017. Potential effects of incorporating fertility control into typical culling regimes in wild pig populations. PLoS ONE 12: e0183441. https://doi.org/10.1371/journal.pone.0183441

Pittiglio, C., Khomenko, S. and Beltran-Alcrudo, D., 2018. Wild boar mapping using population-density statistics: from polygons to high resolution raster maps. PLoS ONE 13: e0193295. https://doi.org/10.1371/journal.pone.0193295

Probst, C., Globig, A., Knoll, B., Conraths, F.J. and Depner, K., 2017. Behaviour of free ranging wild boar towards their dead fellows: potential implications for the transmission of African swine fever. Royal Society Open Science 4: 170054-170054. https://doi.org/10.1098/rsos.170054

Quy, R.J., Massei, G., Lambert, M.S., Coats, J., Miller, L.A. and Cowan, D.P., 2014. Effects of a GnRH vaccine on the movement and activity of free-living wild boar (*Sus scrofa*). Wildlife Research 41: 185-193.

Rosell, C., Navàs, F. and Romero, S., 2012. Reproduction of wild boar in a cropland and coastal wetland area: implications for management. Animal Biodiversity and Conservation 35: 209-217.

Rosell, C., Pericas, B., Colomer, J. and Navàs, F.B.P.C., 2019. Guide to measures for reducing the damage caused by wild mammals in rural areas, urban areas and infrastructures. Barcelona Provincial Council, Barcelona, Spain.

Rossi, S., Staubach, C., Blome, S., Guberti, V., Thulke, H.-H., Vos, A., Koenen, F. and Le Potier, M.-F., 2015. Controlling of CSFV in European wild boar using oral vaccination: a review. Frontiers in Microbiology 6: 1141. https://doi.org/10.3389/fmicb.2015.01141

Ruiz-Fons, F., 2017. A review of the current status of relevant zoonotic pathogens in wild swine (*Sus scrofa*) populations: changes modulating the risk of transmission to humans. Transboundary and Emerging Diseases 64: 68-88.

Sáenz-de-Santa-María, A. and Tellería, J.L., 2015. Wildlife-vehicle collisions in Spain. European Journal of Wildlife Research 61: 399-406. https://doi.org/10.1007/s10344-015-0907-7

Schulz, K., Conraths, F.J., Blome, S., Staubach, C. and Sauter-Louis, C., 2019. African swine fever: fast and furious or slow and steady? Viruses 11: 866. https://doi.org/10.3390/v11090866

Scillitani, L., Monaco, A. and Toso, S., 2010. Do intensive drive hunts affect wild boar (*Sus scrofa*) spatial behaviour in Italy? Some evidences and management implications. European Journal of Wildlife Research 56: 307-318. https://doi.org/10.1007/s10344-009-0314-z

Servanty, S., Gaillard, J.-M., Ronchi, F., Focardi, S., Baubet, E. and Gimenez, O., 2011. Influence of harvesting pressure on demographic tactics: implication for wildlife management. Journal of Applied Ecology 48: 835-843. https://doi.org/10.1111/j.1365-2664.2011.02017.x

Servanty, S., Gaillard, J.-M., Toïgo, C., Serge, B. and Baubet, E., 2009. Pulsed resources and climate-induced variation in the reproductive traits of wild boar under high hunting pressure. Journal of Animal Ecology 78: 1278-90. https://doi.org/10.1111/j.1365-2656.2009.01579.x

Snow, N.P., Foster, J.A., Kinsey, J.C., Humphrys, S.T., Staples, L.D., Hewitt, D.G. and Vercauteren, K.C., 2017. Development of toxic bait to control invasive wild pigs and reduce damage. Wildlife Society Bulletin 41: 256-263. https://doi.org/10.1002/wsb.775

Snow, N.P., Halseth, J.M., Lavelle, M.J., Hanson, T.E., Blass, C.R., Foster, J.A., Humphrys, S.T., Staples, L.D., Hewitt, D.G. and VerCauteren, K.C., 2016. Bait preference of free-ranging feral swine for delivery of a novel toxicant. PLoS ONE 11: e0146712. https://doi.org/10.1371/journal.pone.0146712

Thurfjell, H., Ball, J.P., Åhlén, P.-A., Kornacher, P., Dettki, H. and Sjöberg K., 2009. Habitat use and spatial patterns of wild boar *Sus scrofa* (L.): agricultural fields and edges. European Journal of Wildlife Research 55: 517-523. https://doi.org/10.1007/s10344-009-0268-1

Tolon, V., Dray, S., Loison, A., Zeileis, A., Fischer, C. and Baubet, E., 2009. Responding to spatial and temporal variations in predation risk: Space use of a game species in a changing landscape of fear. Canadian Journal of Zoology 87: 1129-1137. https://doi.org/10.1139/Z09-101

Touzot, L., Schermer, É., Venner, S., Delzon, S., Rousset, C., Baubet, É., Gaillard, J.-M. and Gamelon, M., 2020. How does increasing mast seeding frequency affect population dynamics of seed consumers? Wild boar as a case study. Ecological Applications 30: e02134. https://doi.org/10.1002/eap.2134

Veeroja, R. and Männil, P., 2014. Population development and reproduction of wild boar (*Sus scrofa*) in Estonia. Wildlife Biology in Practice 10: 17-21. https://doi.org/10.2461/wbp.2014.un.3

Welander, J., 2000. Spatial and temporal dynamics of wild boar (*Sus scrofa*) rooting in a mosaic landscape. Journal of Zoology 252: 263-271.

Wilson, C.J., 2014. The establishment and distribution of feral wild boar (*Sus scrofa*) in England. Wildlife Biology in Practice 10: 1-6. https://doi.org/10.1046/j.1365-2907.2003.00016.x

Woodroffe, R., Hedges, S. and Durant, S.M., 2014. To fence or not to fence. Science 344: 46-48. https://doi.org/10.1126/science.1246251

World Organization for Animal Health (OIE). 2019. Zoning and compartmentalisation. In: Terrestrial animal health code. World Animal Health, Paris, France. Available at: https://www.oie.int/fileadmin/home/eng/health_standards/tahc/current/chapitre_zoning_compartment.pdf.

9. African swine fever epidemiology, surveillance and control

A. Viltrop[1], F. Boinas[2], K. Depner[3], F. Jori[4,5], D. Kolbasov[6], A. Laddomada[7], K. Ståhl[8] and E. Chenais[8*]

[1]Estonian University of Life Science, Institute of Veterinary Medicine and Animal Sciences, 51014 Tartu, Estonia; [2]CIISA - Centro de Investigação Interdisciplinar em Sanidade Animal, Faculdade de Medicina Veterinária, Universidade de Lisboa, Avenida da Universidade Técnica, 1300-477 Lisboa, Portugal; [3]Friedrich-Loeffler-Institut (FLI), Federal Research Institute for Animal Health, 17493 Greifswald-Insel Riems, Germany; [4]CIRAD, UMR ASTRE (Animal, Health, Territories, Risks and Ecosystems), 34398 Montpellier, France; [5]ASTRE, Univ Montpellier, CIRAD, INRAE, 34398 Montpellier, France; [6]Federal Research Center for Virology and Microbiology, 601125 Vladimir region, Russia; [7]Istituto Zooprofilattico Sperimentale della Sardegna, 07100 Sassari, Italy; [8]National Veterinary Institute, 751 89 Uppsala, Sweden; erika.chenais@sva.se

Abstract

The introduction of genotype II African swine fever (ASF) virus (ASFV) into the Caucasus in 2007 resulted in unprecedented disease propagation via slow geographical expansion through wild boar populations, short- and long-distance human-mediated translocations, and incursions into naïve wild boar and domestic pig populations. The disease is now widespread in eastern and central Europe as well as in Asia, including China. The global dimension of the current epidemic shows that all countries need to be prepared for an introduction. In its natural habitat in Africa, ASFV is maintained within an ancient cycle between soft argasid ticks and the common warthog. Once introduced to the domestic pig population, direct and indirect virus transmission occurs with or without involvement of the tick vector in the pig-tick and domestic pig epidemiological cycles respectively. In the domestic pig cycle, human activities involving pigs or pig derived products are the dominating driver of virus transmission. ASF epidemiology in the presence of wild boar and northern European climates has proved to have specific characteristics, described in the wild boar-habitat epidemiological cycle. In this cycle wild boar carcasses and the resulting contamination of the environment play key roles in virus persistence. In both the wild boar-habitat and the domestic pig epidemiological cycle, fully implemented biosecurity is the key for stopping virus transmission and controlling the disease. Positive examples from the Czech

This publication is based upon work from COST Action CA15116, ASF-STOP, supported by COST (European Cooperation in Science and Technology).
www.cost.eu

COST EUROPEAN COOPERATION IN SCIENCE & TECHNOLOGY

Funded by the Horizon 2020 Framework Programme of the European Union

DOI 10.3920/978-90-8686-910-7_9, © A. Viltrop *et al.*, 2021

Republic and Belgium show that control and eradication of ASF from the wild boar-habitat cycle can be achieved. Both these cases, as well as the example of Sardinia, where ASFV genotype I now seem very close to eradication after more than 40 years presence, further underline the importance of involving, engaging and understanding all stakeholders in the value chains from farm and forest to fork in order to accomplish ASF control and eradication.

Keywords: ASF, transmission, wild boar, genotype II, epidemic pig disease

9.1 Introduction

First described in 1921 as a disease affecting domestic pigs, African swine fever (ASF) was restricted to the African continent until 1957 when it showed up in Lisbon, Portugal (Manso Ribeiro and Rosa Azevedo, 1961; Montgomery, 1921). After a global tour during the decades that followed, ASF was again restricted to Africa from the late 1990s, with the exception of the island of Sardinia, Italy, where it had persisted since 1978. The introduction of an eastern African swine fever virus (ASFV) genotype II into the Caucasus in 2007 (Rowlands *et al.*, 2008) was therefore unexpected and resulted in unprecedented disease propagation via slow geographical expansion through wild boar populations, and long-distance, human-mediated translocations and incursions into naïve wild boar and domestic pig populations. The disease is now widespread in eastern and central Europe and, since 2018, in China and many other countries in Asia (Chenais *et al.*, 2019a; Dixon *et al.*, 2019; Gogin *et al.*, 2013; Zhou *et al.*, 2018). This epidemic (from now on in this chapter referred to as the 'current epidemic') involves both domestic pigs and wild boar, although in some parts of Europe the infection is maintained in wild boar populations independently of domestic pigs (Chenais *et al.*, 2018). To date, 12 European Union (EU) countries have reported cases in wild boar or outbreaks in domestic pigs caused by ASFV genotype II. Of the member countries affected so far (September 2020), the Czech Republic has achieved control and eradication (Charvátová *et al.*, 2019), and in Belgium the infection appears to be under control and eradication to be close (Dellicour *et al.*, 2020). In the Baltic states, downward trends are observed regarding outbreaks in domestic pigs and cases in wild boar (Oļševskis *et al.*, 2020; Schulz *et al.*, 2019b).

In its natural habitat in southern and eastern Africa ASFV exists in an ancient cycle between the biological vector, soft argasid ticks of the *Ornithodoros moubata* complex, and its natural mammalian host, the common warthog (*Phacochoerus africanus*) (see the sections on arthropod vectors and susceptible suids respectively) (Wilkinson, 1984). In rare spill-over events the disease can be transmitted to domestic pigs via the ticks. Once introduced to the domestic pig population, direct and indirect virus transmission occur with or without involvement of the tick vector in the pig-tick and domestic pig epidemiological cycles respectively (Penrith and Vosloo, 2009; Plowright, 1981). Human activities in the domestic pig value chain involving pigs or pig derived products are the dominating driver of virus transmission in Africa as well as globally (Mulumba-Mfumu *et al.*, 2019; Penrith *et al.*, 2019).

In this chapter an overview of ASF epidemiology is given, with focus on the characteristics of the current epidemic, brief orientations of some historic and current local epidemic patterns in Europe, and how ASF is controlled and surveyed in domestic pigs in EU today.

9.2 Susceptibility of Suidae

All African species of wild pigs are often considered to be naturally resistant to ASFV. However, the level of scientific information confirming this is highly variable, and only in exceptional cases have ASFV infections been confirmed in representative population surveys or experimental infections (Table 9.1) (Jori and Bastos, 2009). Most information is available concerning the common warthog (*Phacochoerus africanus*). This is particularly true for warthog populations in Southern and East Africa which cohabit with ticks of the *Ornithodoros moubata* complex, allowing for the occurrence of the well-studied warthog-tick sylvatic cycle (Plowright, 1981). Warthogs get infected as young piglets when bitten by infected soft ticks sharing their burrows. The virus becomes localised in peripheral lymph nodes and adult warthogs do not excrete sufficient virus to be able to transmit the disease directly (Plowright, 1981; Thomson, 1985). In other areas of the African continent, such as West and Central Africa, the sylvatic cycle has not been confirmed to occur, and information confirming resistance or circulation of the virus in natural populations of warthogs from those areas is almost inexistent (Jori *et al.*, 2013). The case of other African wild pig populations is similar: only one species present in Southern Africa, the Southern bush pig (*Potamochoerus larvatus*), has been proven to be naturally resistant by experimental infection (Oura *et al.*, 1998). For all the other wild pig populations, including other bushpigs (*Potamochoerus porcus*) or populations of giant forest hogs (*Hylochoerus meinertzhageni*), data on

Table 9.1. Proven, suspected or unknown susceptibility to African swine fever for different members of the *Suidae* and *Tayasuidae* families.

Family	Genus	Species[1]	Susceptibility	Continent	Reference
Suidae	*Phacochoerus*	*P. africanus*	Proven resistant	Africa	Thomson, 1985
		P. aethiopicus	Unknown	Africa	
	Potamochoerus	*P. larvatus*	Proven resistant	Africa	Montgomery, 1921
		P. porcus	Unknown	Africa	
	Hylochoerus	*H. meinertzhageni*	Suspected resistant	Africa	Heuschele and Coggins, 1965
	Sus	*S. scrofa*	Yes	Eurasia	
		*S. verrucosus**	Unknown	Asia	
		*S. cebifrons**	Unknown	Asia	
		*S. celebensis**	Unknown	Asia	
		S. barbatus	Unknown	Asia	
		*S. oliveri**	Unknown	Asia	
		S. ahoenobarbus	Unknown	Asia	
		S. philippensis	Unknown	Asia	
		*Porcula salvania**	Unknown	India	
	Babyrousa	*B. babyrussa*	Unknown	Asia	
		*B. togeanensis**	Unknown	Asia	
Tayasuidae	*Tayassu*	*Tayassu pecari*	Suspected resistant	America	Fowler, 1996
	Pecari	*Pecari tajacu*	Suspected resistant	America	Fowler, 1996
	Catagonus	*C. wagneri**	Suspected resistant	America	Fowler, 1996

[1] * means endangered species.

the circulation of ASFV that could provide an indication of potential susceptibility or resistance are not available. Episodes of mass mortality due to ASF have never been reported for any species of wild African suidae, supporting the hypothesis of widespread natural resistance.

The resistant status of African suids towards ASF infection contrasts with the high susceptibility to ASFV seen in European wild boar (*Sus scrofa*, from now on in this chapter referred to as 'wild boar'), but is unsurprising as they belong to the same species as domestic pigs. This susceptibility was noticed already during the first incursion of ASFV genotype I in Europe in the 1960s. On that occasion, however, mortalities were limited in space and time and the virus disappeared from wild boar populations after a few weeks if there were no further contacts with domestic pigs (Pérez *et al.*, 1998). *Tayasuidae* (peccaries and javelinas) are reported to be resistant to ASFV and during the incursion of the virus in Brazil, no cases in those species were ever reported (Fowler, 1996).

With ASF becoming endemic in new regions, any country with pig production is at risk for introduction. This risk is particularly high for Asia, which maintains the largest population of domestic pigs in the world, mainly kept under low biosecurity (Dixon *et al.*, 2019). In this scenario, the potential role of wild pigs in the dissemination of the disease, and its consequences, needs to be considered. Many Asian countries hold large populations of wild boar, but also important remaining populations of endangered wild pig species, see Table 9.1 (Meijaard *et al.*, 2011). In Asia, the Suidae family comprises 12 wild species (including several subspecies of wild boar). Except for a critically endangered population of pygmy hog (*Porcula salvanius*) in the Indian subcontinent, all the other eight species of Suidae (*Sus* spp.) and three species of *Babyrousa* spp. are distributed across different islands in Indonesia and the Philippines. Since ASF is spreading very quickly across this region, the risk of these populations being exposed to ASFV is high. Their susceptibility for ASFV has never been studied, but considering the high tropism of the virus for pig cells and the fact that they are immunologically naïve, the probability of the virus being fatal to these populations is high. Outbreaks of ASF anywhere outside its presently known geographical range should be closely monitored to anticipate whether ASF could add further pressure to already highly threatened endemic pig species.

9.3 Epidemiological parameters

9.3.1 Transmission

Infected domestic pigs and wild boar excrete the virus with all body fluids and excretions including oronasal fluids, faeces and urine. The virus excretion starts about two days before onset of clinical signs. The virus load is particularly large in blood of infected animals, thus the haemorrhages and sometimes bloody diarrhoea caused by the infection result in extensive contamination of the environment.

Transmission through direct contact in domestic pigs and wild boar has been repeatedly demonstrated in animal experiments and in field observations. It is thought that direct transmission usually occurs oronasally. It has also been demonstrated that oronasal infection of pigs usually requires a relatively high virus dose to be successful (~100 haemadsorbing units

(HAU)) (Olesen *et al.*, 2017). In an experiment with wild boar Pietschman and co-workers (2015) demonstrated that only weak and runty animals could be directly infected with very low doses (<10 HAU) of ASFV by the oronasal route. Nevertheless, direct contact seems to play an important role in transmission of the virus within wild boar sounders (family groups). Wild boar sounders are, however, territorial and tend to avoid other sounders. Therefore, physical contacts between discrete sounders are scarce and not believed to contribute to the spatial spread of the virus as much as infected carcasses (slow, local spread) and human mediated spread (fast, long distance spread) (Lange *et al.*, 2018). The maintenance of ASF in European wild boar populations is believed to be mainly driven by contacts of susceptible animals with infected carcasses and contaminated environment (Chenais *et al.*, 2018). Direct contacts of domestic pigs with infected wild boar have not played a major role in disease transmission during the current epidemic. Such contacts may be of more importance in areas where free range pig keeping is practised and direct contacts between wild boar and domestic pigs thus are more likely to occur (EFSA, 2018, 2020). Airborne transmission has been shown to occur only over short distances such as between pigs kept in the same barn (Olesen *et al.*, 2017).

Attempts to infect pigs indirectly through contaminated environment have not been very successful. In a recent experiment, infection of naïve contact pigs by ASFV contaminated environment could not be demonstrated (Eble *et al.*, 2019). Olesen and co-workers (2018) demonstrated transmission through the environment to sentinel pigs introduced into contaminated pens one day after removing the infected pigs but not after three or more days. At the same time, reports from field outbreak investigations in affected countries indicate that indirect transmission via contaminated fomites (vehicles, clothes, equipment and various materials) is the predominating identified plausible route of introduction of the infection into pig farms, particularly for larger commercial farms (EFSA, 2020; Nurmoja *et al.*, 2018; Oļševskis *et al.*, 2016; Zani *et al.*, 2019). Transmission via pig feed (grain, fresh forage) has also been suspected. In small farms, the feed is often prepared on site and pigs are frequently fed with freshly harvested forage. The latter has been suggested to constitute an important route of introduction of the virus into farms in areas where wild boar are affected by ASF (Boklund *et al.*, 2020). In larger farms in Estonia, it has been suspected that cereal feeds may have become contaminated in the process of milling and mixing of feed on farm (Nurmoja *et al.*, 2018). Contaminated bedding material could potentially also carry the virus into farms. In general, it has been suggested that contamination of the environment surrounding pig farms with ASFV increases the risk of a farm becoming infected, indicating once again the importance of indirect transmission of the virus (Boklund *et al.*, 2020; Nurmoja *et al.*, 2018). In an Estonian study the risk of a pig farm to become infected was positively associated with the number of affected wild boar in the surrounding area (Nurmoja *et al.*, 2018). In a study by Boklund and co-workers (2020) in Romania close proximity to outbreaks in domestic farms was a risk factor in commercial and backyard farms. In backyard farms, wild boar abundance around the farm, number of domestic outbreaks within two kilometres around farms and short distance to wild boar cases were also significant risk factors (Boklund *et al.*, 2020). The conclusion made by the authors is that these significant risk factors should be understood as proxies for a high level of virus contamination in the environment.

Translocation of ASFV through live animal movement has been rare during the current epidemic while long distance introductions to new countries and territories on several occasions have been associated with people bringing along contaminated pork or wild boar products, and food waste being deliberately or unintentionally fed to domestic pigs or wild boar (Chenais *et al.*, 2019a). Once introduced, if the virus gets established in domestic pig or wild boar populations the probability of indirect transmission via contaminated environment and fomites increases.

Currently, there is no evidence that ASFV persists long-term in a latent state in animals surviving the infection, or that survivors play any epidemiological role, either in the current epidemic, or in previous ones (Ståhl *et al.*, 2019). Some animals with a chronic form of the disease may, however, excrete the virus for prolonged periods (up to two months). In animals recovering from the disease (true survivors) virus excretion has been shown to last up to 70 days. Longer excretion periods seem to be associated with less virulent strains. However, in an experimental study with the moderately virulent Malta/78 isolate, no transmission from donor to in-contact pigs occurred more than 30 days after they had developed pyrexia (Wilkinson *et al.*, 1983).

9.3.2 Arthropod vectors

The natural reservoir and biological vector of ASFV in South and East Africa is the soft tick *O. moubata*. In the ASF-epidemic on the Iberian Peninsula *Ornithodoros erraticus* acted as a biological vector. Other species from the *Ornithodoros* genus have experimentally been proven to be competent vectors, but never found to play an epidemiological role in the field. *Ornithodoros* spp. ticks are common in Africa, the Middle East (from Turkey to the north to western Iran), and some areas of Southern Europe (EFSA AHAW Panel, 2015). Hard ticks cannot act as biological vectors for ASFV, but they feed on wild boar as well as domestic pigs, and have been shown to contain the virus after feeding on infected animals (Olesen *et al.*, 2020). The transmission potential of hard ticks is unknown; however, it is known that they do not feed again for a long time after having had a blood meal. The potential for other bloodsucking arthropods or insects to serve as mechanical vectors has been widely discussed. ASFV has been found in hog lice (*Haematopinus suis*) collected from experimentally infected pigs. Hog lice normally spend their entire brief lives on the same pig and would only be transferred to another pig by close contact (Bonnet *et al.*, 2020). It is unlikely that they have any epidemiological significance. Stable flies (*Stomoxys calicitrans*) have been shown to transmit ASFV mechanically (via bites) and via ingestion in experimental settings. Evidence from the field, however, is lacking. In flies collected from infected farms in Lithuania the virus could not be detected whereas trace amounts of virus DNA could be detected in insects collected from an outbreak farm in Estonia (Olesen *et al.*, 2020). Furthermore, horse flies that are common in forest environments could be detected in low numbers in non-affected pig farms in Estonia (Tummeleht *et al.*, 2020). Nevertheless, the role, if any, of blood-feeding insects in the transmission of ASFV in field conditions and in the introduction of the virus into pig farms still needs to be elucidated.

9.3.3 Incubation period, morbidity and mortality

Clinical manifestation, morbidity and mortality in domestic pigs and wild boar caused by ASFV genotypes I and II are similar, as demonstrated in numerous animal experiments (Blome *et al.*, 2013). In general, the incubation period for ASF is considered to be between 5 to 15 days. In experimental studies with genotype I and II viruses it ranged between 1 and 33 days, depending mainly on route and dose of infection (Dórea *et al.*, 2017). The incubation period tends to be shorter after intradermal and intramuscular inoculation compared to oronasal or oral infections.

Higher doses of virus mostly result in shorter incubation periods compared to lower doses (Dórea *et al.*, 2017).

The morbidity among pigs experimentally infected with virulent strains of the ASF genotype I or II viruses has been 100% (Dórea *et al.*, 2017). However, the severity of the disease caused by the infection has been somewhat variable depending on the virus strain. Strains classified as moderately virulent and causing subacute disease in most of the experimental pigs have been discovered in the Baltic states. There are also examples of attenuation of field virus strains resulting in asymptomatic infections in challenged domestic pigs (Gallardo *et al.*, 2019; Zani *et al.*, 2018).

Similarly to the morbidity, the mortality among infected animals (the case-fatality rate or lethality) is dependent on the virulence of the virus strain. The highly virulent strains dominating in the field cause a case-fatality rate approaching 100%. In challenge experiments conducted with highly virulent strains since 2007 only single animals have survived and recovered. However, virus strains showing reduced virulence or attenuation have caused case-fatality rates ranging from 0 to 50% in animal experiments (Gallardo *et al.*, 2018; Zani *et al.*, 2018). The morbidity and mortality observed in the field among infected pigs and wild boar are not identical to what is measured in animal experiments. In the field, the infection status of the animals is generally not known, and the morbidity and mortality are calculated based on the number of animals in the group under consideration, i.e. the population at risk (PAR). Thus, the morbidity and mortality estimates are dependent on how the PAR is defined and how large it is. If the PAR is defined as all pigs of a large farm (stable) or unit, the morbidity and mortality estimates may result in very low numbers (less than 1%), whereas the mortality and morbidity in the pen of pigs where the disease was discovered, or in a small back yard farm, may be similar to that observed in experimental studies (up to 100%). Morbidity and mortality estimates are also dependent on the time period that elapsed since the start of the outbreak. Due to accumulation of the cases in time, the morbidity and mortality estimates will increase with time. Therefore, if the disease has been detected in a very early stage of an outbreak in a larger farm, the observed morbidity and mortality of the disease may be very low even if the outbreak is caused by a virulent strain of the virus. Consequently, morbidity and mortality estimates observed in the field should be interpreted with caution and strictly in the context of the outbreak. Generally, morbidity and mortality levels observed in larger farms often do not exceed the normal morbidity and mortality levels of the specific farm until several weeks have passed since the introduction of the virus. During the current epidemic, mortality and morbidity rates reported from outbreaks in Estonia have been in the range of 29.7 to 100% in small back yard farms and 0.04 to 25% in commercial farms (Nurmoja *et al.*, 2018). In conclusion: the first indication of the introduction of ASFV on a farm is not always high mortality or morbidity, and the absence of these signs should not lead farmers and veterinarians to exclude the possibility of ASF. It has to be emphasised that ASF, however, can cause high morbidity and mortality among pigs even of larger herds after prolonged spread of the virus (for several weeks) within the herd or in case of simultaneous infection of pigs from the same source like contaminated feed (swill) or water.

Estimating morbidity and mortality among wild boar populations in the field is even more difficult as the number of animals in the PAR is not exactly known. Therefore, such estimates are largely based either on expert opinion or mathematical models. Field observations indicate that the morbidity and mortality within sounders are high, as several wild boar carcasses have been found together. Nevertheless, occurrence of seropositive wild boar indicates that the case fatality rate among wild boar in the field is not 100%, and that a fraction of animals survives the infection and the disease.

9.3.4 Contagiousness

Field evidence and animal experiments have demonstrated that in most cases ASFV spreads slowly within a pig herd, staying in one pen or part of the stable for weeks. The transmission is accelerated when bleeding occurs in sick pigs and a large amount of virus is released into the environment with the blood. The spread of the virus across a larger pig herd may take weeks or months, depending on how the pigs are separated in the building. Along with the spread of the infection in a herd the morbidity and mortality gradually increase. ASF should thus be considered a moderately contagious disease (Depner *et al.*, 2020; Schulz *et al.*, 2019a).

Transmission studies conducted in recent years, involving various virus strains and experimental conditions, have resulted in rather similar transmission parameters, all indicating moderate or low contagiousness of the infection. In an experiment with the highly virulent strain of ASFV genotype II basic reproduction ratio (R_0) within pens ranged from 5.0 to 6.1 and between pens it was found to be 0.5 (Pietschmann *et al.*, 2015). These estimates coincide with the results obtained by Guinat and co-workers (2016) who estimated the pig-to-pig R_0 for the Georgia 2007/1 ASFV strain using data obtained from another challenge experiment. The models showed that the pig-to-pig R_0 was 5.0 (95% CI 2.4-9.1) and between pen 2.7 (95% CI 0.7-5.2) (Guinat *et al.*, 2016). Nielsen and co-workers (2017) recalculated these transmission parameters taking into account that during the challenge experiment animals were only tested every other day, ending up with similar point estimates for parameters but somewhat different confidence intervals (Nielsen *et al.*, 2017).

There have been several attempts to estimate R_0 for ASFV genotype II from field data. Based on outbreak data in domestic pig herds from Russia during the period 2007-2010 the R_0 was estimated to range from 8 to 11 within farms and from 2 to 3 between farms (Gulenkin *et al.*, 2011). Using field data on cases in wild boar R_0 was estimated at 1.58 (95% CI 1.13-3.77) in Russia (Iglesias *et al.*, 2016), 1.95 for Czech Republic, and 1.65 for Belgium (Marcon *et al.*, 2020). Estimates from field data include uncertainties requiring special attention in the statistical analysis, such as unknown infection dates of index cases, and are further influenced by local conditions affecting contact rates between animals and animal groups (farms or herds) such as herd characteristics (size and production type), management practices in domestic pigs and population density of wild boar. Also, intervention measures may influence the estimates (isolation of domestic herds or removal of wild boar carcasses). Therefore, R_0 estimates obtained from field data should be considered in their particular context. Nevertheless, the estimates obtained in different countries are notably similar and comparable with those obtained in experimental conditions.

9.3.5 Transmission patterns

As mentioned in the section above, within domestic pig herds an ASF outbreak develops relatively slowly in the initial stage. It may take two or more weeks until the mortality and morbidity notably exceed the normal levels of the herd. In wild boar the spread follows a pattern of a propagating epidemic. There are two examples where the development of the disease in a wild boar population has been well recorded within a restricted area: in Czech Republic in the Zlin region, where the affected area was ~60 km^2, and in southern Belgium where the disease was spreading in an area of ~1,600 km^2. In the Zlin region, the epidemic had two peaks before it was eradicated, the first after approximately four weeks following the first detection of the disease and the second after five months. In Belgium, the epidemic developed very slowly reaching its single peak after little more than four months following the first detection (Charvátová *et al.*, 2019; Dellicour *et al.*, 2020) (Figure 9.1).

The temporal evolution of ASF in wild boar populations has been studied based on the Estonian disease notification data (PCR or antibody positive) at different spatial resolutions (EFSA, 2017). The smoothed temporal trend at county level (the average area of a county is 3,000 km^2) indicated the first peak in notified ASF cases around six months after the first case was reported, and a gradual reduction of the number of cases over the following two years. At around 30 months after the first reported case a second, smaller peak could be observed, but the number of cases subsequently decreased rapidly (Figure 9.2).

The detection of ASF cases in wild boar and outbreaks in domestic pig farms in the current epidemic has not been evenly distributed over the year. Outbreaks in domestic pigs have exhibited strong seasonality, with highest incidence during the summer months in all European countries in Europe where domestic pigs have been affected. In wild boar, summer and winter peaks in numbers of detected ASF cases can be observed (Figure 9.3).

The winter peak can be explained with the main hunting season lasting from autumn to early spring, with most hunting being performed in the winter months. In areas under restriction due to ASF all hunted animals are tested for ASF. More wild boars are thus tested during the hunting season. Likewise, more carcasses are detected during the hunting season when hunters are more present in the forests. Reasons for the summer peak could be the increase in population density after the breeding season in spring, and animals moving closer to farmlands to feed on the fields and thus becoming easier to detect. The observed number of cases may, however, be biased due to different reporting frequency in different seasons. Therefore, trends in proportions of positive findings are more informative for assessment of seasonal trends. The analysis of trends

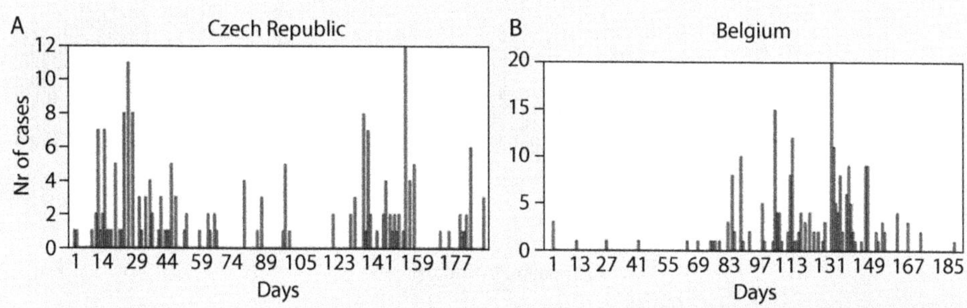

Figure 9.1. Number of infected wild boar carcasses found in the Zlin area, Czech Republic, and in Virton Forest, Belgium, during the ASF epidemic. Adapted from Marcon *et al.* (2020).

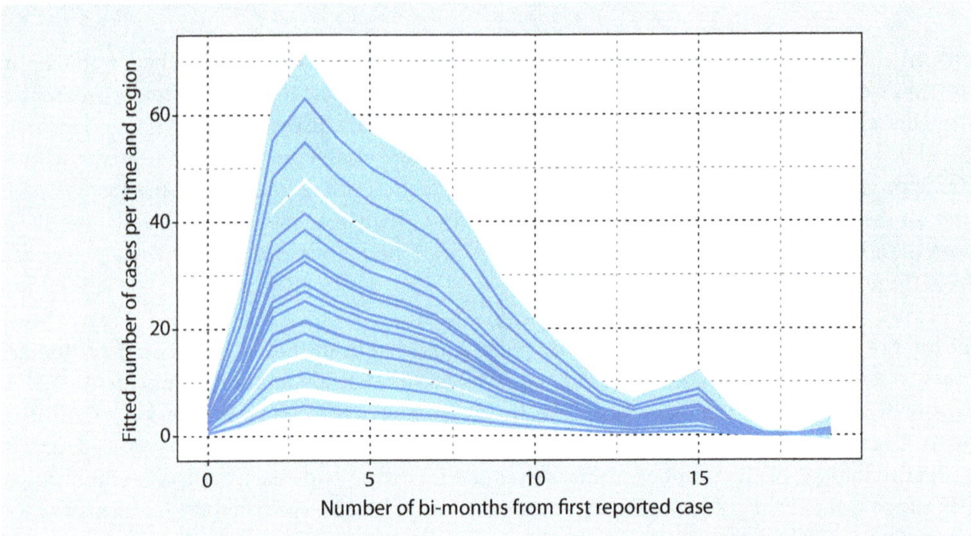

Figure 2. Smoothed 38 months temporal trends of ASF cases (PCR or antibody positive) in wild boar per county during the period of January 2014 to August 2017 in Estonia based on official surveillance data. Starting point for each trend line is the date of the first reported case in a county. Bi-months refers to a time interval of two months.

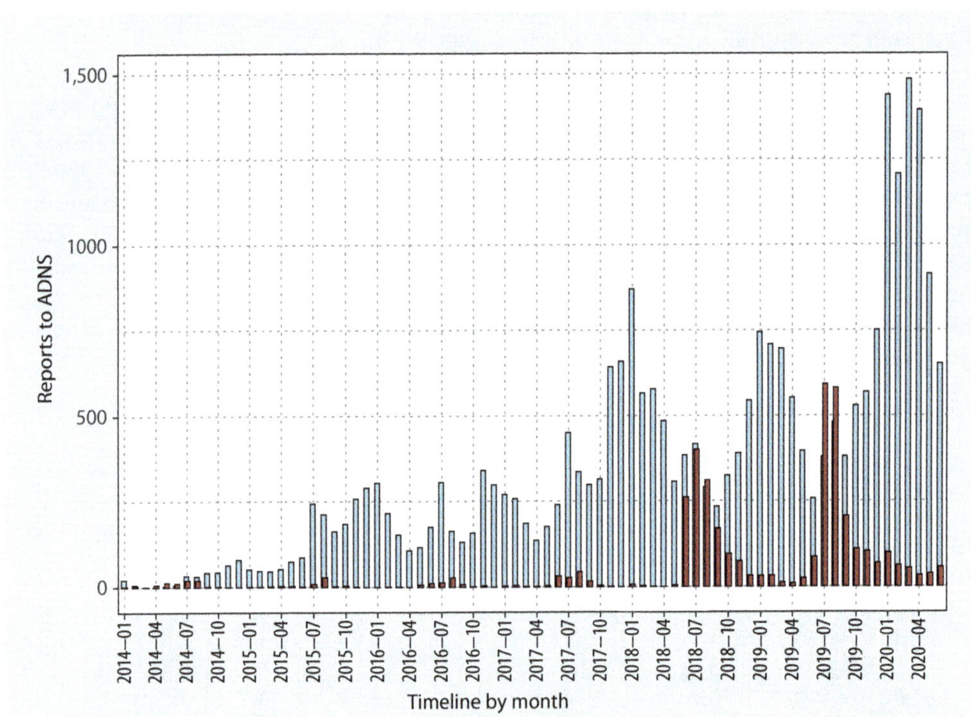

Figure 9.3. Notifications of outbreaks in domestic pigs (red bars) and cases in wild boar (blue bars) in the European Union. Data extracted from the Animal Disease Notification System (ADNS) from January 2014 until June 2020.

in proportions of positive cases, including samples only from affected areas, performed by EFSA (2020) has revealed more or less similar seasonal variation among wild boar found dead in Estonia and Latvia, where both summer and winter peaks could be observed. In contrast, in Lithuania and Poland the summer peaks were not evident among animals found dead. Among hunted wild boar the prevalence levels have been steady in all four countries, with significant drops in proportion of positive cases only for spring months. Data from other affected countries (Romania, Hungary, Czech Republic and Belgium) do not show the same trend. In these countries the occurrence of ASF was generally lower in summer months with a single peak observed in winter.

Based on these observations it may be concluded that there is a seasonal variation in incidence of ASF among wild boar, dependent on environmental and climatic conditions as well as ecology and management. In cooler climates the decomposition of wild boar carcasses and inactivation of the virus in the environment takes longer, increasing the infectious period for these virus sources, which in turn increases the probability of effective contacts for susceptible animals (Probst *et al.*, 2020). The wild boar hunting season may differ between countries, but in general most wild boar hunting is done in autumn and winter. If an infected wild boar is hunted, blood from the shot wound, body fluids spilled during dressing, as well as hunting remains left in the forest can contaminate the environment with large amounts of virus. Additionally, hunting in infected areas may cause dispersal of wild boar and push infection into new susceptible populations. Furthermore, the mating season of wild boar lasts from October to January in most of Europe, coinciding with the main hunting season. During that period males get into contact with females and fighting between males may take place, increasing the probability of effective contacts between infected and susceptible animals. Furthermore, food availability may have an effect on contacts between sounders. If natural food is scarce it may lead to wild boar roaming in larger areas in search of food. If supplementary feeding is provided it may cause aggregation of wild boar at feeding sites. These consequences may each increase the probability of contacts and possibly fights between sounders that in turn may contribute to the higher incidence of ASF among wild boar seen during the winter (EFSA AHAW Panel, 2018). Some of the factors that may increase the disease incidence among wild boar in summer are mentioned above, like increase of the population density after the farrowing in spring. After the birth of the new generation, the sub-adult females may disperse, leaving the maternal group to form new sounders, resulting in increased probability of direct contacts. In addition, in the summer field crops ripen, and wild boar move to feed on fields, making contacts between sounders more likely. Lastly, blood sucking insects, although not convincingly proven, may play a role in the incidence increase by mechanically transmitting the virus to susceptible host.

The factors that potentially enhance virus spread in wild boar populations during the summer season may also be the reasons for the increase in incidence of outbreaks in domestic pig farms during the same time period. Pig farm activities are seasonal, and those related to the summer season may lead to more likely and frequent contacts with infected wild boar or the environment contaminated by them. For example, during the warmer season of the year, freshly cut forage can be fed to pigs; pigs having outdoor access are let out; crops are harvested and farm vehicles and machinery move frequently between the farm and potentially contaminated fields. The role of blood sucking flying insects in the transmission has not been confirmed, see Section 9.3.2 on arthropod vectors.

9.3.6 Speed of disease propagation in wild boar populations

The speed of propagation of ASF in wild boar populations has been assessed using different analytical methods. Using network analysis the median speed of the local spread of the virus (excluding likely human mediated long distance translocation events) was estimated to be between 2.9 and 11.7 km/year in Belgium, the Czech Republic, Estonia, Hungary, Latvia, Lithuania and Poland (EFSA, 2020). Marcon and co-workers (2020) estimated the infection wavefront velocity in the Belgium wild boar epidemic using an interpolation procedure that resulted in an overall wavefront velocity estimate of 0.39 km/week (20.3 km/year) ranging from 0.1 to 1 km/week (5.2-52 km/year) in different parts of the infected area. Niine and co-workers (unpublished data) estimated the average linear speed of expansion of the infected area (including likely human mediated spread) in Estonia during the ascending phase of the epidemic in 2014-2017, using the difference in the radii of the infected area in consecutive weeks as a measure of speed. The average speed of expansion of the infected area was found to be 0.64 km/week (33.1 km/year) ranging from 0.32 to 0.84 km/week (17.1-44.0 km/year) in different years. Large seasonal variation in speed of propagation could be observed in Estonia with notable winter and summer peaks.

The speed of the infection spread is not equal in all directions of the infected areas. The expansion of the affected area follows the area of wild boar suitable habitat and depends on wild boar density. However, other ecological and anthropogenic factors likely have effect on the direction and speed of the spread of the disease. In Belgium, the main direction of the spread of the infection was towards the north and the west. Marcon *et al.* (2020) demonstrated that artificial barriers like roads and fences had significant restricting effect on the spread of the infection in the wild boar population. The speed of expansion of areas with infected wild boar has been different in different EU countries, seeming to be more rapid in the northernmost countries (Estonia and Latvia) and less rapid in Lithuania, Poland and Hungary. Reasons for this still need to be clarified.

9.4 The role of wild boar in African swine fever epidemiology

The role of wild boar in ASF epidemiology varies between regions. Wild boars are highly susceptible to ASFV infection in both natural and experimental infections. Sick wild boars excrete the virus in the same quantities as domestic pigs and the transmission parameters established in experimental conditions for wild boar are similar to those verified for domestic pigs.

During the ASFV genotype I epidemic in Europe from the 1950s to '90s, wild boar was not considered to play any major role in spreading the virus to domestic pigs or in maintaining the virus locally. ASFV tended to persist in the wild boar population only when the virus was circulating in domestic pigs in the same area. As soon as the virus was eradicated from domestic pigs it also disappeared from the wild boar population. It was found that wild boar could transmit the virus mainly to free range domestic pigs through direct contacts or contacts with infected carcasses (Costard *et al.*, 2013; Jori and Bastos, 2009). A similar pattern has also been observed on the island of Sardinia, where the ASF genotype I virus has persisted in limited areas in free range domestic 'brado' pig herds (FRP) and among wild boar in the same areas (Laddomada *et al.*, 2019). However, the depopulation of the FRP-herds has led to a strong decline of ASFV in the local wild boar population (see Section 9.5.2 on Sardinia).

At the start of the current epidemic in 2007 (see below) the disease spread mainly among domestic pig farms with low biosecurity. Incidental virus spill-overs to wild boar populations were observed, but it was presumed that the epidemic would follow the pattern of former European epidemics and spontaneously fade out from the local wild boar population following disease eradication from the domestic pigs, due to high case fatality rate and the absence of long-term carriers (Costard *et al.*, 2013). Nevertheless, already in 2008 the first concerns regarding a more significant role of wild boar in dissemination and persistence of the virus in domestic pig and wild boar populations were expressed based on field observations in the Caucasus and Russia (Beltrán-Alcrudo *et al.*, 2008).

Soon after the ASF incursion into the Baltic States and Poland in 2014, it became evident that the infection can survive locally for a long period in wild boar populations independently of outbreaks in domestic pigs. Wild boar surveillance data have shown that the disease moves in the form of a slow epidemic wave (~1-5 km/per month) through local wild boar populations, killing most of the animals in the area. However, the disease does not disappear completely in the back of the epidemic wave but continues to spread in the affected area with low incidence for several years (Figure 9.2), even without disease transmission in domestic pigs. This indicates the existence of specific ecological conditions enabling such a long term circulation of ASFV in wild boar populations at least in northern temperate climates, and has led to the suggestion to add a distinct cycle, the 'wild boar-habitat cycle', to the list of three previous transmission cycles of ASF including 'sylvatic cycle' in Africa, 'tick to pig cycle' in areas where competent biological vectors can transmit the virus, and the 'domestic pig cycle' (Figure 9.4) (Chenais *et al.*, 2018).

Figure 9.4. The four epidemiologic cycles of African swine fever and main transmission agents. (1) Sylvatic cycle: the common warthog (*Phacochoerus africanuus*) and soft ticks of *Ornithodoros* spp. The role of the bushpig (*Potamochoerus larvatus*) in the sylvatic cycle remains unclear. (2) The tick-pig cycle: soft ticks and domestic pigs (*Sus scrofa domesticus*). (3) The domestic cycle: domestic pigs and pig-derived products (pork, blood, fat, lard, bones, bone marrow, hides). (4) The wild boar-habitat cycle: wild boar (*S. scrofa*), pig- and wild boar-derived products and carcasses, and the habitat. The figure is reproduced from Chenais *et al.* (2018) and published with permission from Emerging Infectious Diseases.

In the wild boar-habitat cycle both direct transmission between infected and susceptible wild boar and indirect transmission through carcasses and contaminated environment in the habitat may occur. Indirect transmission, particularly through infected carcasses, seems to play a key role in the cycle. Direct transmission seems less important for several reasons: physical contacts between sounders are not frequent, particularly after the epidemic has depleted an area of most of the wild boar; mathematical simulation of disease transmission suggests that if direct transmission between sounders would occur frequently, the disease spread in wild boar populations would be much faster than what is actually observed (Lange *et al.*, 2018); less virulent or attenuated strains of ASFV seem to disappear fairly quickly from the wild boar population (Zani *et al.*, 2018), indicating that if the virus is not generating enough carcasses it is not able to persist in the wild boar population. In this regard the combination of high case-fatality rate, long-term virus persistence in animal carcasses and the environment, as well as the relatively low contagiousness preventing complete depopulation of the host population, seem to interact in a way that maximises both local persistence and constant geographical spread (Figure 9.5) (Depner *et al.*, 2020).

Cool and moist climates favour environmental persistence of the virus and indirect transmission. As mentioned, spread of ASFV in wild boar populations poses a great risk for domestic pigs in the area. Outbreak investigations have revealed that direct contacts between wild boar and domestic pigs have been rather exceptional during the present epidemic, and that in most cases the introduction of the virus to pig herds has occurred via indirect transmission routes (Boklund *et al.*, 2020; EFSA AHAW Panel, 2018; Nurmoja *et al.*, 2018; Oļševskis *et al.*, 2016). Unexpectedly, most outbreaks in domestic pigs occur during the warmest season of the year, not when the conditions in the environment are generally cold and humid, i.e. optimal for virus persistence. As discussed above, despite the unfavourable weather conditions, sharing the environment (like fields in harvest season where the virus load is high due to increased wild boar population density and ASF incidence among wild boars), apparently creates an effective link to pass the virus between wild boar and domestic pigs.

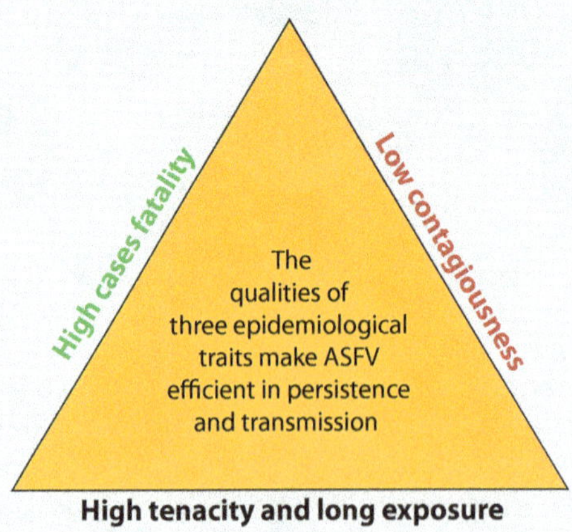

Figure 9.5. The persistency triangle illustrating how epidemiological characteristics make eradication of ASF in wild boar population challenging (Depner *et al.*, 2020). © World Organisation for Animal Health (OIE).

In summary, wild boar has played an important role in spreading ASFV across, and maintaining it in, large territories during the present epidemic. However, it has to be emphasised that this spread has been strongly supported by human activities. In case of long-distance spread, wild boar has been the victim and not the cause of virus propagation. Spread of ASFV in wild boar populations poses a risk for infection in domestic pig herds, but also in this situation human mediated transmission tends to play the key role. ASFV can persist in the wild boar-habitat cycle without continuous new introductions, but it seems possible to break the cycle. Evidence from Estonia and Latvia suggests that if the wild boar population density is kept low enough (with hunting efforts) for a sufficiently long period after the epidemic wave has passed, the infection is likely to die out (Oļševskis *et al.*, 2020; Schulz *et al.*, 2020).

9.5 Historic and present African swine fever epidemics in Europe

9.5.1 African swine fever in the Iberian Peninsula (1957-1994)

ASF occurred for the first time ever outside Africa in Portugal, near Lisbon, in 1957. It was concluded that the infection originated from feeding pigs with uncooked swill from planes originating from Angola, the swill containing meat from infected pigs (Manso Ribeiro *et al.*, 1958).

The eradication measures implemented by the Portuguese Veterinary Authorities during the first outbreak comprised strict quarantines and slaughter of affected herds, affected piggeries were cleaned and disinfected and a long delay was imposed before repopulation (Boinas, 1986). Control of pig movements in a 'sanitary control zone' was imposed around the first outbreak but, due to intense commercial pressures, control was not complete and new outbreaks appeared in other districts. In spite of this, by 1958 the Veterinary Services had achieved control of the disease and there was no further evidence of ASF in Portugal for about two years. It was clear that the key measures for controlling and eradicating the disease were early identification of infected herds, isolation, restriction of pig movement to markets and to other farms, and stamping out infected populations. Additionally, disinfection and sanitation of the pig housing before restocking prevented re-infection. Two years later, in 1960, the disease reoccurred in Portugal (Manso Ribeiro and Rosa Azevedo, 1961). Analysis of the virus genomes of the outbreaks in 1957 and 1960, although both from genotype I, showed large differences and corroborated the hypothesis of two separate virus introductions (Wilkinson *et al.*, 1993). From Portugal, the disease spread to Spain where the original outbreak was diagnosed close to the Spanish border with Portugal, in 1960 (Polo Jover and Sanchez Botija, 1961). The same set of sanitary measures used in the 1957-58 eradication programme was re-imposed in Portugal when ASF reappeared in 1960, but this time they were not successful. In both Iberian countries ASF became established as an endemic disease affecting herds throughout the territories. Eradication was only achieved more than three decades later (1993 in Portugal and 1994 in Spain) after the enforcement of stricter sanitary measures, extensive surveys for detection of infected animals, slaughter of infected herds with depopulation and fair compensation, implementation of surveillance and protection zones, herd census, increased biosecurity of farms, enhanced animal movement control and increased awareness and collaboration of pig producers (Arias and Sánchez-Vizcaíno, 2002; Boinas, 1994).

During the 1957/58 outbreaks in Portugal, and those in Spain and Portugal in the early 1960s, infection caused a very acute clinical disease leading to death of virtually 100% of pigs within seven days after the onset of clinical signs (Manso Ribeiro and Rosa Azevedo, 1961; Polo Jover and Sanchez Botija, 1961). As the disease was established in the two countries, sub-acute, chronic and sub-clinical forms of the disease became more frequent (Sanchez Botija, 1982; Vigario and Caiado, 1989). This change has been attributed both to spontaneous decrease of virus pathogenicity and a newly developed live attenuated virus vaccine that was used in the extensive Iberian pig production areas in the southern regions of Portugal in the early 1960s. The reversion of virulence of the vaccine virus isolate, and its subsequent spread to the pig population lead to the development of carrier pigs (Manso Ribeiro *et al.*, 1958). The vaccine was withdrawn after a very short time.

In the traditional old pigsties used in Iberian pig production systems a haematophagous soft tick, *Ornithodoros erraticus*, can be found (Boinas *et al.*, 2011). The tick can harbour ASFV for up to five years and is thus a possible source of disease when feeding on susceptible pigs. Tick presence was associated with repeated outbreaks in farms in Spain (Pérez-Sánchez *et al.*, 1994), and its presence was considered the most probable origin of a sporadic outbreak in the south of Portugal in 1999 when some infested premises were repopulated with pigs after the country had been declared ASF free (Boinas *et al.*, 2011). These regions also had a higher wild boar density and greater opportunity of contact between them and free ranging pigs in production systems with generally low biosecurity. On the Iberian Peninsula in the 1980s, wild boar was not considered a risk factor for disease transmission, supported by epidemiological surveys reporting wild boar to be responsible for only 5 to 6% of the ASF outbreaks in Portugal and Spain (Ordas *et al.*, 1983; Perestrelo Vieira, 1993). Due to these specific risk factors for ASF, the south-western portion of the Iberian Peninsula, where extensive pig production predominates, were the last remaining ASFV infected areas before the eradication. In addition, in the final stages of the eradication, these areas were the object of a joint 'Coordinated ASF Eradication Programme' between Portuguese and Spanish Veterinary Authorities (Arias and Sánchez-Vizcaíno, 2002; Boinas, 1994).

9.5.2 African swine fever in Sardinia

ASFV genotype I was introduced into south Sardinia in 1978 via contaminated food waste from the Iberian Peninsula. It rapidly spread to several areas of the island, mainly due to uncontrolled movements of domestic pigs. ASF found ideal conditions to become endemic in the 'brado' FRP population, kept in the inner mountainous areas of central Sardinia, where these pigs interact closely with wild boar. Disease prevention in high biosecurity farms has almost always been successful, while ASF outbreaks frequently occurred in poor-biosecurity backyard farms regularly exposed to contact with FRPs. Evidence supports a hypothesis that, compared to FRPs, wild boar played a secondary role as virus source (Laddomada *et al.*, 2019). Indeed, a large amount of data in the last forty years suggests that, in Sardinia, ASFV was not able to persist in wild boar populations alone for more than a few years if this population was not re-infected by FRPs or domestic pigs. Any role for *Ornithodoros* spp. ticks in ASFV transmission in Sardinia has been excluded.

Until recently, any attempt to eradicate the disease encountered strong resistance by the FRP-keepers, who considered this traditional way of keeping pigs as a part of their cultural identity

(Cappai *et al.*, 2018). As from 2015, a new ASF eradication plan (EPASF-15-20) has been implemented, fully empowered by the Sardinian Regional Government, and supported by the National Government. Based to a large extent on conventional veterinary measures adapted to the local condition, the new strategy favoured financial incentives for good husbandry practices and biosecurity more than compensation to affected farmers. Furthermore, the EPASF-15-20 took into account the socio-economic and cultural aspects associated with ASF occurrence (Cappai *et al.*, 2018; Loi *et al.*, 2019a,b). Veterinary controls were strengthened all along the pig production chain in an increasingly rigorous manner, particularly in backyard pig farms. More stringent rules were applied to hunting, including safe disposal of wild boar offal. Control measures were accompanied by very intensive activities of education, awareness and communication, targeted to farmers, hunters and the local population. Open-air, double fenced pig farms were authorised and subsidised, as an alternative to keeping FRPs.

As an integrated part of the EPASF-15-20, almost 5,000 FRPs were culled from November 2015 to February 2020. Testing of these pigs confirmed that ASFV was endemic in this population located in a few municipalities in central Sardinia, with very high prevalence of virus- and seropositive animals (Laddomada *et al.*, 2019). As a consequence of the intensive control measures and of the progressive reduction in the density and number of the FRPs, the overall epidemiological situation has largely improved also in wild boar. At present (July 2020) the ASF situation can be summarised as follows:

- Domestic pig farms. The last outbreak (detected by means of RT-PCR) occurred in September 2018. On a few later occasions (the latest in November 2019), seropositive pigs were detected in pig farms or at slaughter, but no virus genome could be detected in these farms or animals.
- FRPs. The last PCR-positive pigs were detected in January 2019. To date, this population has been reduced to 100-200 pigs, divided into small groups. This very sparse population no longer seems to play any role in ASFV transmission.
- Wild boar. The disease control measures applied in the last five years have been associated with a clear decrease in ASFV and ASF antibodies detection during both passive and active surveillance. This favourable trend became even clearer after the FRP-population was reduced to negligible numbers. The last PCR-positive wild boar was detected in April 2019 during passive surveillance, while absence of ASFV and continuous, strong decline of seroprevalence has been confirmed in over 6,000 wild boars tested during the hunting season November 2019-January 2020.

Although the EPASF-15-20 has led to very positive results, it cannot be excluded that ASF still occurs at very low level in wild boar populations in some remote areas. It is thus necessary to continue and intensify ASF surveillance. The favourable trend indicates that complete eradication will most likely be achieved in the near future, provided that the current disease control measures continue to apply.

9.5.3 African swine fever in Europe outside EU

9.5.3.1 Caucasus

The first signs of ASFV returning to mainland Europe appeared in March 2007, in the seaport of Poti, Georgia (Rowlands *et al.*, 2008). According to official reports, the virus was introduced via ship waste containing pork products, with the first cases occurring in pig farms close to the port (from now on in this section, 'reported' means officially reported to the World Organisation for Animal Health (OIE). Reports are available at www.oie.int). The causative agent was identified as ASFV genotype II, otherwise circulating in Mozambique, Madagascar and Zambia (Rowlands *et al.*, 2008). The majority of pigs in the Caucasus are kept by poor subsistence farmers in backyard systems with low general biosecurity levels, with free-range management being widely practised. Slaughter is carried out on the premises, even in larger commercial farms. These factors, in combination with active transboundary movements and markets, facilitated quick spread of the infection (25-35 km per month) after the first introduction in 2007 (Beltrán-Alcrudo *et al.*, 2018). Delayed diagnosis and control measures led to unmanageable propagation of the disease. From June to August 2007, Georgia reported 58 outbreaks in domestic pigs. According to OIE, they were resolved completely in January 2008, but the virus re-emerged in Georgia twice (in 2010 and 2011), resulting in outbreaks in different areas (Vepkhvadze *et al.*, 2017). The situation regarding ASF in wild boar in Georgia is unknown (Vepkhvadze *et al.*, 2017). After Georgia, the virus was detected in Armenia in early August 2007. It is believed that the virus was introduced by transboundary movement of infected pigs and wild boar. Similarly, to Georgia, there were three waves of ASF epidemic in Armenia, in 2007, 2010 and 2011. The latter outbreaks affected the wild boar population and domestic pigs close to the Georgian border (Sanchez-Vizcaino *et al.*, 2013). In 2019 the presence of the virus in Armenian territory was still suspected despite of lack of official reports of outbreaks.

In Azerbaijan the majority of the population is Muslim, consequently the country has very low numbers of domestic pigs. Nevertheless, in late January 2008, the presence of ASF was officially confirmed in a village in north-western Azerbaijan, about 180 km east of the Georgian border. It is the only reported outbreak from the country. In June 2011 Azerbaijan self-declared freedom from ASF to the OIE, while media reported some more ASF cases in the Nagorno-Karabakh (a region with a political status of disputed territory) located in the Lesser Caucasus mountains from 2007 to 2013 (Grigoryan, 2013).

In December 2008 the virus had covered about 500 km and reached wild boar populations in north-western Iran. ASFV isolates from Iran expressed 100% similarity with those from Georgia, possibly brought into Iran by infected wild boar crossing the Aras River from Armenia (Rahimi *et al.*, 2010).

Repeated outbreaks in 2007-2019 in the north and south Caucasus, including a case in Kabardino-Balkarian Republic in 2019, suggest that the virus is still circulating in the Caucasus region. The exact mechanism of virus persistence is still unknown. Involvement of *Ornithodoros* spp. has not been proven, but cannot be excluded.

9.5.3.2 Russia

It took the virus about nine months to pass through Georgia and cross the Caucasus mountains reaching Russia. The very first case of ASF in Russia was registered in wild boar in November 2007. From the wild boar population, the disease entered the domestic pig population in the Republic of North Ossetia-Alania and spread further northward, affecting both domestic pigs and wild boar. Since its introduction in 2007, the ASF epidemic in Russia is characterised by unpredictable long-distance jumps followed by local epidemics. Delayed intervention strategies and inappropriate financial compensation have led to underreporting by small-scale farmers and inappropriate disposal of dead or infected pigs (including illegal selling of apparently healthy pigs) increasing the transmission (FAO, 2013). A key epidemiological role of backyard farms is supported by several statistical models (Korennoy *et al.*, 2014; Vergne *et al.*, 2016, 2017), as well as by observations of seasonality of ASF outbreaks in both domestic pigs and wild boar, coinciding with periods of increased economic activity in this sector. Therefore, the main factors influencing the progressive spread of ASF in Russia were illegal movement of contaminated pork products, large number of backyard holdings, swill feeding and free-range management practices, as well as poor preparedness of regional veterinary services. These factors all contributed to the wide distribution of the disease from Kaliningrad on the Baltic Sea to the Russian Far East. To achieve control of the disease, the majority of small farms with low biosecurity were gradually eliminated.

Since the introduction in 2007 the biological properties of the virus have not changed substantially. Both field observations and laboratory studies indicate that circulating viruses remain highly virulent, manifesting clinically mainly as the acute form with short incubation period (three to five days) and the course of the disease lasting around ten days (Belyanin *et al.*, 2011). In most cases clinical symptoms are nonspecific (anorexia, depression, affected breathing with wheezing, high temperature, in some cases cyanotic discoloration, paresis of the hind limbs, abortion and rhinitis), complicating disease diagnosis.

9.5.3.3 Belarus

So far two ASF outbreaks in domestic pigs have been notified in Belarus, both in 2013. No cases in wild boar have been reported. There are some indirect signs that this might not be a true representation of the ASF situation in Belarus. According to data from the National Statistic Committee of the Republic of Belarus, the number of domestic pigs decreased significantly in 2014-2015 (Federal Service for Veterinary and Phytosanitary Surveillance in Russia, 2018) and again in 2019-2020. Likewise, information collected from open sources in 2013-2014 by the Federal Service for Veterinary and Phytosanitary Surveillance in Russia reveals many cases of ASFV PCR positive pork products imported to Russia from Belarus in 2014-2018 (Federal Service for Veterinary and Phytosanitary Surveillance in Russia, 2020). In addition, a decision of the government of Belarus to significantly reduce the wild boar population indicates possible concerns about disease transmission in this species.

9.5.3.4 Ukraine

The first ASF outbreak in Ukraine was notified in July 2012 in the Zaporozhye region on the Black Sea coast. Two years later a wild boar was found dead on the riverside on the border with Russia. At the end of 2016, several outbreaks affecting domestic pigs and wild boar occurred in the southern part of Ukraine, bordering Hungary. In 2018, ASF had affected domestic and wild boar populations of all Ukrainian regions.

8.5.3.5 Moldova

The first ASF outbreaks in Moldova were registered in small backyard farms in the northern parts on the border with Ukraine in September 2016. Feeding swill containing leftovers from infected pork and pork products originating from Ukraine has been hypothesised as the introduction route. Currently, most of the outbreaks in domestic and wild pigs are concentrated along the border with Romania.

The emergence of ASF from Georgia and the Caucasus region can to some extent be explained by the domestic pig husbandry systems in the affected countries with many small-scale farms with pigs frequently roaming free, the common socio-economic relationships between the countries in the region based on shared history and culture, as well as the neglected status of veterinary service systems in the post-Soviet era.

9.5.4 African swine fever in the EU 2014-present

On January 24th 2014, ASFV was confirmed in two wild boar carcasses found by the Lithuanian veterinary authorities. Three weeks later, on February 17th and 18th, two cases in wild boar were confirmed in Poland. These four cases of wild boar found dead represented the start of what would become the second major epidemic of ASF within the EU, approximately 20 years after the end of the first. During the following months ASF also appeared in Latvia (by the end of June) and in Estonia (September). In Lithuania, Poland and Latvia the virus was first detected in wild boar found dead not far from the national borders with Belarus, suggesting multiple separate initial introductions of ASFV to the EU, most likely through transboundary movements of infected wild boar. At the time, Belarus had only notified two outbreaks of ASF, both in backyard pigs and far from the border to the EU. However, it has been speculated that the disease was also present in the Belarussian wild boar population, and that depopulation campaigns were carried out in an effort to control the disease. Intensive hunting is known to lead to dispersion of wild boar, and such a depopulation campaign may thus have facilitated the progressive geographical spread of the disease (EFSA AHAW Panel, 2014; Pejsak *et al.*, 2014).

The first years of the epidemic in the EU were characterised by slow geographical expansion of the disease within the affected member countries, dominated by cases in wild boar, and only sporadic outbreaks in domestic pig farms. In contrast to what was initially predicted, the epidemic thus seemed to be driven by wild boar, and ASFV circulation to be maintained within the affected wild boar population independently from outbreaks in domestic pigs (Chenais *et al.*, 2019a). In this new epidemiological scenario, described as the wild boar-habitat cycle, infected wild boar carcasses and the virus contaminated environment are believed to constitute the long-term source of the virus needed to maintain the virus circulation over time (Chenais *et al.*, 2018).

Whereas local disease expansion occurred through natural movements of infected wild boar, longer distance translocations of the virus, locally or regionally within affected countries or to more distant and previously unaffected parts of the EU, were most likely related to human activities (EFSA, 2018). ASFV contaminated meat products originating from an affected area and left in the environment in reach for hungry wild boar are often mentioned as a likely source of virus for naïve populations, although alternative routes of introduction are also possible, and the true route rarely known. The incursions of ASFV to the wild boar population around the city of Zlín, in the eastern parts of the Czech Republic, during the summer of 2017, and to the wild boar population in the province of Luxembourg in southern Belgium during the summer of 2018, are both believed to be the result of such human mediated spread. On both occasions, the disease was confirmed very unexpectedly in wild boar found dead far away from the nearest reported case, more than 400 km and 800 km, respectively. Similarly, the incursions of the virus to Romania in 2017, to Hungary and Bulgaria in 2018, to western Poland in 2019 and to Greece in 2020 were associated with human activities.

In spite of extensive measures implemented to control the disease in the affected countries and to prevent further spread, the geographical expansion within the EU has continued. After the first long distance translocation of the virus, to the Czech Republic in 2017, at least one new country has become affected each year: the Czech Republic and Romania in 2017, Hungary, Bulgaria and Belgium in 2018, Slovakia in 2019, and Greece and Germany in 2020. The total number of reported wild boar cases has increased from year to year, from around 250 in 2014, to almost 4,000 in 2017 and almost 6,500 in 2019. However, at the same time the number of outbreaks in domestic pigs has been limited in most of the affected countries, and any extensive secondary spread within the pig sector has been prevented. Romania constitutes the exception. After a second incursion of the virus during summer 2018 in the area around the Danube delta, a dramatic epidemiological evolution emerged. A first outbreak was confirmed in a backyard farm in Tulcea county on June 10th, and on June 12th a first case in wild boar was reported (EFSA, 2018). Four months later the number of reported outbreaks in domestic pigs had reached almost 1000, whereas less than 100 cases in wild boar had been confirmed. Since then, the disease has spread widely within the country, mainly affecting backyard holders, but also a number of large and very large commercial farms. The observed epidemiological pattern remains different from the other affected countries in the EU, with outbreaks among domestic pigs by far outnumbering cases in wild boar (3,211 in pigs compared to 1,422 in wild boar by 1 July 2020). Most likely this observed pattern is associated with the characteristics of the pig sector in Romania, with a very large proportion of the more than four million pigs kept in backyard farms with low levels of biosecurity, and with disease spread between farms driven by human activities. Thus, the situation in Romania reflects the domestic pig epidemiological cycle of ASF, rather than the wild boar-habitat cycle that dominates in most of the other affected countries in the EU.

To date 12 EU countries have reported cases of ASF in wild boar or outbreaks in domestic pigs within the current epidemic. The economic consequences have been vast (see Chapter 7), and the situation is not yet under control. During the last few years, however, positive signs in the development of the epidemic have appeared in some of the affected countries. In the Czech Republic, early detection and timely implementation of relevant control measures, including e.g. restricted access to the area, intensive surveillance, hunting and fencing, allowed the epidemic to be contained within a small area in the District of Zlín, where it was first introduced. Through enhanced biosecurity requirements on domestic pig farms, outbreaks in domestic pigs could be prevented. The last ASFV positive wild boar, out of a total of 230, was found less than 10 months after the first and, one year later in March 2019, the country was again declared free from ASF (Charvátová *et al.*, 2019). As in the Czech Republic, the ASF epidemic in Belgium was initiated by a long-distance translocation of the virus and a localised incursion into the wild boar population. After confirmation of ASFV in two wild boar carcasses found in the province of Luxembourg, located only 12 km from the border with France, a control strategy based on the successful Czech experience was implemented. And just as in the Czech Republic, the strategy seems to have worked. The epidemic has been contained within a rather limited area, and without spill over to domestic pigs. The last fresh positive case was found in August 2019, 11 months after the first (Dellicour *et al.*, 2020). Thus, Belgium also seems to be well on the way towards regaining freedom. Likewise, in Estonia and Latvia positive trends have been observed. In Latvia the numbers of outbreaks in domestic pigs as well as cases in wild boar are decreasing markedly, especially in the eastern parts of the country that was infected first (EFSA, 2020; Oļševskis *et al.*, 2020; Schulz *et al.*, 2020). In Estonia no cases had been detected for 19 months then two positive cases in wild boar were detected in August 2020 in one hunting ground in the western parts.

9.6 Regulatory framework for prevention and control of African swine fever in the EU

ASF is one of the five diseases listed in the new animal health law of the EU (EC, 2016), considered to be highly transmissible with risk for transboundary spread. The other diseases are foot and mouth disease (FMD), classical swine fever (CSF), African horse sickness and highly pathogenic avian influenza (HPAI). These five diseases all pose risks to animal health and have the potential to cause major economic losses (with HPAI having additional public health implications). Control of ASF follows the common concept for controlling infectious diseases as stated in Directive 92/119/EEC (EC, 1992). Roughly, as soon as the presence of the disease is suspected, immediate actions have to be taken to confirm or exclude ASF. Once the disease is confirmed, the infected holding has to be depopulated. Further spread must be prevented by tracing contact holdings, monitoring movements of animals and potentially contaminated products, and by establishing surveillance and protection zones around the outbreak. The CSF directive (Council Directive 2001/89/EC; EC, 2001) has been used as a model for drafting the ASF directive (Council Directive 2002/60/EC; EC, 2002) and the ASF measures currently in place in the EU thus follow the measures foreseen to control and eradicate CSF. The measures are based on the assumption that ASF, like CSF, is a highly contagious disease, which rapidly spreads from pig to pig or from farm to farm. The aim of all activities is to eradicate the disease inside the affected area(s) and, at the same time, allow trade and movement of animals and animal products outside the restricted regions so that 'business as usual' can continue. In summary, the measures should not only lead to the eradication of the disease, but also protect the pig industry outside the restricted regions, keeping trade and business going.

The measures foreseen in the EU legislation set a minimum standard. National authorities can implement additional and more severe measures if deemed necessary, as can the European Commission. This was the case during the current epidemic.

9.6.1 Control of African swine fever in domestic pigs

Suspicion of ASF is raised if a pig exhibits clinical signs, shows post-mortem lesions or reactions to laboratory tests that indicate the possible presence of ASFV. In such a case, the holding has to be placed under official surveillance until the ASF situation is clarified. The most important measures in case of suspicion include to count all the pigs in the various categories on the holding and compile a list of the number of pigs already sick, dead or likely to be infected in each category, and further to construct a map of the holding for the epidemiological investigations. All pigs shall be restricted to their living quarters and no pigs or pig products should leave the holding until the final results exclude the presence of ASF. In addition, the movement of persons and vehicles to or from the farm should be restricted and appropriate means of disinfection used at the entrance and exit of stables.

Upon confirmation of ASF immediate measures have to be taken in the affected holding. All pigs on the holding have to be euthanised without delay and a sufficient number of samples taken for further epidemiological investigations, in particular for tracing the virus introduction and for estimating the high-risk period (HRP, the likely length of time that ASF has been present on the farm prior to the notification). Further, an epidemiological inquiry should be conducted to estimate the HRP, determine the possible origin of the virus, trace possible contact holdings that could have been infected from the same source and determine if vectors (e.g. soft ticks) or feral pigs (wild boar) caused the infection. Live pigs, slaughtered pigs, meat, meat products, semen, ova or embryos, etc. which left the holding during the HRP should be traced and cleaning and disinfection of the holding (see Chapter 11 for details on the procedures) should be performed.

A protection zone with a radius of at least three km and a surveillance zone with a radius of at least ten km has to be established around the outbreak site. When establishing zones, the authorities must take account of the results of the epidemiological inquiry, the geographical situation (particularly natural or artificial boundaries), the location and proximity of holdings, patterns of movements and trade in pigs, availability of slaughterhouses and facilities for processing carcasses as well as the facilities and personnel available to control movement of pigs within the zones. The latter is particularly important if the pigs to be euthanised have to be moved away from their holding of origin. The measures in the restricted zones primarily aim to identify further holdings that might be infected with ASFV. A census of all holdings has to be carried out as soon as possible and pigs on all holdings examined clinically. All dead or diseased pigs have to be examined for ASF. Furthermore, random samples for laboratory examination have to be taken from all holdings in the protection zone, and in case of suspicion, from holdings in the surveillance zone. A second important set of measures aim to prevent a possible virus escape from the restricted areas by implementing a standstill policy. Movement and transport of pigs is prohibited and pigs may not be moved from the holding in which they are kept. Also, other domestic animals may not enter or leave the holdings (in the surveillance zone only during the first seven days of restriction). Trucks or vehicles are not allowed to leave the zones without being cleaned and disinfected. Persons entering or leaving pig holdings have to comply with appropriate hygiene measures to reduce the risk of ASFV transmission.

The measures can be lifted only if the sampling and testing programme has been completed and the presence of ASF has been ruled out. However, restrictions cannot be lifted earlier than 30 days after cleaning and disinfection have been completed in the infected holding in the protection zone, and 20 days in the surveillance zone. Restocking of pig holdings which were infected with ASFV can take place not earlier than 40 days after cleaning and disinfection have been completed. It is advisable to introduce sentinel pigs before repopulating the farm.

9.6.2 Surveillance of African swine fever in domestic pigs

The main strategic aims of surveillance in domestic pigs are early detection of potentially infected holdings, and to prove freedom from disease in a previously infected region or country. Surveillance is compulsory in the protection and surveillance zones around outbreak holdings, as well as in holdings located in areas that are under restrictions due to the presence of ASF in wild boar.

Nowadays surveillance is mainly based on passive surveillance, testing sick and dead animals for the presence of ASFV. This has proved to be the most efficient strategy (Danzetta *et al.*, 2020). The passive surveillance approach is based on the very high case fatality rate of ASF (>90%), which means that almost all infected animals will become sick and die. Contagiousness, on the other hand, is relatively low and only few animals in a holding are affected at the beginning of an infection (initial low mortality) (Chenais *et al.*, 2019a; Schulz *et al.*, 2019a). Seropositive animals are found only during an advanced stage of the epidemic. Therefore, active surveillance based on random sampling for serological testing is not recommended anymore for early detection of ASF. Nevertheless, the EU legislation on ASF diagnosis (Decision 2003/422; EC, 2003) has not been updated and random serological testing of holdings within the protection and surveillance zones is still requested for lifting restrictions.

However, in areas under restriction due to ASF in wild boar, it is recommended to conduct passive surveillance and to sample diseased animals (animals showing clinical signs resembling ASF, e.g. fever or haemorrhagic lesions), dead animals (at least the first two deaths each week in each production unit with post weaning pigs or pigs older than two months) as well as animals with ante or post-mortem signs that raise suspicion at home slaughtering (EC, 2014, 2015).

In regions affected by ASF in wild boar, inspections of domestic pig holdings should take place at least once per year through interviews with farmers by an assigned veterinarian who also performs a census of the pigs, checks their identification and assesses the biosecurity of the farm while observing and, if necessary, examining the pigs.

9.6.3 Control and eradication of ASF in wild boar

The practical applications of controlling ASF in wild boar are described in detail in Chapter 8. In 2021, the European Council Directive 2002/60/EC (EC, 2002), which has so far been applied in case of ASF in wild boar in the EU, will be replaced by Regulation 2016/429 and its delegated act 2020/687 (EC, 2014, 2020). This legislation confirms the main elements of Directive 2002/60/EC, including the establishment of an expert group that shall assist the authorities in developing an eradication plan with the major objective of preventing virus transmission to domestic pigs and achieve ASF eradication from the affected wild boar population. The current epidemic has highlighted that containment and eradication of ASF from wild boar populations may be a major challenge and that the disease may persist in an affected area for several years. Under these circumstances it is not yet clear when disease prevention and control measures, including trade restrictions concerning domestic pig farms, can be safely lifted in an area where ASF appears to have been successfully controlled and eradicated from wild boar.

9.6.4 Conclusions and recommendations

In the new animal health law of the EU (Regulation 2016/429; EC, 2016) disease-specific rules for prevention and control apply for the listed diseases. Furthermore, preventive and control measures should be 'tailor made' in order to address every disease's unique epidemiological profile, consequences and distribution within the EU. In the past, ASF was often described as a highly contagious disease with high mortality affecting large number of pigs within an epidemiological unit. However, analyses of the domestic pig outbreaks in the current epidemic, as well as experimental studies, have revealed that the contagiousness is rather low, and that under field conditions ASFV transmission between animals can be a slow process (Chenais *et al.*, 2019a; Schulz *et al.*, 2019a). Consequently, ASF control and eradication measures need a different approach compared to highly contagious diseases, such as FMD or CSF (Figure 9.6).

Spread of ASFV in domestic pigs is mainly facilitated by human activities and insufficient farm biosecurity. Therefore, epidemiological tracing of contact farms is paramount for identifying secondary infections and stopping disease spread. A holding located outside a restricted area but linked to an infected farm through human activities can be under higher risk than a holding with good biosecurity within a protection or surveillance zone. Based on the experiences gained during the last years a correction of disease control measures, particularly concerning tracing and detecting of potentially infected holdings is needed. Detailed epidemiological farm investigations combined with a surveillance scheme based on enhanced passive surveillance would be advisable. In particular, updated guidelines are needed for sampling procedures in holdings where pigs

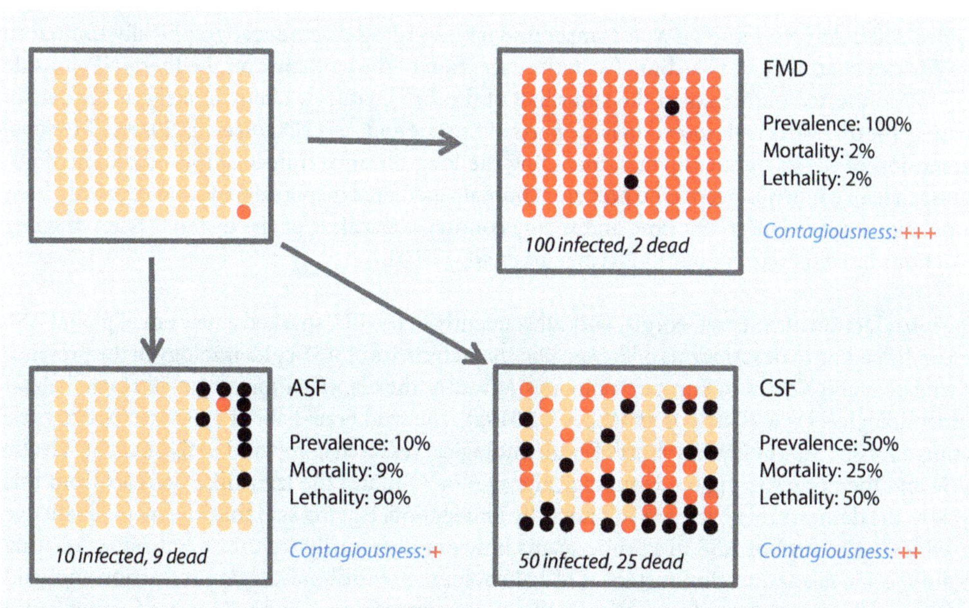

Figure 9.6. Exemplified disease spread of three major pig diseases highlighting the epidemiological differences between foot and mouth disease (FMD), classical swine fever (CSF) and African swine fever (ASF). Red dot = infected animal, black dot = dead animal.

are euthanised following confirmation of the disease; in holdings located in the protection and surveillance zones before lifting the restrictions; when pigs are euthanised as a prevention measure on a suspect holding; before authorisation is given to move pigs from holdings located in protection or surveillance zones and in case these pigs are slaughtered or euthanised; and in holdings being repopulated.

In this context, it is also advisable to redefine the size of the restriction zone, for example the effect of the surveillance zone on the prevention of ASF spread should be evaluated. Random sampling of holdings for serological screening within the restricted areas, as it is stated in the present EU legislation (Decision 2003/422/EC; EC, 2003), will not provide certainty of virus freedom. Apart from wasting human and laboratory resources, the effect of such a measure might instead be a false sense of security.

The better early detection surveillance schemes for detecting potentially infected holdings are implemented, the earlier infected farms are found, meaning that only one or a few animals will be virus positive on detection. This leads to a paradoxical situation regarding the acceptance of the depopulation measures, which have to follow without any delay. Due to this dilemma, alternative culling schemes for large farms with only few infected animals should be developed. For that to be possible, good management, strict internal biosecurity measures and intelligent farm surveillance schemes need to be in place.

9.7 Final remarks

Before 2007, ASF was known as a transboundary emerging disease, geographically limited to the African continent and Sardinia (with the exception of the epidemic on the Iberian Peninsula and the connected outbreaks in the latter half of the 20th century). Unfortunately, at the date of writing, the disease is widespread in large parts of Europe and Asia (Dixon *et al.*, 2019). The global dimension of the current epidemic, including the long distance translocations and incursions, shows that all countries need to be prepared: human-mediated dispersal to domestic or wild boar populations can occur at any time and to any country, regardless of the distance from ongoing infections in wild boar populations (Chenais *et al.*, 2019a).

The introduction of ASF to Georgia, and subsequently to the EU, sparked a new era of global ASF research, leading to descriptions of the specific characteristics of ASF epidemiology in the presence of wild boar and Northern European climates, as well as the identification of the wild boar-habitat epidemiological cycle (Chenais *et al.*, 2018, 2019a). The wild boar-habitat epidemiological cycle is thus almost a hundred years short of epidemiological research compared to the ancient sylvatic cycle and the domestic pig epidemiological cycles of ASF, and the transmission risks from this cycle to the domestic pig cycle are still not fully understood. Having said that, the major challenge in achieving control of ASF in Europe seems now not necessarily technical, but rather relating to the specific needs and circumstances of stakeholders in the domestic pig production and wild boar value chains in affected areas. Especially in the parts of Europe where pig production is still dominated by smallholder systems involving mostly poor farmers, low-cost control options fully adapted to the local context and highly accepted by the end users are required to achieve control (Chenais *et al.*, 2019b). National legislation as well as EU regulations set out clear solutions for controlling ASF. However, if these rules are not implemented at the local level where the disease is transmitted during the daily activities of people, they are of no value for disease control and eradication. Stakeholders in the pig production and wild boar value chains are largely aware of these legislations and regulations, but implementation is despite that far from fully executed. It is becoming clear that epidemiological knowledge alone is not sufficient to control ASF, and that understanding of local sociocultural, economic and political dimensions, as well as individual keys to effective communication is equally important (Chenais *et al.*, 2019a; Jori *et al.*, 2020; Loi *et al.*, 2019b). In both the wild boar-habitat and the domestic pig epidemiological cycle of ASF, fully implemented biosecurity is the key for stopping virus transmission and controlling the disease. The positive examples from Sardinia, the Czech Republic and Belgium show that control and finally also eradication of ASF can be achieved, but also that to reach this goal, all stakeholders in the value chains from farmer and forests to fork need to be involved, engaged and understood.

Acknowledgements

This publication is based on work from 'Understanding and combating African swine fever in Europe (ASF-STOP COST action 15116)' supported by COST (European Cooperation in Science and Technology). The authors kindly acknowledge José Cortinas Abrahantes for Figure 9.2 and Linda Ernholm for Figure 9.3.

References

Arias, M. and Sánchez-Vizcaíno, J., 2002. African swine fever eradication: the Spanish model. In: Morilla, A., Yoon, K. and Zimmerman, J. (eds.) Trends in emerging viral infections of swine. Iowa State Press, Ames, IA, USA, pp. 133-139.

Beltrán-Alcrudo, D., Kukielka, E.A., De Groot, N., Dietze, K., Sokhadze, M. and Martinez-Lopez, B., 2018. Descriptive and multivariate analysis of the pig sector in Georgia and its implications for disease transmission. PloS One 13: e0202800.

Beltrán-Alcrudo, D., Lubroth, J., Depner, K. and De La Rocque, S., 2008. African swine fever in the Caucasus. FAO Empres Watch 1.

Belyanin, S., Vasiliev, A., Kolbasov, D., Tsybanov, S., Balyshev, V. and Kurinnov, V., 2011. ASFV virulence isolates. Veterinary Medicine of the Kuban 5.

Blome, S., Gabriel, C. and Beer, M., 2013. Pathogenesis of African swine fever in domestic pigs and European wild boar. Virus Research 173: 122-130. https://doi.org/10.1016/j.virusres.2012.10.026

Boinas, F., 1986. Information system to facilitate the eradication of African swine fever and control of pig diseases in Portugal. University of Reading, Reading, UK.

Boinas, F.J.S., 1994. The role of *Ornithodoros erraticus* in the epidemiology of African swine fever in Portugal. University of Reading, Reading, UK.

Boinas, F.S., Wilson, A.J., Hutchings, G.H., Martins, C. and Dixon, L.J., 2011. The persistence of African swine fever virus in field-infected *Ornithodoros erraticus* during the ASF endemic period in Portugal. PloS One 6: e20383.

Boklund, A., Dhollander, S., Vasile, T.C., Abrahantes, J., Bøtner, A., Gogin, A., Villeta, L.G., Gortázar, C., More, S. and Papanikolaou, A., 2020. Risk factors for African swine fever incursion in Romanian domestic farms during 2019. Scientific Reports 10: 10215.

Bonnet, S.I., Bouhsira, E., De Regge, N., Fite, J., Etoré, F., Garigliany, M.-M., Jori, F., Lempereur, L., Le Potier, M.-F. and Quillery, E., 2020. Putative role of arthropod vectors in African swine fever virus transmission in relation to their bio-ecological properties. Viruses 12: 778.

Cappai, S., Rolesu, S., Coccollone, A., Laddomada, A. and Loi, F., 2018. Evaluation of biological and socio-economic factors related to persistence of African swine fever in Sardinia. Preventive Veterinary Medicine 152: 1-11.

Charvátová, P., Wallo, R., Jarosil, T. and Šatrán, P., 2019. How ASF was eradicated in the Czech Republic. Available at: https://www.pigprogress.net/Health/Articles/2019/6/How-ASF-was-eradicated-in-the-Czech-Republic-429472E/.

Chenais, E., Depner, K., Guberti, V., Dietze, K., Viltrop, A. and Stahl, K., 2019a. Epidemiological considerations on African swine fever in Europe 2014-2018. Porcine Health Management 5: 6. https://doi.org/10.1186/s40813-018-0109-2

Chenais, E., Lewerin, S.S., Boqvist, S., Stahl, K., Alike, S., Nokorach, B. and Emanuelson, U., 2019b. Smallholders' perceptions on biosecurity and disease control in relation to African swine fever in an endemically infected area in Northern Uganda. BMC Veterinary Research 15: 279. https://doi.org/10.1186/s12917-019-2005-7

Chenais, E., Ståhl, K., Guberti, V. and Depner, K., 2018. Identification of wild boar-habitat epidemiologic cycle in African swine fever epizootic. Emerging Infectious Diseases 24: 810.

Costard, S., Mur, L., Lubroth, J., Sanchez-Vizcaino, J.M. and Pfeiffer, D.U., 2013. Epidemiology of African swine fever virus. Virus Research 173: 191-197. https://doi.org/10.1016/j.virusres.2012.10.030

Danzetta, M.L., Marenzoni, M.L., Iannetti, S., Tizzani, P., Calistri, P. and Feliziani, F., 2020. African swine fever: lessons to learn from past eradication experiences. A systematic review. Frontiers in Veterinary Science 7: 296.

Dellicour, S., Desmecht, D., Paternostre, J., Malengreaux, C., Licoppe, A., Gilbert, M. and Linden, A., 2020. Unravelling the dispersal dynamics and ecological drivers of the African swine fever outbreak in Belgium. Journal of Applied Ecology 57: 1619-1629.

Depner, K., Dietze, K., Globig, A., Zani, L., Mettenleiter, T. and Chenais, E., 2020. African swine fever and the dilemma of a relatively low contagiousness. OIE Buletin 02.

Dixon, L.K., Stahl, K., Jori, F., Vial, L. and Pfeiffer, D.U., 2019. African swine fever epidemiology and control. Annual Review of Animal Biosciences 8: 211-246.

Dórea, F.C., Swanenburg, M., van Roermund, H., Horigan, V., de Vos, C., Gale, P., Lilja, T., Comin, A., Bahuon, C. and Zientara, S., 2017. Data collection for risk assessments on animal health (Acronym: DACRAH). EFSA Supporting Publications 14: 1171E.

Eble, P., Hagenaars, T., Weesendorp, E., Quak, S., Moonen-Leusen, H. and Loeffen, W., 2019. Transmission of African swine fever virus via carrier (survivor) pigs does occur. Veterinary Microbiology 237: 108345.

European Commission (EC), 1992. Council Directive 92/119/EEC of 17 December 1992 introducing general Community measures for the control of certain animal diseases and specific measures relating to swine vesicular disease. Official Journal of the European Union L 62: 69-85.

European Commission (EC), 2001. Council Directive 2001/89/EC of 23 October 2001 on Community measures for the control of classical swine fever. Official Journal of the European Union L 316: 5-35. Available at: https://eur-lex.europa.eu/legal-content/EN/ALL/?uri=celex%3A32001L0089.

European Commission (EC), 2002. Council Directive 2002/60/EC of 27 June 2002 laying down specific provisions for the control of African swine fever and amending Directive 92/119/EEC as regards Teschen disease and African swine fever. Official Journal of the European Union L 192: 27-46. Available at: https://eur-lex.europa.eu/legal-content/EN/TXT/?uri=celex%3A32002L0060.

European Commission (EC), 2003. 2003/422/EC: Commission Decision of 26 May 2003 approving an African swine fever diagnostic manual. Official Journal of the European Union L 143: 35-49. Available at: https://eur-lex.europa.eu/legal-content/EN/ALL/?uri=CELEX%3A32003D0422.

European Commission (EC), 2014. 2014/709/EU: Commission implementing Decision of 9 October 2014 concerning animal health control measures relating to African swine fever in certain Member States and repealing Implementing Decision 2014/178/EU (notified under document C(2014) 7222). Official Journal of the European Union L 295: 63-78. Available at: https://eur-lex.europa.eu/legal-content/EN/TXT/?uri=CELEX%3A32014D0709.

European Commission (EC), 2015. Strategic approach to the management of African Swine Fever for the EU. Working document. SANTE/7113/2015 rev 12. European Union DG Sante, Brussels, Belgium. Available at: https://ec.europa.eu/food/sites/food/files/animals/docs/ad_control-measures_asf_wrk-doc-sante-2015-7113.pdf.

European Commission (EC), 2016. Regulation (EU) 2016/429 of the European Parliament and of the Council of 9 March 2016 on transmissible animal diseases and amending and repealing certain acts in the area of animal health ('Animal Health Law'). Official Journal of the European Union L 84: 1-208. Available at: https://eur-lex.europa.eu/legal-content/EN/TXT/?uri=uriserv%3AOJ.L_.2016.084.01.0001.01.ENG.

European Commission (EC), 2020. Commission Delegated Regulation (EU) 2020/687 of 17 December 2019 supplementing Regulation (EU) 2016/429 of the European Parliament and the Council, as regards rules for the prevention and control of certain listed diseases. Official Journal of the European Union L 174: 64-139. Available at: https://eur-lex.europa.eu/legal-content/EN/TXT/?uri=uriserv%3AOJ.L_.2020.174.01.0064.01.ENG&toc=OJ%3AL%3A2020%3A174%3ATOC.

European Food Safety Authority (EFSA), 2017. Epidemiological analyses of African swine fever in the Baltic States and Poland: (Update September 2016-September 2017). EFSA Journal 15: 5068. https://doi.org/10.2903/j.efsa.2017.5068

European Food Safety Authority (EFSA), 2018. Epidemiological analyses of African swine fever in the European Union (November 2017 until November 2018). EFSA Journal 16: 5494. https://doi.org/10.2903/j.efsa.2018.5494

European Food Safety Authority (EFSA), 2020. Epidemiological analyses of African swine fever in the European Union (November 2018 to October 2019). EFSA Journal 18: 5996. https://doi.org/10.2903/j.efsa.2020.5996

European Food Safety Authority Panel on Animal Health and Welfare (EFSA AHAW Panel), 2014. Scientific opinion on African swine fever. EFSA Journal 12: 3628. https://doi.org/10.2903/j.efsa.2014.3628

European Food Safety Authority Panel on Animal Health and Welfare (EFSA AHAW Panel), 2015. Scientific opinion on African swine fever. EFSA Journal 13: 4163. https://doi.org/10.2903/j.efsa.2015.4163

European Food Safety Authority Panel on Animal Health and Welfare (EFSA AHAW Panel), 2018. African swine fever in wild boar. EFSA Journal 16: 5344. https://doi.org/10.2903/j.efsa.2018.5344

Federal Service for Veterinary and Phytosanitary Surveillance in Russia, 2018. The Rosselkhoznadzor is forced to introduce a temporary ban on the supply of live pigs and pig products from the territory of the Republic of Belarus. Available at: https://fsvps.gov.ru/fsvps/asf/news/25955.html.

Federal Service for Veterinary and Phytosanitary Surveillance in Russia, 2020. Available at: https://fsvps.gov.ru/fsvps/importExport/belarus/restrictions.html.

Food and Agriculture Organization of the United nations (FAO), 2013. Empres watch. African swine fever in the Russian Federation: risk factors for Europe and beyond. Available at: http://www.fao.org/docrep/018/aq240e/aq240e.pdf.

Fowler, M., 1996. Husbandry and diseases of captive wild swine and peccaries. Revue Scientifique et Technique (International Office of Epizootics) 15: 141-154.

Gallardo, C., Nurmoja, I., Soler, A., Delicado, V., Simón, A., Martin, E., Perez, C., Nieto, R. and Arias, M., 2018. Evolution in Europe of African swine fever genotype II viruses from highly to moderately virulent. Veterinary Microbiology 219: 70-79.

Gallardo, C., Soler, A., Rodze, I., Nieto, R., Cano-Gómez, C., Fernandez-Pinero, J. and Arias, M., 2019. Attenuated and non-haemadsorbing (non-HAD) genotype II African swine fever virus (ASFV) isolated in Europe, Latvia 2017. Transboundary and Emerging Diseases 66: 1399-1404.

Gogin, A., Gerasimov, V., Malogolovkin, A. and Kolbasov, D., 2013. African swine fever in the North Caucasus region and the Russian Federation in years 2007-2012. Virus Research 173: 198-203. https://doi.org/10.1016/j.virusres.2012.12.007

Grigoryan, A., 2013. Head of Veterinary Service of Nagorno-Karabakh: spread of ASF was prevented. Available at: https://www.kavkaz-uzel.eu/articles/219762/.

Guinat, C., Gubbins, S., Vergne, T., Gonzales, J., Dixon, L. and Pfeiffer, D., 2016. Experimental pig-to-pig transmission dynamics for African swine fever virus, Georgia 2007/1 strain. Epidemiology and Infection 144: 25-34.

Gulenkin, V., Korennoy, F., Karaulov, A. and Dudnikov, S., 2011. Cartographical analysis of African swine fever outbreaks in the territory of the Russian Federation and computer modeling of the basic reproduction ratio. Preventive Veterinary Medicine 102: 167-174.

Heuschele, W.P and Coggins, L., 1965. Isolation of African swine fever from a giant forest hog. Bulletin of Epizootic Diseases of Africa 13: 255-256.

Iglesias, I., Munoz, M., Montes, F., Perez, A., Gogin, A., Kolbasov, D. and De la Torre, A., 2016. Reproductive ratio for the local spread of African swine fever in wild boars in the Russian Federation. Transboundary and Emerging Diseases 63: e237-e245.

Jori, F. and Bastos, A.D., 2009. Role of wild suids in the epidemiology of African swine fever. EcoHealth 6: 296-310. https://doi.org/10.1007/s10393-009-0248-7

Jori, F., Chenais, E., Boinas, F., Busauskas, P., Dholllander, S., Fleischmann, L., Olsevskis, E., Rijks, J., Schulz, K. and Thulke, H., 2020. Application of the World Café method to discuss the efficiency of African swine fever control strategies in European wild boar (*Sus scrofa*) populations. Preventive Veterinary Medicine 185: 105178.

Jori, F., Vial, L., Penrith, M.L., Perez-Sanchez, R., Etter, E., Albina, E., Michaud, V. and Roger, F., 2013. Review of the sylvatic cycle of African swine fever in sub-Saharan Africa and the Indian ocean. Virus Research 173: 212-227. https://doi.org/10.1016/j.virusres.2012.10.005

Korennoy, F., Gulenkin, V., Malone, J., Mores, C., Dudnikov, S. and Stevenson, M., 2014. Spatio-temporal modeling of the African swine fever epidemic in the Russian Federation, 2007-2012. Spatial and Spatio-temporal Epidemiology 11: 135-141.

Laddomada, A., Rolesu, S., Loi, F., Cappai, S., Oggiano, A., Madrau, M.P., Sanna, M.L., Pilo, G., Bandino, E. and Brundu, D., 2019. Surveillance and control of African swine fever in free-ranging pigs in Sardinia. Transboundary and Emerging Diseases 66: 1114-1119.

Lange, M., Guberti, V. and Thulke, H.H., 2018. Understanding ASF spread and emergency control concepts in wild boar populations using individual-based modelling and spatio-temporal surveillance data. EFSA Supporting Publications 15: 1521E.

Loi, F., Cappai, S., Coccollone, A. and Rolesu, S., 2019a. Standardized risk analysis approach aimed to evaluate the last African swine fever eradication program performance, in Sardinia. Frontiers in Veterinary Science 6: 299.

Loi, F., Laddomada, A., Coccollone, A., Marrocu, E., Piseddu, T., Masala, G., Bandino, E., Cappai, S. and Rolesu, S., 2019b. Socio-economic factors as indicators for various animal diseases in Sardinia. PloS One 14: e0217367.

Manso Ribeiro, J. and Rosa Azevedo, J., 1961. Reapparition de la peste porcine africaine au Portugal. Bulletin de l'Office International des Epizooties 55: 88-106.
Manso Ribeiro, J., Rosa Azevedo, J., Teixeira, M., Braco Forte, M., Rodrigues Ribeiro, A., Oliveira e Noronha, F., Grave Pereira, C. and Dias Vigario, J., 1958. An atypical strain of swine fever virus in Portugal. Bulletin de l'Office International des Epizooties 50: 516-534.
Marcon, A., Linden, A., Satran, P., Gervasi, V., Licoppe, A. and Guberti, V., 2020. R0 Estimation for the African swine fever epidemics in wild boar of Czech Republic and Belgium. Veterinary Sciences 7: 2.
Meijaard, E., d'Huart, J. and Oliver, W., 2011. Family Suidae (pigs). In: Wilson, DE and Mittermeier, R. (eds.) Handbook of the mammals of the world. Lynx Edicions, Barcelona, Spain, pp. 248-291.
Montgomery, E., 1921. On a form of swine fever occuring in British East Africa (Kenya colony). Journal of Comparative Pathology and Therapeutics 24: 159-191 (part I), 243-269 (part II).
Mulumba-Mfumu, L.K., Saegerman, C., Dixon, L.K., Madimba, K.C., Kazadi, E., Mukalakata, N.T., Oura, C.A., Chenais, E., Masembe, C. and Ståhl, K., 2019. African swine fever: update on Eastern, Central and Southern Africa. Transboundary and Emerging Diseases 66: 1462-1480.
Nielsen, J., Larsen, T., Halasa, T. and Christiansen, L.E., 2017. Estimation of the transmission dynamics of African swine fever virus within a swine house. Epidemiology and Infection 145: 2787-2796.
Nurmoja, I., Mõtus, K., Kristian, M., Niine, T., Schulz, K., Depner, K. and Viltrop, A., 2018. Epidemiological analysis of the 2015-2017 African swine fever outbreaks in Estonia. Preventive Veterinary Medicine 181: 104556
Olesen, A.S., Belsham, G.J., Bruun Rasmussen, T., Lohse, L., Bødker, R., Halasa, T., Boklund, A. and Bøtner, A., 2020. Potential routes for indirect transmission of African swine fever virus into domestic pig herds. Transboundary and Emerging Diseases 67: 1472-1484.
Olesen, A.S., Lohse, L., Boklund, A., Halasa, T., Belsham, G., Rasmussen, T. and Bøtner, A., 2018. Short time window for transmissibility of African swine fever virus from a contaminated environment. Transboundary and Emerging Diseases 65: 1024-1032.
Olesen, A.S., Lohse, L., Boklund, A., Halasa, T., Gallardo, C., Pejsak, Z., Belsham, G.J., Rasmussen, T.B. and Bøtner, A., 2017. Transmission of African swine fever virus from infected pigs by direct contact and aerosol routes. Veterinary Microbiology 211: 92-102.
Oļševskis, E., Guberti, V., Seržants, M., Westergaard, J., Gallardo, C., Rodze, I. and Depner, K., 2016. African swine fever virus introduction into the EU in 2014: experience of Latvia. Research in Veterinary Science 105: 28-30.
Oļševskis, E., Schulz, K., Staubach, C., Seržants, M., Lamberga, K., Pūle, D., Ozoliņš, J., Conraths, F.J. and Sauter-Louis, C., 2020. African swine fever in Latvian wild boar – a step closer to elimination. Transboundary and Emerging Diseases 67: 2615-2629.
Ordas, A., Sanchez-Botija, C., Bruyel, V. and Olias, J., 1983. African swine fever. The current situation in Spain. In: Wilkinson, P.J. (ed.) Coordination of Agricultural research. African swine fever. Eur 8466. Office for Official Publications of the European Communities, Luxembourg, Luxembourg, pp. 7-11.
Oura, C., Powell, P., Anderson, E. and Parkhouse, R., 1998. The pathogenesis of African swine fever in the resistant bushpig. Journal of General Virology 79: 1439-1443.
Pejsak, Z., Truszczyński, M., Kozak, E. and Markowska-Daniel, I., 2014. Epidemiological analysis of two first cases of African swine fever in wild boars in Poland. Medycyna Weterynaryjna 70: 369-372.
Penrith, M.L. and Vosloo, W., 2009. Review of African swine fever: transmission, spread and control. Journal of the South African Veterinary Association 80: 58-62.
Penrith, M.L., Bastos, A.D., Etter, E.M. and Beltrán-Alcrudo, D., 2019. Epidemiology of African swine fever in Africa today: sylvatic cycle versus socio-economic imperatives. Transboundary and Emerging Diseases 66: 672-686.

Perestrelo Vieira, R., 1993. Evolution of African swine fever in Portugal. Coordination of agricultural research. In: Galo, A. (ed.) Coordination of Agricultural research. African swine fever. Office for Official Publications of the European Communities, Luxembourg, Luxembourg, pp. 43-51.

Pérez, J., Fernández, A., Sierra, M., Herraez, P., Fernández, A. and de las Mulas, J.M., 1998. Serological and immunohistochemical study of African swine fever in wild boar in Spain. Veterinary Record 143: 136-139.

Pérez-Sánchez, R., Astigarraga, A., Oleaga-Perez, A. and Encinas-Grandes, A., 1994. Relationship between the persistence of African swine fever and the distribution of *Ornithodoros erraticus* in the province of Salamanca, Spain. Veterinary Record 135: 207-209.

Pietschmann, J., Guinat, C., Beer, M., Pronin, V., Tauscher, K., Petrov, A., Keil, G. and Blome, S., 2015. Course and transmission characteristics of oral low-dose infection of domestic pigs and European wild boar with a Caucasian African swine fever virus isolate. Archives of Virology 160: 1657-1667.

Plowright, W., 1981. African swine fever. In: Williams, E.S. and Barker, I.K. (eds.) Infectious diseases of wild mammals. Iowa State University Press, Ames, pp. 178-190.

Polo Jover, F. and Sanchez Botija, C., 1961. African swine fever in Spain. Bulletin de l'Office International des Epizooties 55: 107-147.

Probst, C., Gethmann, J., Amendt, J., Lutz, L., Teifke, J.P. and Conraths, F.J., 2020. Estimating the postmortem interval of wild boar carcasses. Veterinary Sciences 7: 6.

Rahimi, P., Sohrabi, A., Ashrafihelan, J., Edalat, R., Alamdari, M., Masoudi, M., Mostofi, S. and Azadmanesh, K., 2010. Emergence of African swine fever virus, northwestern Iran. Emerging Infectious Diseases 16: 1946.

Rowlands, R.J., Michaud, V., Heath, L., Hutchings, G., Oura, C., Vosloo, W., Dwarka, R., Onashvili, T., Albina, E. and Dixon, L.K., 2008. African swine fever virus isolate, Georgia, 2007. Emerging Infectious Diseases 14: 1870-1874. https://doi.org/10.3201/eid1412.080591

Sanchez Botija, C., 1982. African swine fever – new developments. Revue Scientifique et Technique 1: 1065-1094.

Sanchez-Vizcaino, J.M., Mur, L. and Martinez-Lopez, B., 2013. African swine fever (ASF): five years around Europe. Veterinary Microbiology 165: 45-50. https://doi.org/10.1016/j.vetmic.2012.11.030

Schulz, K., Conraths, F.J., Blome, S., Staubach, C. and Sauter-Louis, C., 2019a. African swine fever: fast and furious or slow and steady? Viruses 11: 866.

Schulz, K., Staubach, C., Blome, S., Nurmoja, I., Viltrop, A., Conraths, F., Kristian, M. and C, S.-L., 2020. How to demonstrate freedom from African swine fever in wild boar – Estonia as an example. Vaccines 8: 336.

Schulz, K., Staubach, C., Blome, S., Viltrop, A., Nurmoja, I., Conraths, F.J. and Sauter-Louis, C., 2019b. Analysis of Estonian surveillance in wild boar suggests a decline in the incidence of African swine fever. Scientific Reports 9: 8490.

Ståhl, K., Sternberg-Lewerin, S., Blome, S., Viltrop, A., Penrith, M.-L. and Chenais, E., 2019. Lack of evidence for long term carriers of African swine fever virus-a systematic review. Virus Research 272: 197725.

Thomson, G.R., 1985. The epidemiology of African swine fever: the role of free-living hosts in Africa. The Onderstepoort Journal of Veterinary Research 52: 201-209.

Tummeleht, L., Jürison, M., Kurina, O., Kirik, H., Jeremejeva, J. and Viltrop, A., 2020. Diversity of Diptera species in Estonian pig farms. Veterinary Sciences 7: 13.

Vepkhvadze, N., Menteshashvili, I., Kokhreidze, M., Goginashvili, K., Tigilauri, T., Mamisashvili, E., Gelashvili, L., Abramishvili, T., Donduashvili, M. and Ghvinjilia, G., 2017. Active surveillance of African swine fever in domestic swine herds in Georgia, 2014. Revue Scientifique et Technique 36: 879-887.

Vergne, T., Gogin, A. and Pfeiffer, D., 2017. Statistical exploration of local transmission routes for African swine fever in pigs in the Russian Federation, 2007-2014. Transboundary and Emerging Diseases 64: 504-512.

Vergne, T., Korennoy, F., Combelles, L., Gogin, A. and Pfeiffer, D.U., 2016. Modelling African swine fever presence and reported abundance in the Russian Federation using national surveillance data from 2007 to 2014. Spatial and Spatio-temporal Epidemiology 19: 70-77.

Vigario, J.D. and Caiado, J., 1989. Situazione epidemiologica in Portogallo. In: Peste Suina Africana. Istituto Zooprofilattico Sperimentale Sardegna, Nuoro, Italy, pp. 123-132.

Wilkinson, P., 1984. The persistence of African swine fever in Africa and the Mediterranean. Preventive Veterinary Medicine 2: 71-82.

Wilkinson, P., Wardley, R. and Williams, S., 1983. Studies in pigs infected with African swine fever virus (Malta/78). CEC/FAO Expert Consultation on African Swine Fever Research. EEC Publication EUR 8466 EN, Sardinia, Italy, pp. 74-84.

Wilkinson, P.J., Dixon, L., Sumption, K., Ekue, F., Hutchings, G., Payne, A. and Boinas, F., 1993. Genetic variation and epidemiology of African swine fever in Europe and Africa. In: International Congress of Virology, Glasgow, UK, p. 223.

Zani, L., Dietze, K., Dimova, Z., Forth, J.H., Denev, D., Depner, K. and Alexandrov, T., 2019. African swine fever in a Bulgarian backyard farm – a case report. Veterinary Sciences 6: 94.

Zani, L., Forth, J.H., Forth, L., Nurmoja, I., Leidenberger, S., Henke, J., Carlson, J., Breidenstein, C., Viltrop, A. and Höper, D., 2018. Deletion at the 5'-end of Estonian ASFV strains associated with an attenuated phenotype. Scientific Reports 8: 6510.

Zhou, X., Li, N., Luo, Y., Liu, Y., Miao, F., Chen, T., Zhang, S., Cao, P., Li, X. and Tian, K., 2018. Emergence of African swine fever in China, 2018. Transboundary and Emerging Diseases 65: 1482.

10. Biosecurity measures against African swine fever in domestic pigs

M. Martínez[1], A. de la Torre[1], J.M. Sánchez-Vizcaíno[2] and S. Bellini[3*]
[1]Animal Health Research Centre, National Institute for Agricultural and Food Research and Technology (INIA-CISA), Carretera Algete-El Casar de Talamanca, Km. 8,1, 28130 Valdeolmos, Madrid, Spain; [2]VISAVET Health Surveillance Centre and Animal Health Department, Veterinary Faculty, Complutense University of Madrid, Av. Puerta de Hierro, 28040 Madrid, Spain; [3]Istituto Zooprofilattico Sperimentale della Lombardia e dell'Emilia Romagna (IZSLER), Via Bianchi, 9 25124 Brescia, Italy; silvia.bellini@izsler.it

Abstract

During the past decade, African swine fever (ASF) has spread from the Caucasus region to eastern Europe and Asia, affecting domestic pig and wild boar populations. In the absence of an effective vaccine or treatment for ASF, the spread of ASF onto domestic pig farms can only be prevented by strict compliance with biosecurity measures. Many ASF outbreak investigations have reported biosecurity shortcomings as a critical element for virus introduction and spread. The structure of the European swine industry makes it necessary to put in place differentiated biosecurity measures, in order to meet the different risk levels of introduction and spread of ASF among the diversified farming systems. Biosecurity programmes are normally applied to commercial holdings. Indeed, non-commercial farming is often seen as a dead end in terms of disease transmission; however, farms that sell animals locally or regionally can play an important role in the spread of disease. This is the reason why all holdings that have access to markets should be included in such biosecurity programmes. Here we summarise the main principles and include information targeted to backyard holdings. If strictly implemented such measures can prove effective in minimising the risk of an ASF outbreak.

Keywords: ASF, pigs, biosecurity

This publication is based upon work from COST Action CA15116, ASF-STOP, supported by COST (European Cooperation in Science and Technology).
www.cost.eu

EUROPEAN COOPERATION IN SCIENCE & TECHNOLOGY

Funded by the Horizon 2020 Framework Programme of the European Union

Laura Iacolina *et al.* (eds.) **Understanding and combatting African swine fever**
DOI 10.3920/978-90-8686-910-7_10,

10.1 Introduction

The global spread of African swine fever (ASF) worldwide has placed a large part of the world's swine population under threat. One of the major challenges for the control of ASF is that there are no safe vaccines or treatment available to provide adequate protection of domestic and wild pigs. Therefore, farm biosecurity and good farming practices are considered the most effective tool for preventing ASF introduction to pig herds.

Biosecurity is one of the key prevention tools at the disposal of operators and others working with animals to prevent the introduction, development and spread of transmissible animal diseases to, from and within an animal population. This implies that biosecurity measures shall take into consideration the epidemiology of the disease, virus resistance in the environment, routes of transmission and excretion as well as the characteristics of the farming systems in place (Bellini *et al.*, 2016) and the health situation of the territory. Furthermore, the measures to adopt should be sufficiently flexible, suit the type of production and the categories of animals involved and take into account the local circumstances and technical developments.

The following are the three main components of biosecurity (FAO/OIE/World Bank, 2010):
- Segregation/separation: the creation and maintenance of barriers to limit the potential opportunities for infected animals and contaminated materials to enter a farm and when properly applied, this step will prevent most contamination and infection.
- Cleaning: objects (e.g. vehicles, equipment) that have to enter (or leave) a farm must be thoroughly cleaned to remove visible dirt. This will also remove most of the pathogens that contaminate the materials.
- Disinfection: when properly applied, disinfection will inactivate any pathogen that is present on materials that have already been thoroughly cleaned.

In pig farming, biosecurity measures are normally targeted to commercial holdings, whilst they are not well defined or easy to implement in other types of holdings where the investment in infrastructures is minimal and few arrangements are made to provide the pigs with housing. The level of professionalism varies according to farm size, productive orientation, available economic resources and motivation or awareness among farmers. In the current ASF epidemiological situation in Europe, many outbreak investigations have reported biosecurity shortcomings and poor early detection as a critical factor for virus introduction and spread. The improvement of biosecurity in the backyard sector represents an important challenge for the farmers, due to the sector's heterogeneity and the costs of its implementation and maintenance. However, considering the current ASF situation, at least all pig holdings that have access to markets should be included in the biosecurity plan. Although non-commercial farms which, by definition, should not sell pigs and products, could be considered a dead end in terms of disease transmission, backyard units that sell animals at local or regional level can play a relevant role in the local spread of ASF, also contributing to the endemicity of the disease.

In the framework of a control strategy, the farms' biosecurity plans have to be verified by veterinary services and because of the virus ability to survive in the environment, the plans shall also include detailed procedures on cleaning and disinfection, which are an essential part of the farm biosecurity plan. See Chapter 11 for details on cleaning and disinfection.

Pig producers need to understand the risk posed by the presence of the disease and they have to adopt all the necessary precautionary measures to protect their own herds. To achieve this, veterinary services shall provide basic information to pig holders through appropriate communication campaigns and by promoting the adoption of preventive measures. The key for changing behaviours and practices lies in people's perception of the level of risk and it is of crucial importance that all the elements of the pig production chain are involved and aware of the pivotal role played by biosecurity in the implementation of the control strategy for the disease. Biosecurity may require some upfront investment; however, the resulting reduction in animal disease should be a positive incentive for operators. It should also be considered that the implementation of the basic requirements confers a barrier that contributes to prevent the spread of other main transmissible diseases as well.

10.2 Relevant measures to prevent the spread of African swine fever in the domestic pig sector

In the European Union (EU), the measures for the control of ASF are reported in Council Directive 2002/60/EC (EC, 2002), which establishes the measures to be adopted in case of ASF occurrence. The eradication measures are applied in combination with strict quarantine and biosecurity measures on domestic pig holdings and animal movement control. Specific regionalisation measures are laid down in Commission Decision 2014/709/EU (EC, 2014), which establishes control measures on the movement of pigs and certain pig products from the areas at risk of infection in order to prevent the spread of ASF to other areas of the EU. These measures also include the application of a specific biosecurity plan, approved by the Veterinary Authorities. This means that biosecurity has a recognised crucial role in preventing the introduction of the virus, managing the outbreaks, stemming the spread of diseases and finally facilitating trade.

Within the COST (European Cooperation in Science and Technology) action: 'Understanding and combating African swine fever in Europe' (ASF-STOP), a systematic review targeting preventive measures to avoid the spread of ASF in the domestic pig sector was conducted with the final aim of developing biosecurity guidelines for the different pig farming systems. The search was mainly addressed to the epidemiological characteristics of the current scenario in the EU and took into account the different pig farming systems present. Indeed, pig production in the EU is highly heterogeneous regarding farm type (industrialised, outdoor or backyard), biosecurity standards and production levels (commercial, own consumption). In most countries, the different pig production systems cohabit, from the simplest system with minimal investment, to large-scale market-oriented enterprises. For more details on the characteristics of the pig sector in the EU, see Chapter 7.

The biosecurity guidelines were developed considering the classification of pig holdings as established in the EU strategic approach for the management of ASF. This document was developed with the aim of establishing harmonised measures in response to the ASF epidemiological situation.

Based on the Working Document of the Directorate General for Health and Food Safety (DGSANTE) (EC, 2015) pig farms are classified as:

A. Non-commercial farms (NCM): farms where pigs are kept only for fattening for own consumption and neither pigs nor any of their products leave the holding.
B. Commercial farms (CM): farms which sell pigs, send pigs to a slaughterhouse or move pig products off the holding.
C. Outdoor farms (OD): farms in which pigs are kept temporarily or permanently outdoors.

In the preparation of the guidelines, as a first step, a systematic review of the documentation (scientific literature, handbooks, national and international regulations, technical guidelines) on the measures to prevent the spread of ASF in the domestic pig sector was conducted. The search was mainly focused on the current epidemiological context. The identified preventive measures (Table 10.1) were classified in four different groups: general preventive measures (measures in common for commercial, non-commercial and outdoor farms), and three groups of suggested measures for each of the identified types of farms. For this purpose, a questionnaire accompanied by general instructions was prepared and experts were asked to assess the relevance of each described preventive measure for the different production systems by answering yes or no to the closed question. A group of experts was invited to participate in an expert opinion session to assess the relevance of the different preventive measures identified in the systematic review. The experts were also encouraged to suggest additional measures if they thought any were missing. Based on the outcome of the assessment, guidelines for commercial, non-commercial and outdoor farms were produced as well as flyers for the pig producers. The results of the experts' assessment on the list of measures are shown in Table 10.2.

The systematic review process yielded 52 pieces of literature from which preventive measures were classified into four groups:

- General preventive measures applicable to all type of farms: 12 measures.
- Preventive measures for commercial farms (CM): additional 8 specific measures.
- Preventive measures for non-commercial farms (NCM): additional 13 specific measures.
- Preventive measures for outdoor farms (OD): additional 4 specific measures.

The complete methodology and results of the study are available in the publication by Jurado *et al.* (2018b).

All experts agreed that these three measures were relevant for ASF prevention on the three types of pig farm:

- Identification of animals and farm records including animal movements.
- Strict enforcement of the ban on swill feeding.
- Containment of pigs, to avoid contact with pigs from other farms, feral pigs or wild boar or their products.

In addition, on commercial farms, these measures were considered relevant by all experts:

- Check ASF-free certificates and health status before acquiring new animals as well as semen, ova or embryos on breeding farms.
- Limited farm visits with proper register and establishment of biosecurity measures regarding footwear and clothing.

Table 10.1. Identified preventive measures from the systematic review described for commercial, non-commercial outdoor and general (commercial, non-commercial and outdoor) pig farms.

ID	Preventive measures	Type of farm[1]
1	ASF-free certificates and health status check before acquiring new animals, semen, ova or embryos on breeding farms	General
2	Limit farm visits. Register and biosecurity (footwear and clothing)	General
3	Farmers/workers and operators education	General
4	Farmers/workers should not have contact with external pigs	General
5	Fences to prevent contacts with external pigs and wild boar	General
6	Removal of carcasses, slaughter residues and food waste	General
7	Discouragement of sharing used equipment between units	General
8	Use of footbaths in entrance of units where animals are held	General
9	Daily health checks for clinical signs and mortality rates	General
10	Cleaning and disinfection protocols for facilities and equipment	General
11	Farm location far from suitable wild boar areas and close to geographical barriers	General
12	Control measures against flies	General
13	Establish clean/dirty areas (including changing rooms and showers)	CM
14	Arrange logistics for entry and exit of animals including protocols for entrance of vehicles, loading areas, role of pig transporters, etc.	CM
15	Cleaning and disinfection protocols for transport vehicles	CM
16	Quarantine period for purchased animals and quarantine rooms	CM
17	Identification of animals and farm records and animal movements	CM
18	Internal audits and evaluations to enforce biosecurity measures	CM
19	Rules for food for staff entering the farm (i.e. restricted to eating rooms or not allowed)	CM
20	Proper disposal of manure and dead animals	CM
21	Strict enforcement of the ban on swill feeding	NCM
22	Containment of pigs, do not allow contact with pigs from other farms, feral pigs or wild boar or their products	NCM
23	Farmers/farm staff should not have hunted in ASF risk areas, allowing a 48 hours interval between hunting and being in contact with domestic pigs	NCM
24	Effective disinfection and cleaning of the surrounding of the holding including its entrance	NCM
25	Veterinary supervision prior and while home slaughtering	NCM
26	Cleaning and disinfection protocols before and after home slaughter	NCM
27	No sows or boars used for mating purposes held on non-commercial farm	NCM
28	No movements between/from non-commercial farms	NCM
29	Avoid use of fresh fodder in areas at risk of exposure to ASF	NCM
30	Promote educational programmes through governmental training programmes and improve access to health services	NCM
31	Treatment and storage (out of reach of wild boars) of grass or grains for at least 30 days or prohibit its use	NCM
32	Avoid the use of straw bedding unless treated to inactivate ASF and stored for at least 90 days	NCM
33	No exchange of feed or bedding with other farms	NCM
34	Banning of free-range management on communal areas or public forests with no biosecurity measure	OD
35	Distance between outdoor farms (at least 1 km) to minimise the risk of ASF introduction through direct and indirect contact	OD
36	In the presence of *Ornithodoros* ticks avoid using traditional pig-housing facilities (usually made of wood and stones where ticks can be hidden)	OD
37	Apply chemical control if ticks are present in traditional pig-housing facilities	OD

[1] CM = commercial; NCM = non-commercial; OD = outdoor.

Table 10.2. Preventive measures against African swine fever (ASF) ranked as relevant (% agreement) as per results of the expert elicitation for each type of pig farm (modified from Table 9.2 of Jurado *et al.*, 2018b).[1]

ID	Preventive measures	CM	NCM	OD
1	ASF-free certificates and health status before trade	100	83	100
2	Farm visits: limited, registered and with biosecurity for footwear and clothing	100	83	92
3	Farmers/workers and operators' education	100	92	100
4	Farmers/workers should not contact external pigs	92	92	92
5	Perimeter fences	92	67	100
6	Appropriate removal of carcasses, residues and waste	100	92	100
7	Discouragement of sharing used equipment	100	83	100
8	Use of footbaths at entrance of animals' premises	75	50	33
9	Daily health checks (clinical signs and mortality rates)	100	58	92
10	Cleaning and disinfection protocols for facilities, vehicles and equipment	100	50	92
11	Farm location	67	58	75
12	Control measures against flies	75	42	17
13	Establishing clean/dirty areas	92	58	67
14	Entry and exit (animals, vehicles, loading areas, role of pig transporters, etc).	100	33	75
15	Cleaning and disinfection protocols for transport vehicles	100	58	83
16	Quarantine for purchased animals and quarantine rooms	100	25	67
17	Identification of animals and farm records including animal movements	100	100	100
18	Internal audits and evaluations to enforce biosecurity measures	92	25	67
19	Rules for food for staff entering the farm (i.e. restricted to eating rooms or not allowed)	100	58	75
20	Proper disposal of manure and dead animals	100	92	92
21	Strict enforcement of the ban on swill feeding	100	100	100
22	Containment of pigs: no contacts allowed with other pigs or their products	100	100	100
23	48 hours interval between hunting and being in contact with domestic pigs in an ASF infected/risk wild boar area, or banning hunting altogether	92	92	92
24	Cleaning and disinfection around the holding including its entrance	58	50	42
25	Ante-mortem vet supervision and during home slaughter	50 (NA:33)	83	67 (NA:16.5)
26	Cleaning and disinfection protocols before and after home slaughter	42 (NA:25)	83	67 (NA:8)
27	No sows or boars for mating purposes on NCM farms	58	67 (NA:8)	42 (NA:8)
28	No movements between/from NCM farms	75 (NA:17)	92	67 (NA:16.5)
29	Avoid use of fresh fodder in high risk ASF areas	97	75	75
30	Training programmes and improved access to health services	92	100	100
31	Treatment and storage (out of reach of wild boars) of grass or grains for at least 30 days or prohibit its use	75	75	67
32	Straw bedding: treatment to inactivate ASF and stored for at least 90 days, or avoid its use	83	83	83
33	No exchange of feed or bedding with other farms	92	92	92
34	Banning of free-range management on communal areas or public forests with no biosecurity measure	67 (NA:16.5)	92	92
35	Distance between outdoor farms (at least 1 km)	16 (NA:42)	33 (NA:25)	67
36	If *Ornithodoros* ticks present, avoid using traditional pig-housing facilities (usually made of wood and stones where ticks can be hidden)	84 (NA:8)	75 (NA:8)	84 (NA:8)
37	Apply chemical control if ticks were present in traditional pig-housing facilities	84 (NA:8)	92 (NA:8)	92 (NA:8)

[1] CM = commercial; NA = not applicable; NCM = non-commercial; OD = outdoor.

- Farmers/workers and operators education.
- Appropriate removal of carcasses, slaughter residues and food waste.
- Discouragement of sharing used equipment between holdings and/or units.
- Daily health checks for clinical signs and mortality rates.
- Cleaning and disinfectant protocols for facilities, vehicles and equipment.
- Logistical arrangement for the entry and exit of animals, including protocols regarding entrance of vehicles, loading areas, role of pig transporters, etc.
- Cleaning and disinfection protocols for transport vehicles.
- Quarantine period for purchased animals and quarantine rooms.
- Rules for food for staff entering the farm (i.e. restricted to eating rooms or not allowed).
- Proper disposal of manure and dead animals.

For non-commercial and outdoor farms, in addition, all experts agreed on the promotion of educational programmes through governmental training programmes and improve access to health services as a relevant measure. For outdoor farms the following additional measures were also relevant for all experts:
- Check ASF-free certificates and health status before acquiring new animals as well as semen, ova or embryos on breeding farms.
- Farmers/workers and operators education.
- Perimeter fences to prevent contacts with external pigs and wild boar.
- Appropriate removal of carcasses, slaughter residues and food waste.
- Discouragement of sharing used equipment between holdings and/or units.

Some experts suggested other measures not included in the list, such as:
- Netting animal facilities.
- Pest control programmes.
- Use of carbonic dioxide traps to check for *Ornithodoros* ticks.
- Change of boots before entering the farm or farm units.

Several experts emphasised the importance of measures already considered in the list, such as:
- Establishing double fencing perimeter on outdoor farms.
- Education of swine veterinarians and farmers, particularly on clinical signs and transmission routes.
- Discourage sharing injection syringes and instruments unless thoroughly disinfected and sterilised between uses.

In conclusion, the three main areas of prevention that were highlighted in this work include:
1. Control of movements into and from the farm.
2. Control of pigs' feed.
3. Improvement of health services and education.

The above reviewed measures, both from the systematic review, the Working Document of DGSANTE 'Strategic Approach to the management of African swine fever for the EU SANTE/7113/2015-Rev 12' (EC, 2015) and the expert opinion elicitation, defined a comprehensive list of criteria and biosecurity measures that should be applied in order to increase the preparedness of each country to face the prominent menace of ASF introduction and/or diffusion in commercial farms, in outdoor farms and in non-commercial farms.

10.3 Minimum biosecurity measures for commercial pig farms

Commercial farms (Figure 10.1) are those farms which sell pigs, send pigs to a slaughterhouse or move pig products off the holding (EC, 2015).

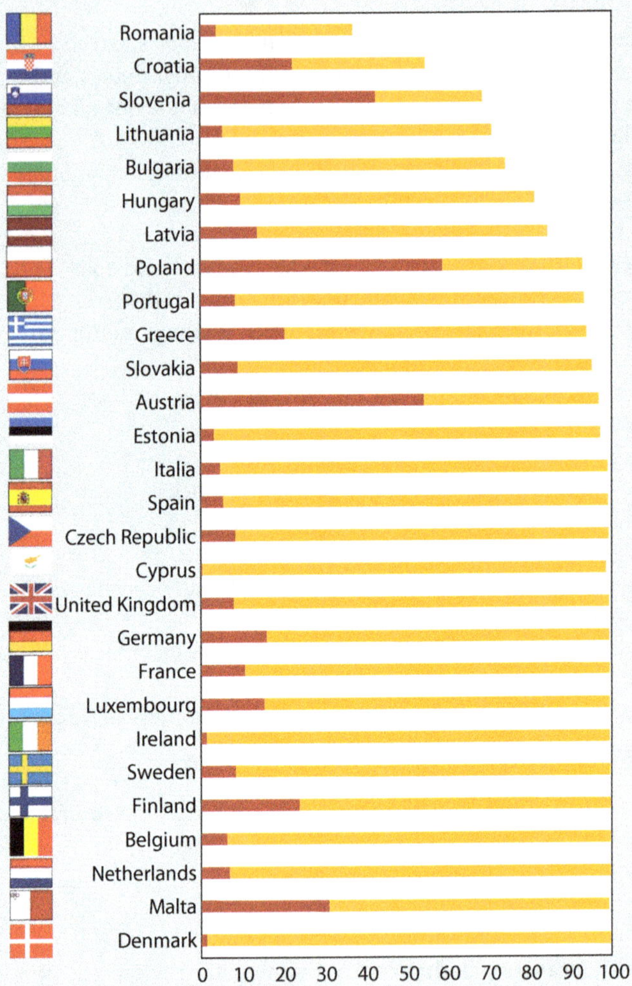

Figure 10.1. Commercial pig farms in the EU. Farms keeping animals in medium-small units of 10-400 pigs (in red) and farms keeping animals in medium-large units of >400 pigs (in orange) as % of country's total farms (EUROSTAT, 2014).

Commercial farms can be divided into multi-site farms which are holdings specialised on one production step (farrowing, nurseries, or finishing) and farrow-to-finish farms which are premises that produce all production steps. The minimum biosecurity measures against ASF in commercial farms are summarised below.

10.3.1 Farm location

Farms should be located far from suitable wild boar areas (EFSA, 2015, Bosch *et al.*, 2017) and other pig farms. The minimum distance between farms depends on the national and local regulations. Farms should maintain distance from frequently used roads, dumping areas and slaughterhouses. Location should consider wind currents that may transport infectious agents.

10.3.2 Structural biosecurity

Cleaning and disinfection procedures for vehicles (see Chapter 11). Drivers should not come into contact with farm workers nor with animals. Deliveries should happen without entering the farm. Decontamination of vehicles should be done before or upon arrival, before entering the farm. Parking areas should avoid cross-contamination of workers' and farm vehicles. Loading and unloading areas should be at least 20 m from animal facilities (Astorga *et al.*, 2016), should be built with materials easy to clean and disinfect and with 1% negative inclination (materials tend to drop outside the farm). Animals that exit the farm should not return. Newly purchased animals should be confined for at least 30 days (Bellini *et al.*, 2016) and be inspected by veterinarians. If applicable, animals may be serologically and virologically tested against targeted diseases. It is recommended to have isolation rooms for infectious/sick animals. Netting against birds and insects (e.g. flies) should be installed. Pest control should be periodically conducted. The presence of domestic animals (e.g. dogs or cats) should be discouraged. Hand-washing facilities and changing rooms are required, showers are needed for higher biosecurity. Footbaths should be at the entrance of every animal unit (see Chapter 11). The farm perimeter should be fenced (double fenced, preferably) and the entrance closed (EFSA, 2014a). Materials easy to clean and disinfect should be used everywhere.

10.3.3 Animal health status

Health status and free-ASF certificates have to be checked before acquiring animals. Semen, ova and embryos should come from approved collection centres as they have been suggested to be a potential source of ASF infection (EC, 2016; Thacker *et al.*, 1984) although this has not been scientifically proven (Blitvich *et al.*, 2020; Maes *et al.*, 2008). Introducing animals lacking a certificate showing their origin should be avoided as untraceable animals are one of the likely causes of ASF entry (EC, 2016).

10.3.4 All-in-all-out system

All-in-all-out in single-site and multi-site farms should be implemented. Cleaning and disinfection protocols should be applied after moving out animals followed by a resting period between batches. All organic material (faeces, feed, bedding materials) should be completely removed.

10.3.5 Good farming practices

Cleaning and disinfection protocols should be established and periodically performed on every farm facility, vehicle and tool. Workers and farm owners should follow the described biosecurity measures and should not have swine herds or pig pets at home. No food should be brought onto animal facilities. Sharing material between animals, especially needles, should be avoided. Any equipment (e.g. veterinarian ultrasound apparatus) that is moved from farm to farm should be cleaned and disinfected before entering the farm. Fresh fodder harvested in areas at risk for ASFV exposure should be avoided (EFSA, 2015). ASF virus may persist in protein-rich environments (FAO, 2001), thus bedding materials should originate from ASF free areas and fields where no slurry has been previously spread and there is no possibility of contamination with ASF infected wild boar. Proper disposal of manure as well as dead animals and other removable material should be ensured. Containers and storage basins should be located far from animal facilities. The Spanish Regulation states that this distance should be at least 100 m (RD, 2000). Vehicles used for the purpose should not enter the farm or even pass near the farm (if possible). Proper management and storage of animal feed to avoid cross-contamination.

10.3.6 Visits

Visits should be discouraged, if visitors enter the farm, they should follow strict biosecurity measures. Visits and visitors should be recorded on the farm record book. Workers and owners should also avoid visiting other pig farms. Veterinarians should follow strict biosecurity protocols as they come into contact with susceptible animals from different farms. Veterinary instruments and equipment should be properly cleaned and disinfected if they are shared between farms.

10.3.7 Awareness and training of farm staff

Training and raising awareness should be conducted and updated according to the level of risk. Both are key factors for the early suspicion and detection of ASF.

10.3.8 Farm records

Animals should be identified and all animal movements recorded. Births and deaths, animal census, entry and exit of animals (live and dead), vehicles, visits, pest control, and cleaning and disinfection procedures should all be properly registered. Protocols for each production phase should be clearly defined, including movement flows within the farm holdings.

Furthermore, in order to have the most complete view on the application of the guidelines mentioned above in Europe, participants of ASF-STOP were encouraged to share their country-specific biosecurity measures applied in commercial farms. Contributions from nine countries were received.

The overall list of biosecurity measures implemented in the various countries participating in ASF-STOP is comprehensive and representative of the needs that emerged from the literature systematic review and the expert opinion advice (Figure 10.2).

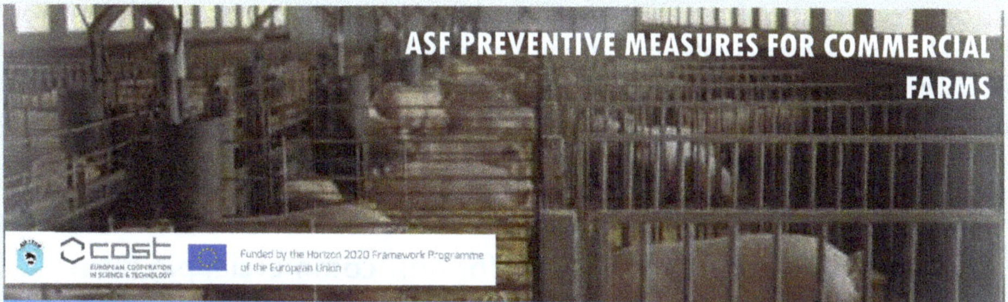

1. FARM LOCATION far from suitable wild boar areas, neighbouring pig farms and frequently used roads. Farm perimeter fenced (double fenced) and entrance closed.
2. STRUCTURAL BIOSECURITY
   - Cleaning and disinfection procedures for vehicles. Drivers should not come in contact with farm workers/animals. Loading and unloading areas placed at least 20 m away from animal facilities within the perimeter of the farm, built with materials easy to clean and disinfect and 1% negative inclination.
   - Quarantine rooms for new animals (14-30 days). Animals inspected by veterinarians (serologically and virologically if possible). Isolation rooms for infectious/sick animals.
   - Nets against birds/insects. Pest control periodically conducted. Domestic animals should be discouraged.
   - Hand-washing facilities/changing rooms/(showers). Footbaths at the entrance of every animal unit. Organic material should be removed from footwear prior to disinfecting.
   - Use of materials in structures or buildings that facilitate cleaning and disinfection procedures.
3. ANIMAL HEALTH STATUS. Health status and free-ASF certificates have to be checked before acquiring those animals, semen, ova and embryos. Buying pigs at local markets and illegally is a high risk practice.
4. ALL-IN-ALL-OUT SYSTEM. Cleaning and disinfection protocols after moving out animals + fallow period. All organic material should be completely removed to maximise the efficacy of disinfection.
5. GOOD FARMING PRACTICES.
   - C&D protocols should be established and periodically performed on every farm facility, vehicle and tool.
   - Workers and farm owners should follow the described biosecurity measures with regard to handwashing, showering, changing of clothes and footwear. They should not have swine herds or pig pets at home.
   - No food should be brought onto animal facilities.
   - Sharing material between animals, especially needles, should be avoided. Any equipment like ultrasound apparatus that usually is owned by the veterinarian and moved from farm to farm should be cleaned and disinfected before entering the farm.
   - Straw and bedding materials should come from ASF free areas where no slurry has been used as fertilizer and there is no possibility of contamination with ASF infected wild boar.
   - Proper disposal of manure and dead animals. Containers located far from animal facilities. Vehicles should access them without entering the farm or even passing near the farm (if possible).
   - Proper management and storage of animal feed to avoid cross-contamination.
6. VISITS should be discouraged. They (e.g. veterinarians) should follow strict biosecurity measures and should be recorded on the farm record book. Veterinary instruments and equipment should be properly cleaned and disinfected. Workers and owners should also avoid visiting other pig farms.
7. AWARENESS AND TRAINING OF FARM STAFF should be raised and updated according to the level of risk.
8. FARM RECORDS (births and deaths; animal census and movements; visits; pest control or C&D procedures) should be registered in a farm record book. Protocols for each production phase should be clearly defined.

Biosecurity measures are based on the latest working document elaborated by the Directorate General for Health and Food Safety, scientific literature systematic review and expert opinion under the COST action ASFSTOP (EU One Health EJP). Funding: EU's Horizon 2020 research and innovation programme (Grant Agreement No 773830).

Figure 10.2. Flyer on African swine fever Biosecurity Guidance in commercial farms distributed by ASF-STOP members. Available at https://www.asf-stop.com/dissemination/

In some countries along with the laws that list the actions to be implemented by farmers, a formal control system was put in place in order to constantly evaluate the overall biosecurity set-up of the farms and their risk level of introduction and spread of ASFV. Making such a system accessible to official veterinarians, farm veterinarians and farmers improves and facilitates cooperation and dialogue between farmers and the Competent Authority, increasing the level of awareness and preparedness needed to face ASFV threats.

10.4 Minimum biosecurity measures for non-commercial pig farms

Non-commercial farms are those farms where pigs are kept only for fattening for own consumption and neither pigs nor any of their products leave the holding (EC, 2015). In the current epidemiological context, backyard holdings with poor biosecurity in place are playing an important role in the spread of ASFV (EFSA, 2014b). They are common in Georgia, the Russian Federation, Poland, Romania and Bulgaria and to a varying extent in the Baltic countries (Figure 10.3). These kinds of farms are characterised by poor farming practices and insufficient biosecurity, which allow animals to have access to kitchen leftovers or even slaughter disposal. Moreover, depending on the country and local practices, pigs are allowed to free range during the day or even scavenge for days or months. Pig slaughter is usually carried out on the farm, although it may be restricted to proper slaughterhouses if there are local regulations (ASFORCE, 2015).

The higher density of pigs in small-scale confined production systems leads to a higher risk of pathogen circulation among herds. In this context, generally pigs are purchased from several sources, such as other small pig producers, from local markets or abroad, and segregation and cleaning and disinfection are often poorly implemented. These are all unsafe practices that increase the risk of spreading pathogens (FAO, 2010). Small-scale pig producers are normally involved in local trade and, in case of disease occurrence, these activities might result in the spread of pathogens locally and eventually also to the wild boar.

Based on the expert opinion elicitation (Jurado *et al.*, 2018b) and the latest working document elaborated by the Directorate General for Health and Food Safety (EFSA, 2014b), there is a set of basic preventive measures applicable also to backyard holdings and, if they are properly and strictly implemented, they can prove effective in minimising the risk of ASFV spread.

10.4.1 General biosecurity measures for non-commercial farms

- Buying pigs from trusted and certified sources (ASFV-free commercial holdings).
- No swill feeding.
- Ensure identification of holdings, pigs and their traceability.
- Workers should not bring food onto the premises.
- The access to pig's stables should be restricted only to people in charge of taking care of the animals.
- Pigs' owners and people in charge of pigs shall avoid visiting other farms.
- Pigs shall be kept confined in stables: pigs should be kept in a way that ensures that there is no direct or indirect contact with other pigs outside the premises nor with wild boar.

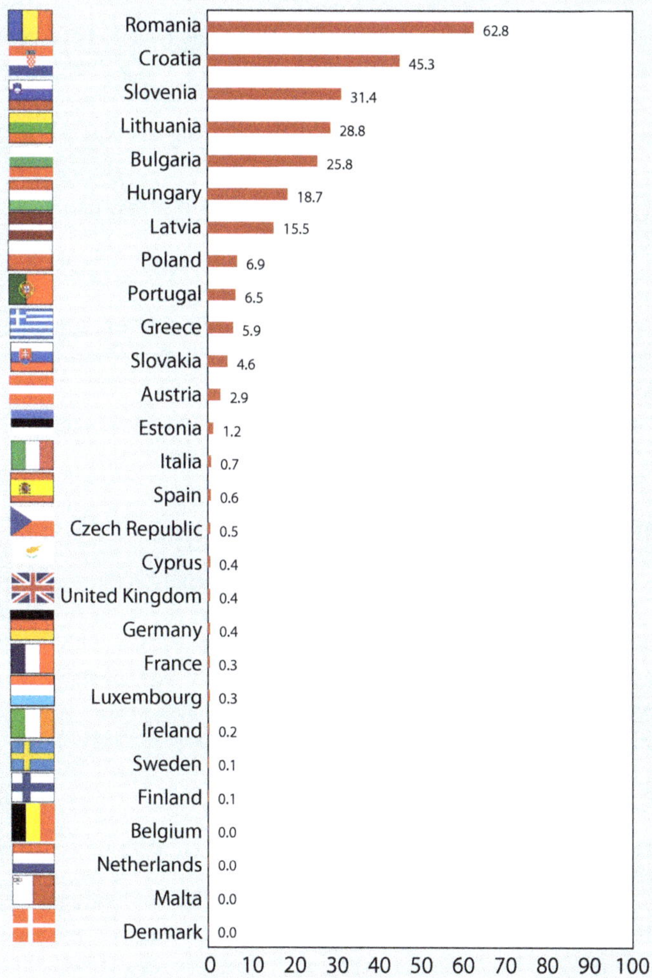

Figure 10.3. Non-commercial pig farms in the EU (farms keeping animals in small units of 10 or fewer pigs (as % of country's total farms) (Adapted from EUROSTAT, 2014).

- Farm buildings should: (1) be built in such a way that no feral pigs or other animals (e.g. dogs) can enter the stable; (2) allow for disinfection facilities (or changing) for footwear at the entrance into the stable.
- Effective disinfectants should be placed at the entrance of the stable.
- No sows or boar used for natural reproduction should be allowed on the holding. In case pigs intended for breeding are kept on the holding, biosecurity should follow the same indications foreseen for commercial farms.
- People working in contact with pigs should wear dedicated clothes and footwear that should be left in the stable after use.
- People working in contact with pigs should wash hands with soap before entering and leaving the premises.

- Proper disposal of dead animals or parts of dead animals to avoid the spread of infected material and attracting wild animals.
- No wild boar or part of it, shall be brought onto the premises.
- No hunting activity should be carried out 48 h prior to being in contact with pigs.

10.4.2 Biosecurity measures if locally harvested grass and straw represent a risk

Commercially traded crops, vegetables, hay and straw have a very low ability to contain and maintain infectious ASFV (EFSA, 2014b). However, if the use of locally harvested grass and straw is considered to represent a risk under local prevailing conditions the following should apply:

- Ban on feeding fresh grass or grains to pigs unless treated to inactivate ASFV or stored (out of reach of wild boar) for at least 30 days before feeding.
- Ban on using straw for bedding of pigs unless treated to inactivate ASFV or stored (out of reach of wild boar) for at least 90 days before use.

10.4.3 Additional biosecurity measures regarding veterinary inspection

All experts involved in the expert opinion elicitation identified as an important preventive measure for non-commercial farms the improvement of access to the farms to veterinarians and veterinary services (Figure 10.4). This allows verification of compliance with the rules established in the eradication strategy and enhances ASFV early detection. Therefore:

- Veterinary services should be informed in case of dead or sick pigs.
- Home slaughter should be carried out only under veterinary supervision.

10.5 Minimum biosecurity measures for outdoor pig holdings

There are a variety of outdoor pig holdings within the EU, with also a wide variety of biosecurity standards. There is organic and traditional farming, as well as outdoor pig raising to elaborate a product with strict quality standards for every part of its process. However, in ASF infected countries and considering the involvement of wild boar in the spread of ASF in the EU, outdoor pig farming is considered a risk practice and it is not allowed in infected areas.

The biosecurity measures to prevent ASF in outdoor farming should consider location and proximity or connectivity to ASF outbreaks, the environment and the characteristics of the pig farms, to tailor them to the specific needs and particularities of each scenario.

Since 1978 ASF has been present in Sardinia and grazing free-range herds are considered a major obstacle to disease control (Jurado *et al.*, 2018a). In the island, raising pigs outdoors is a traditional husbandry practice, mainly practised in mountainous and hilly areas where the disease is more difficult to control. At present, keeping pigs outdoors is forbidden in Sardinia.

In the past epidemic of ASF in Europe that started in the 1960s, there were areas in the Iberian Peninsula that remained endemic for 30 years. In these areas, an ASF cycle was maintained between outdoor pig farming, wild boar and a poor infrastructure in pig farm buildings that allowed the persistence of ASF virus through the involvement of an *Ornithodoros erraticus* tick.

Figure 10.4. Flyer on African swine fever Biosecurity Guidance in non-commercial pig-production systems distributed by ASF-STOP members (https://www.asf-stop.com/dissemination/).

Experimentally it has been shown that such ticks are able to remain infective for more than five years after the removal of infectious hosts (Boinas *et al.*, 2011). At the time of endemicity in Spain and Portugal (1960s-1980s), there were a number of traditionally constructed pig shelters that favoured the presence of *O. erraticus* in cracks in adobe or between the stones of walls (Wilson *et al.*, 2013). To date, no tick involvement has been evidenced in the ongoing epidemic in Eastern Europe.

The current EU legislation recommends to restock as soon as 40 days after an ASF outbreak in the absence of vectors, or not before six years if vector involvement in ASF transmission is suspected (Council Directive 2002/60/CE; EC, 2002). In a worst-case scenario, the connection of outdoor swine production with low levels of biosecurity can constitute a significant risk factor

Table 10.3. Biosecurity measures against ASF for outdoor farms.

To avoid or minimise the risk of direct ASF transmission:
• No swill feeding or feeding with catering waste • Location of the farm as far away as possible from areas with wild boar or areas with other pigs • Location as far away as possible from the main roads • Fences to avoid direct transmission from potentially infected wild boar or other feral/free-ranging suids. Possibility of double fencing of outdoor holdings if necessary (if they are located close to a wild boar suitable habitat) or other solution which excludes the possibility of contact with wildlife • Control of feed/water points to avoid accessibility by other animals (i.e. further fencing; equipment exclusive for swine). Higher surveillance at times of the year when resources are scarce in nature • Entry of new animals, same biosecurity measures as indoor commercial pigs (including quarantine) • Proper disposal of dead animals or parts of dead animals to avoid the spread of infected material or attracting wild animals. • Avoid use of items that increase the contact among pigs, or management practices that would induce fights • Frequent health training and awareness raising should be conducted and updated according to the level of risk. Both are key factors for the early suspicion and detection of ASF control
To avoid or minimise the risk of indirect ASF transmission:
• Infrastructure of existing buildings that avoid tick breeding • Acaricides and insecticides, to avoid tick biting and other insects that could be mechanical vectors • Cleaning and disinfection of vehicles and equipment, clearing of weeds • Waste and residue management (including carcasses removal and slurry) • Treatment of any bedding material. Avoid use of freshly mowed green mass (grass and grains) and straw • Workers' biosecurity practices: avoid visiting other pig farms, hunting, etc.; clothing; hygiene (hand washing); food management • Control of visits • Control of presence of other people (tourists and campers) in the area • No sharing of feed, equipment etc. • Avoid entry of vehicles to the place where animals are kept • Biosecurity requirements controlled in each visited holding • Safe and controlled disposal of dead animals, in rendering plant • No other avoidable animal species in outdoor holdings

Measures to reduce the risk of ASF perpetuation intra-farm

1. NO SWILL/CATERING WASTE FEEDING
2. ENTRY OF NEW ANIMALS, same biosecurity measures as indoor commercial pigs (including quarantine)
3. Safe and controlled DISPOSAL OF DEAD ANIMALS in rendering plant + WASTE/ RESIDUES MANAGEMENT (including slurry) to avoid the spread of infected material or to attract wild animals
4. AVOID USE of items that increase the contact among pigs, or management practices that would induce fights
5. Daily visit to check health status
6. Appropriate infrastructure to avoid tick breeding
7. Use of acaricides and insecticides
8. No other animals species in outdoor holding

Measures to reduce the risk of ASF entrance to the farm

1. FARM LOCATION far away of areas with wild boar or other pigs, and main roads
2. FENCES to avoid direct transmission from potential infected wild boar/feral/free-ranging suids. Double fencing in outdoor holdings if located close to wild boar suitable habitat or other solution that excludes possibility of contact with free wildlife.
3. CONTROL OF FEED/WATER POINTS to avoid accessibility by other animals. Higher surveillance at times of the year when resources are scarce in nature.
4. C&D of vehicles, equipment, weeds, etc
5. TREATMENT OF ANY BEDDING MATERIAL. Avoid usage of freshly mowed green mass (grass and grains) and straw
6. WORKERS' BIOSECURITY PRACTICES: avoid visiting other pig farms, hunting, etc; clothing; hygiene (hand washing); food management
7. CONTROL of visits and other people presence
8. No sharing of feeding, equipment etc.
9. Avoid entry of vehicles to the place where animals are kept
10. BIOSECURITY requirements controlled in each visited holding

Adapted from the COST action ASFSTOP (EU One Health EJP) WG3 Report (Objective 2, Year 3), based on expert opinion. This programme has received funding from the EU's Horizon 2020 research and innovation programme (Grant Agreement No 773830).

Figure 10.5. Flyer on African swine fever Biosecurity Guidance in outdoor pig-production systems distributed by ASF-STOP members (available at https://www.asf-stop.com/dissemination/).

for further spread to commercial pig holdings. Awareness campaigns and training have a pivotal role in contingency planning.

A compilation of minimum biosecurity measures for outdoor pig holdings to avoid or minimise the risk of direct or indirect transmission of ASF is provided in Table 10.3 and Figure 10.5.

Acknowledgements

This publication is based on work from 'Understanding and combating African swine fever in Europe (ASF-STOP COST action 15116)' supported by COST (European Cooperation in Science and Technology).

References

ASFORCE, 2015: Guidelines for the cost-effective prevention and control of African swine fever. ASFORCE project, European Union's Seventh Framework Programme Project Consortium on Targeted research effort on African swine fever. Available at: http://asforce.org/sites/default/files/ASFORCE%20 Guidelines%20ASF.pdf.

Astorga, R.J., Tarradas, C., Argüello, C. and Luque I., 2016. Bioseguridad en las granjas porcinas: bioseguridad relacionada con la estructura y diseño en la granja. Suis 131: 32-36.

Bellini, S., Rutili, D. and Guberti, V., 2016. Preventive measures aimed at minimizing the risk of African swine fever virus spread in pig farming systems. Acta Veterinaria Scandinavica 58: 82. https://doi.org/10.1186/s13028-016-0264-x

Blitvich, B.J., Magalhaes, T., Laredo-Tiscareño, S.V. and Foy, B.D., 2020. Sexual transmission of arboviruses: a systematic review. Viruses 12: 933. https://doi.org/10.3390/v12090933

Boinas, F., Wilson, A.J., Hutchings, G.H., Martins, C. and Dixon, L.J., 2011. The persistence of African swine fever virus in field infected *Ornithodoros erraticus* during the ASF endemic period in Portugal. PLOS One 6: e20383. https://doi.org/10.1371/journal.pone.0020383

Bosch, J., Iglesias, I., Muñoz, M.J. and de la Torre, A., 2017. A cartographic tool for managing African swine fever in Eurasia: mapping wild boar distribution based on the quality of available habitats. Transboundary and Emerging Diseases 64: 1720-1733

European Commission (EC), 2002. Council Directive 2002/60/EC of 27 June 2002 laying down specific provisions for the control of African swine fever and amending Directive 92/119/EEC as regards Teschen disease and African swine fever. Official Journal of the European Union L 192: 27-46. Available at: https://eur-lex.europa.eu/legal-content/EN/TXT/?uri=celex%3A32002L0060.

European Commission (EC), 2014. 2014/709/EU: Commission implementing Decision of 9 October 2014 concerning animal health control measures relating to African swine fever in certain Member States and repealing Implementing Decision 2014/178/EU (notified under document C(2014) 7222). Official Journal of the European Union L 295: 63-78. Available at: https://eur-lex.europa.eu/legal-content/EN/TXT/?uri=CELEX%3A32014D0709.

European Commission (EC), 2015. Strategic approach to the management of African swine fever for the EU. SANTE/7113/2015-Rev 12. European Union Directorate General for Health and Food Safety, Brussels, Belgium. Available at: http://ec.europa.eu/food/sites/food/files/animals/docs/ad_control-measures_asf_wrk-doc-sante-2015-7113.pdf.

European Commission (EC), 2016. Animal health and welfare meetings. Available at http://ec.europa.eu/food/animals/health/regulatory_committee_en#20140304-05.

European Food Safety Authority (EFSA), 2014a. Evaluation of possible mitigation measures to prevent introduction and spread of African swine fever virus through wild boar. EFSA Journal 12: 3616

European Food Safety Authority (EFSA), 2014b. Scientific review on African swine fever. Panel on Animal Health and Welfare. EFSA Journal 12: 3628.

European Food Safety Authority (EFSA), 2015. Scientific opinion on African swine fever. Panel on Animal Health and Welfare. EFSA Journal 13: 4163. https://doi.org/10.2903/j.efsa.2015.4163.

Eurostat, 2014. Statistics in focus, pig farming in the European Union: considerable variations from one Member State to another. Eurostat.

Food and Agriculture Organization of the United Nations (FAO), 2001. Manual on the preparation of African Swine Fever contingency plans. FAO, Rome, Italy.

Food and Agriculture Organization of the United Nations (FAO)/World Organisation for Animal Health (OIE)/World Bank, 2010. Good practices for biosecurity in the pig sector – Issues and options in developing and transition countries. FAO Animal Production and Health Paper, 169. Available at: http://www.fao.org/3/i1435e/i1435e00.htm

Jurado, C., Fernández-Carrión, E., Mur, L., Rolesu, S., Laddomada, A. and Sánchez-Vizcaíno, J.M., 2018a. Why is African swine fever still present in Sardinia? Transboundary and Emerging Diseases 65: 557-566. https://doi.org/10.111/tbed.12740

Jurado, C., Martínez-Avilés, M., de la Torre, A., Štukelj, M., Cardoso de Carvalho Ferreira, H., Cerioli, M., Sánchez-Vizcaíno, J.M. and Bellini, S., 2018b. Relevant measures to prevent the spread of African swine fever in the European Union domestic pig sector. Frontiers in Veterinary Science 5: 77. https://doi.org/10.3389/fvets.2018.00077

Maes, D., Nauwynck, H., Rijsselaere, T., Mateusen, B., Vyt, P., de Kruif, A., and Van Soom, A., 2008. Diseases in swine transmitted by artificial insemination: an overview. Theriogenology 70: 1337-1345. https://doi.org/10.1016/j.theriogenology.2008.06.018

Royal Decree (RD), 2000. Basic rules for the management of pig farms [in Spanish]. RD 324/2000. Available at https://www.boe.es/diario_boe/txt.php?id=BOE-A-2000-4447.

Thacker, B., Larsen, R., Joo, H.S. and Leman, A., 1984: Swine diseases transmissible with artificial insemination. Journal of American Veterinary Medicine Association 185: 511-516.

Wilson, A.J., Ribeiro, R. and Boinas, F., 2013. Use of a Bayesian network model to identify factors associated with the presence of the tick *Ornithodorus erraticus* on pig farms in southern Portugal. Preventive Veterinary Medicine 110: 45-53.

11. Cleaning and disinfection in the domestic pig sector

M. Štukelj[1*], J. Prodanov-Radulović[2] and S. Bellini[3]

[1]University of Ljubljana, Veterinary Faculty, Gerbičeva ulica 60, 1000 Ljubljana, Slovenia; [2]Scientific Veterinary Institute 'Novi Sad', Rumenacki put 20, 21000 Novi Sad, Serbia; [3]Istituto Zooprofilattico Sperimentale della Lombardia e dell'Emilia Romagna, Via A. Bianchi 7/9, 25124 Brescia, Italy; marina.stukelj@vf.uni-lj.si

Abstract

African swine fever (ASF) is one of the most important diseases in pigs and since there are no effective vaccines, cleaning and disinfection remains one of the most effective tools to prevent the spread of the ASF virus (ASFV) in pig holdings. Indeed, cleaning and disinfection procedures are fundamental for pathogen inactivation, to prevent the spread of the disease and to facilitate the repopulation after an outbreak. The choice of disinfectants and of procedures for disinfection should take into consideration the nature of the premises, vehicles and objects to be treated. Disinfectants should further be officially authorised by the veterinary service and the conditions for their use strictly followed. This chapter describes the cleaning and disinfection protocols to apply in pig holdings to prevent the spread of ASFV. Currently available classification of cleaning agents and products is presented in detail as well as different factors that may affect the disinfectant's efficacy.

Keywords: cleaning and disinfection, ASFV, pig holdings

This publication is based upon work from COST Action CA15116, ASF-STOP, supported by COST (European Cooperation in Science and Technology).
www.cost.eu

EUROPEAN COOPERATION IN SCIENCE & TECHNOLOGY

Funded by the Horizon 2020 Framework Programme of the European Union

DOI 10.3920/978-90-8686-910-7_11, © M. Štukelj *et al.*, 2021

11.1 Introduction

African swine fever virus (ASFV) is a highly pathogenic enveloped virus that affects domestic and wild pigs (*Sus scrofa*). Susceptible suids can be infected by direct or indirect contact with infectious animals or their fluids, ingestion of contaminated animal feed, pork or pig products, or contact with contaminated surfaces or fomites (clothing, footwear, vehicles, farming tools, etc.) acting as mechanical vectors of the virus. Indeed, ASFV can persist for weeks in blood, faeces and urine excreted in the environment by infected pigs (Montgomery, 1921; Haas *et al.*, 1995; Turner and Williams 1999). In the southern and eastern parts of Africa and during the outbreak on the Iberian Peninsula, bites from infected soft ticks belonging to *Ornithodoros* genus were another source of ASFV infection. ASFV has a remarkable ability to survive for long periods in a protein rich environment and to remain stable (Table 11.1). The capacity for environmental resistance of ASFV is relevant for the local persistence and geographical spread of the virus. Regular and thorough cleaning and disinfection (C&D) is also an important routine for eradicating endemic diseases, and to reduce the risk of introducing new pathogens into the holding. Indeed, C&D procedures are an essential part of any farm biosecurity plan and in the framework of a control strategy for an epidemic disease they should be designed under the supervision of the Veterinary Authority, which should check the correct implementation.

11.2 African swine fever virus characteristics

The ASFV is an enveloped double-stranded DNA virus resistant to low temperatures. The virus can be inactivated by pH <3.9 or >11.5 in serum-free medium. Serum increases the resistance of the virus, e.g. at pH 13.4 resistance lasts up to 21 hours without serum, and 7 days with serum (Table 11.1) and it is resistant to pH changes that occur during meat maturation. This means that meat from pigs slaughtered in the infective stages of ASF or that die spontaneously of the disease provides a source of infection. ASFV persists in certain tissues for up to six months and can infect susceptible animals fed such meat (see Chapter 9 for details on epidemiology). Therefore, after an African swine fever (ASF) outbreak and before starting the disinfection procedures, elimination of all potential sources of residual virus is indispensable.

Table 11.1. Resistance of African swine fever virus to physical and chemical action (OIE Technical Disease Card for African swine fever, last update June 2019).

Temperature	Highly resistant to low temperatures. Heat inactivated by 56 °C/70 minutes; 60 °C/20 minutes.
pH	Inactivated by pH <3.9 or >11.5 in serum-free medium. Serum increases the resistance of the virus, e.g. at pH 13.4 – resistance lasts up to 21 hours without serum, and 7 days with serum.
Chemicals/disinfectants	Susceptible to ether and chloroform. Inactivated by 8/1000 sodium hydroxide (30 minutes), hypochlorites between 0.03% and 0.5% chlorine (30 minutes), 3/1000 formalin (30 minutes), 3% ortho-phenylphenol (30 minutes) and iodine compounds. Note: disinfectant activity may vary depending on the pH, time of storage and organic content.
Survival	Remains viable for long periods in blood, faeces and tissues; infected, uncooked or undercooked pork products. Can multiply in vectors (*Ornithodoros* sp.).

ASFV has been found to sustain infectivity also for long periods in different matrices, as summarised in Table 11.2 (EFSA, 2014).

Some experiments using different ASFV isolates showed that faeces, urine and oral fluids collected from ASFV infected pigs can contain infective virus (Davies *et al.*, 2017). It was demonstrated that ASFV can remain infective in slurry up to 112 days at 4 °C, and up to 84 days at 17 °C (unpublished results, reviewed by Haas *et al.*, 1995), and can be inactivated after 15 min at 60 °C, one hour at 50 °C and 4 h at 40 °C. According to a recent experimental study carried out using Georgia 2007/1 isolate, it appears that shedding of ASFV in excretions such as oral fluids, urine and faeces begins at approximately the same time as the onset of fever (Davies *et al.*, 2017). In this study, the level of viraemia was not related to the production of infectious excretions (Davies *et al.*, 2017). The same study demonstrated that the half-life of the virus in urine was longer than the half-life in faeces and oral fluids, suggesting that urine was the most stable medium for virus survival. This implies that faeces, urine and oral fluids can easily contaminate water, the premises and the environment. Therefore, if not properly treated, ASFV may persist in the environment and become a possible source of infection. Indeed, even a small amount of infective material can lead to transmission via oro-nasal route (Davies *et al.*, 2017; Gallardo *et al.*, 2013).

11.3 Cleaning and disinfection protocol

The general protocol to apply on pig farms for C&D is similar for all diseases. However, the choice of specific disinfectant depends on the pathogen and its resistance capacity. The C&D protocol described in this chapter follows a general scheme to apply in a systematic manner (FAD PReP and USDA, 2018) to prevent the introduction of pathogens (including ASFV) into the holding as well as in case of an ASF outbreak. Indeed, EU Council Directive 2002/60/EC (EC, 2002) establishes that in case of ASF occurrence, the destruction of carcasses shall be followed by thorough C&D of all premises, vehicles and equipment and this is required to allow the reintroduction of pigs to the farm after the outbreak. The C&D activity must be documented

Table 11.2. African swine fever virus stability in different matrices (summarised in EFSA, 2014).

Material	Stability	References
Chilled meat	105-165 days	Kowalenko *et al.*, 1965
Frozen meat	>1000 days	Adkin *et al.*, 2004
Bone marrow	180 days	Kowalenko *et al.*, 1965
Dry-cured meat	140 days	McKercher *et al.*, 1987; Mebus *et al.*, 1993, 1997
Meat and pork fat salted (22-27 °C)	16 days	Kolbasov *et al.*, 2011
Blood	70 days wooden surface	USDA, 2007
	105 days putrefied blood	
Faeces	60-100 days	Haas *et al.*, 1995
	11 days ~21 °C	
Liquid manure	84 days 17°	Haas *et al.*, 1995
Nasal aerosol	5-19 minutes	De Carvalho *et al.*, 2014

in the holding or vehicle register and, where official approval is required, be certified by the supervising official veterinarian. Adequate safety measures, depending on the chemical used (see below), must be taken during the implementation of the disinfection processes and all team members must be equipped with appropriate protective clothing, boots, hats, visors, gloves and respirators (FAD PreP and NAHEMS, 2014). The buildings, barns, equipment, clothing and footwear that pigs have come in contact with should be cleaned and disinfected (Capretti *et al.*, 2009, Levis and Baker, 2011). Cleaning and disinfection should be performed from clean to dirty areas of the premises and from up to down. It is important to include the entire stable in the C&D process, including ceiling, walls, floor, water pipes, ventilation system, manure removal system and all present equipment, to minimise potential contamination of previously cleaned areas (Van Immerseel *et al.*, 2018). All removable equipment, such as feeders, drinkers and any other loose equipment should be cleaned and disinfected separately (Capretti *et al.*, 2009; OIE, 1995).

A complete C&D protocol consists of seven steps: dry cleaning, wet cleaning, rinsing, drying, disinfection, drying and testing the efficiency of the procedure (FAD PreP and NAHEMS, 2014; Ramirez, 2009; Van Immerseel *et al.*, 2018).

11.3.1 Classification of cleaning agents

No single cleaning agent is able to perform all the functions necessary for cleaning. Therefore, a cleaning solution or detergent is made up of a range of components:

- *Water*: is the main ingredient of all 'wet' cleaning systems and provides the cheapest transport medium for rinsing and dispersing dirt (Holah, 1995). Water used for cleaning must be of microbiologically acceptable quality, maintained at the desired temperature and applied abundantly (OIE, 1995). Water has dissolving powers and can remove soluble ionic compounds (e.g. salts and sugars). Moreover, it helps to emulsify fats at temperatures above their melting point and it can be used as an abrasive agent in pressure-jet cleaning. Water under pressure must be used with caution as it can produce pockmarks in concrete and fractures in mortar, tiles and masonry grouting. Organic material harbouring pathogens may accumulate in such pockmarks and fractures (Holah, 1995; OIE, 1995).
- *Surfactants* (surface-active or wetting agents): are compounds that lower the surface tension (or interfacial tension) between two liquids, between a gas and a liquid, or between a liquid and a solid. Surfactants may act as detergents, wetting agents, emulsifiers, foaming agents, or dispersants. Surfactants are classified as anionic, non-ionic, cationic and amphoteric (Holah, 1995).

  Anionic surfactants are dissociated in water into an amphiphilic anion and a cation, which is in general an alkaline metal (Na^+, K^+) or a quaternary ammonium. They are the most commonly used surfactants, are good detergents, emulsifiers and have good foaming properties. They can be used in association with phenolic disinfectants, which include alkylbenzene sulfonates (detergents), soaps (fatty acids), lauryl sulphate (foaming agent), di-alkyl sulfosuccinate (wetting agent) and lignosulphonates (dispersants) (Capretti *et al.*, 2009).

  Non-ionic surfactants do not produce ions in aqueous solution because their hydrophilic group is of a non-dissociable type, such as alcohol, phenol, ether, ester or amide. Consequently, they are compatible with other types of surfactant and are excellent candidates to enter complex mixtures, as found in many commercial products. They are much less sensitive to electrolytes, particularly divalent cations, than ionic surfactants, and can be used with high salinity or hard water. Non-ionic surfactants are good detergents and emulsifiers but do not have good foaming properties. These compounds include ethoxylated linear alcohols, ethoxylated alkyl phenols, fatty acid esters, amine and amine derivates (Capretti *et al.*, 2009).

Cationic surfactants are dissociated in water into an amphiphilic cation and an anion, most often of the halogen type. A very large proportion of this class corresponds to nitrogen compounds such as fatty amine salts and quaternary ammonium compounds, with one or several long chains of the alkyl type, often deriving from natural fatty acids. These surfactants are not good detergents, but many cationic surfactants are bactericides (Capretti *et al.*, 2009).

Amphoteric surfactants are molecules that exhibit both anionic and cationic dissociations. This is the case of synthetic products like betaines or sulfobetaines and natural substances such as amino acids and phospholipids. Some amphoteric surfactants are not sensitive to pH, whereas others are cationic at low pH and anionic at high pH, manifesting amphoteric behaviour at intermediate pH. These compounds are good detergents and have biocidal properties (Capretti *et al.*, 2009).

- *Alkalis:* are useful cleaning agents, as they are inexpensive and able to break down proteins (through the action of hydroxyl ions) as well as to saponify fats. At higher concentrations, alkalis may also be microbicidal. They are able to kill most bacteria and viruses at a pH greater than 9. However, alkalis are not effective against spores or non-enveloped viruses (Missouri Department of Agriculture, 2008). The main disadvantages of alkalis are their potential to precipitate hard water ions, the formation of foam with soaps and their corrosiveness (Holah, 1995; OIE, 1995). Moreover, alkalis are generally difficult to rinse and often require follow up with acid rinsing. The main alkaline compounds used for their cleaning properties are sodium hydroxide (caustic soda), potassium hydroxide and sodium carbonate (washing soda) (OIE, 1995).

  Sodium hydroxide (caustic soda) is highly corrosive and irritating to the skin, eyes and mucous membranes of animals and humans; contact could produce severe burns. Thus, when applying sodium hydroxide solutions, workers should wear waterproof coats and hats, boots, goggles or other protective devices for eyes (CEREP and COVEPI, 2004; OIE, 1995). Due to its corrosive properties, caustic soda must only be used on resistant materials (e.g. walls) and users should be aware that it will remove paint (CEREP and COVEPI, 2004; OIE, 1995). Aluminium surfaces should be protected from contact with caustic soda (OIE, 1995). Sodium hydroxide should be used only when there is absolute certainty that the environment will not be negatively affected. In case of an ASF outbreak, it can be sprinkled on carcasses of dead animals and for the treatment of slurry (CEREP and COVEPI, 2004).

  Potassium hydroxide is effective against ASFV but, like caustic soda, is very corrosive and must be handled with care. It has good grease- and debris-removing properties (CEREP and COVEPI, 2004; OIE, 1995).

  Sodium carbonate (washing soda) has limited microbicidal activity, so it is mainly used in formulations together with other chemicals to increase their grease-removing properties, enhance penetration capacity and raise pH (OIE, 2019). In case of an ASF outbreak, it can be used on farm equipment and stables (CEREP and COVEPI, 2004).
- *Acids:* have minor detergent properties, although they are very useful in solubilising carbonate, mineral scales (including hard water salts) and proteinaceous deposits (Holah, 1995). These agents are commonly used for cleaning procedures, are widely available and relatively inexpensive.

Inorganic acids (e.g. nitric, hydrochloric, sulphuric, phosphoric and sulphamic acids) are used as cleaners for removing lime scale, milk stones, etc. They all have microbicidal properties due to their low pH levels but are generally slow acting. Inorganic acids are efficient cleaners but have strict limitations due to their corrosiveness to both skin and materials (OIE, 1995).
Organic acids (e.g. citric and acetic acids) are less toxic and less corrosive than inorganic acids. They are characterised by bactericidal, virucidal and fungicidal properties, but are generally slow acting (OIE, 1995).

- *Sequestering agents* (chelating agents): are used to prevent the precipitation of mineral ions by forming soluble complexes with these ions. The primary use of these agents is in the control of hard water ions. Moreover, they are added to surfactants to aid dispersion capacity and rinsability (Holah, 1995).

Detergents are chemical products used to disperse and remove soil and organic materials from surfaces by reducing surface tension and increasing the penetrating ability of water. This improves disinfectants' ability to reach and destroy microorganisms within or beneath the organic material. Soaps and detergents are cleaning agents and as surfactants (surface active agents) their chemical structure is amphiphilic (consists of a hydrophilic and lipophilic portion), thus providing both water-soluble and oil-soluble properties. This surfactant action of soaps and detergents can also aid in inactivating bacteria and disrupting the lipid envelope of some viruses.

The choice of the most suitable detergent agent should take into account several factors such as: the type of surface (e.g. material composition, porosity) (Table 11.3), the characteristics of the cleaning solution (water concentration, water hardness and temperature), nature and solubility of organic material to be removed, the efficacy and practicability under farm conditions and compatibility with the disinfectant selected (Capretti *et al.*, 2009; Holah, 1995; Marriott *et al.*, 2018). Cleaning products must be chosen to ensure compatibility with the disinfectant selected (FAD PreP and NAHEMS, 2014) (Table 11.4).

Table 11.3. Characteristics of the materials and detergents recommended in African swine fever control (modified from Marriott *et al.*, 2018).

Material	Characteristics	Recommended detergents
Ferrous metal	Easily damaged by rust due to the use of acid and alkaline chlorine detergents	Neutral detergents
Aluminium	Damaged by acidic and strongly alkaline detergents	Neutral or weakly alkaline detergents
Concrete	Damaged by acid detergents	Alkaline detergents
Glass	Damaged by strongly alkaline detergents	Neutral or weakly alkaline detergents
Paints and resins	Damaged by alkaline detergents	Acid detergents
Rubber	Damaged by strong acids. It must not be porous or spongy	Alkaline detergents
Stainless steel	Smooth, non-porous, easy to clean, resistant to corrosion, high temperature oxidation	Indifferently acidic or alkaline detergents

Table 11.4. Examples of detergents and disinfectants not to be used in combination.

Disinfectants	Detergents	Cause	Reference
Quaternary ammonium compounds (QACs)	alkalis	alkaline detergents may react chemically with QACs and destroy their antimicrobial properties	Holah, 1992
Hypochlorites	acids detergents	if these compounds are mixed, the resultant reaction releases toxic chlorine gas	RUMA, 2013
Phenols	soaps based on tallow, tall oil or oleic acids	these detergents are able to markedly decrease the activity of phenol compounds	RUMA, 2013
Chlorexidine	alkalis	alkalin detergents may interfere with disinfectant action of chlorexidine	RUMA, 2013

11.3.2 Dry cleaning to remove all organic material

The presence of organic material inactivates some disinfectants or protects microorganisms from their action; chlorine-based disinfectants are especially problematic in this regard. Indeed, the active ingredient in the bleach, chlorine, is relatively quickly inactivated by organic debris such as manure. Therefore, removing organic debris and dirt (dry cleaning) before disinfection is the most important step in the C&D process. It removes over 90% of microorganisms when properly performed, and improves the disinfection efficacy (FAD PreP and USDA, 2018).

The stables need to be empty (without animals) bêfore starting C&D; easy-to-dismantle equipment should also be removed and cleaned and disinfected outside the stable (Van Immerseel *et al.*, 2018). Dry cleaning involves the removal of any gross contamination and organic material (e.g. soil, manure, feed and bedding) from production areas or equipment. All organic waste should be removed using adequate and clean equipment, such as shovels, scrapers and brushes (Capretti *et al.*, 2009; Levis and Baker, 2011). If necessary, sills and floor should be hand scraped to remove any caked-on manure, food, debris and organic material; the same treatment should be applied to all permanently attached equipment. All floors, light fixtures, fan blades and louvers must be cleaned (FAD PReP and USDA, 2018). Solid waste material must be taken away and disposed of appropriately. In case of C&D in connection with an ASF outbreak, any items, materials or products that cannot be disinfected should be destroyed (e.g. burned or buried). Handling of waste materials and disposal should be conducted in a way to minimise further spread of pathogens, contamination of soil, air or water. Air blowers should not be used for dry cleaning due to the risk of spreading microorganisms.

Moistening the area or items with water may be helpful to control dust and minimise aerosolization of pathogens (pre-soaking) (OIE, 1995). To ease the cleaning process, it is recommended to pre-soak for at least 2 hours, preferably 6-24 hours (OIE, 1995; Waddilove, 2010). For this, pressure washers with a high-water flow but low pressure are recommended (OIE, 1995). This loosens any organic material left after dry cleaning and turns it into a solution (Van Immerseel *et al.*, 2018). The next washing phase can then be performed using less water and accumulating less sewage (Capretti *et al.*, 2009; Levis and Baker, 2011).

11.3.3 Wet cleaning

The cleaning solution can contain different components, such as water, alkaline or acid detergents. The efficacy of wet cleaning depends on the cleaning agents used and therefore these are discussed in some detail before describing the cleaning process.

Following the removal of gross contamination by dry cleaning and soaking, areas or items shall be washed with detergent. Prior to washing and applying the detergent, all electrical equipment should be turned off and removed, or covered tightly with plastic sheets. Washing can be performed by scraping, scrubbing and flushing with high-pressure washers with a flow rate of at least 20 litres per minute and pressure between 130 and 200 bar. High pressure water is very effective in removing heavy accumulation of urine and faeces. Whenever possible, hot water (32-54 °C) should be used. Hot water is far more effective than cold water at inactivating pathogens (FAD PReP and USDA, 2018). To prevent environmental pollution, drainage should be controlled (RUMA, 2013). The complete removal of the dirt residues can be obtained by rubbing the irregular surfaces with metal brushes. Deep cracks, crevices, pits, pores or other surface irregularities should be given special attention to dislodge accumulated grime (FAD PreP and NAHEMS, 2014). Steam cleaners may also be used and are particularly effective where compacted soilage is to be removed (OIE, 1995). Hot water and steam can be effective for cleaning cracks, crevices, and the inside of pipes where pathogens are likely to linger (FAD PreP and NAHEMS, 2014).

The washing process and the use of detergents help to further reduce the number of microorganisms and remove any oil, grease, or exudates that may inhibit the action of disinfectants (FAD PreP and NAHEMS, 2014; Waddilove, 2010). It has been reported that cleaning may lead to a two to six log reduction of microorganisms (Holah, 1992). The detergent should ideally be safe, non-tainting, non-corrosive, stable and not harmful to the environment. The choice of the detergent depends on the nature of the material to be removed (Holah, 1995). The detergent can either be previously mixed with water or directly applied onto the surfaces and equipment according to the instructions on the label. The application of foam is generally more effective. Whenever possible, foam should be allowed to sit on surfaces for the period recommended by the manufacturer. A systematic approach to spraying should be used, starting at the back of the facility and working toward the front, spraying the ceiling first, then the walls and finally the floor, in order to avoid splashing previously cleaned areas with dirty water (FAD PReP and USDA, 2018; OIE, 1995; Waddilove, 2010).

It is necessary to allow sufficient contact time for the detergent to be effective (20-60 minutes) and during this time surfaces should be monitored to make sure they do not dry off. If surfaces are drying, they should be sprayed with detergents again (Capretti *et al.*, 2009).

11.3.4 Rinsing to remove residues of detergents

After washing, all surfaces should be rinsed in order to remove detergents and all traces of material used in the cleaning process because residues can neutralise or inactivate some chemical disinfectants (Table 11.4) (OIE, 1995). Rinsing should be carried out using a high volume of cold water at low pressure. When the rinsing process is completed, surfaces should be carefully inspected to ensure they are visibly clean (Hurnik, 2005).

11.3.5 Drying surfaces to avoid dilution of the disinfectant

Whenever possible, surfaces should dry completely (if possible, overnight) before application of a disinfectant (OIE, 1995; RUMA, 2013). Indeed, excess moisture, especially on porous surfaces, may dilute and reduce the efficacy of the disinfectant and harm the equipment. In cool or cold weather, drying can be accomplished by heating the building and increasing air circulation with auxiliary blowers. In hot weather, drying may be accomplished with blowers or fans. In confined areas or on equipment where air circulation from fans is not enough, the use of high-pressure air from a compressor or high-volume blowers can aid in the removal of excess moisture, so that drying can take place. If highly infectious or zoonotic pathogens are suspected, high-pressure systems should be avoided or used with caution to avoid inadvertent spread of pathogens (Waddilove, 2010).

11.3.6 Disinfection

Disinfection can be accomplished using physical (e.g. heat, ultraviolet light) or chemical (e.g. disinfectant) means; in certain instances, a combination of methods may be needed. Heat is the best disinfectant, but it is rarely used in pig stables. However, for open surfaces, the use of hot water or steam is uneconomical, hazardous or impossible; consequently, disinfection is performed using chemical biocides (Holah, 1995; Hurnik, 2005).

In areas affected or at risk of introducing ASF, the disinfectants to choose for the control of the disease must be effective against the ASFV and approved by the official veterinarian. In case of ASF outbreak, the procedures to follow are described in CD 2002/60/EC (EC, 2002):

- *Application of the disinfectant.* Once surfaces are completely cleaned, rinsed, and dried, the disinfection can be implemented. The application form of disinfectants influences the time of contact between the chemical and surface-bound microorganisms (OIE, 1995). Application methods may involve spraying, fogging or misting, immersing, or wiping or mop-on methods. In the first case, disinfectants must be capable of ensuring microbicidal effect in 5 minutes or less, as within this time frame the disinfectant will have run off non-horizontal surfaces. According to Van Immerseel *et al.* (2018) two main ways of disinfectant application can be distinguished: wet (surface) and dry disinfection. Surface disinfection is often carried out with high pressure. In the case of dry disinfection (thermal fogging), a highly concentrated disinfectant is heated and subsequently converted to a fog by a fogger. Disinfection applied in a fumigation stage may ensure elimination of pathogens in difficult-to-reach zones and can be performed where it is possible to seal the building completely.
- *Fogging or aerial disinfection.* Aerial contaminants generated in the pig buildings are classified into particulates, gases and airborne microorganisms, which may significantly impact human and animal health. Dust or particulate matter represent important aerial contaminants in piggeries since they can combine with inorganic compounds, gases, bacteria, and viable endotoxins on the surface of dust particles, which can become potentially hazardous agents. This risk may affect not only animals but also piggery workers, who may show signs of immune system activation, as well as overt respiratory disease (Costa *et al.*, 2014). For this reason, after

surface disinfection, a widespread environmental disinfection procedure is recommended by 'fogging' the premises with the disinfectant suitably diluted according to the manufacturer's instructions. This procedure must be done in the absence of pigs, and the premises must be kept closed after the treatment (Capretti *et al.*, 2009).
The amount of disinfectant necessary varies considerably. It is considered that for a polished, non-porous floor, 100 ml/m^3 of disinfectant/chemical is sufficient (FAO, 2010). It is very important to properly calculate the amount of solution needed to disinfect an animal facility (Van Immerseel *et al.*, 2018).

- *Contact time.* After the application, disinfectants should be left on the surface to act for an appropriate contact time according to the manufacturer's instructions on the label. In some cases, the chemical disinfectant may need to be reapplied to keep the surface wet for the required time (FAD PReP and USDA, 2018; Waddilove, 2010). In case of ASF, disinfectants must remain on the surfaces for at least 24 hours (EC, 2002). Contact time can be increased by applying the disinfectant as a foam or gel, in which case contact time of 10-15 minutes (foam) and >15 minutes (gel) are possible (OIE, 1995). It is recommended to apply the disinfectant uniformly using low pressure/low volume methods, using adequate and approved equipment (Capretti *et al.*, 2009).
- *Temperature.* All disinfectants, whether they are sprays, foams, aerosols or fumigants, work best at temperatures above 18.3 °C, whereas temperatures for chlorine and iodine-based disinfectants should not exceed 43.3 °C (FAD PReP and USDA, 2018).

11.3.6.1 Disinfectants effective against African swine fever virus

When choosing the disinfectant one must take into consideration different aspects such as the type of surfaces to be treated, the spectrum of activity, the efficacy and practicability under farm conditions (e.g. ease of handling, risk of corrosion of the equipment, temperature stability), safety for the staff, animals and the environment, costs, risk to store, etc. (FAO, 2010; Missouri Department of Agriculture, 2008). The preparation of disinfectant solutions must be performed by qualified operators strictly following the manufacturer's instructions (concentration, contact time, pH, temperature) (FAO, 2010). It is important to keep solutions clean and freshly made. Mixing disinfectants is inadvisable, as the potency of each may be nullified or a dangerous reaction may be caused, releasing heat or dangerous gases (CEREP and COVEPI, 2004). There are no indications in the literature regarding the ideal disinfectant against ASFV, but every country has approved and/or authorised a list of biocides effective against ASFV and thus, only authorised biocides should be used and applied according to the producer's instructions (Juszkiewicz *et al.*, 2019). However, on the OIE website, a list of effective disinfectants against ASFV is provided (OIE, 2019). General knowledge of and experience in the use of disinfectants against enveloped viruses (CEREP, 2019; OIE, 2019) have shown that the following chemical compounds are effective in inactivation of ASFV:

- formaldehyde 1%;
- sodium hypochlorite (0.03 to 0.0075%);
- caustic soda solution 2%;
- glutaraldehyde, formic;
- sodium or calcium hydroxide 1% (effective at virus inactivation in slurry at 4 °C);
- phenols – lysol, lysephoform, and creolin;
- chemical compounds based on lipid solvents;

- multi-constituent compounds – sodium chloride, potassium peroxymonosulfate, lysoformin, desoform, octyldodeceth-20 (OD-20) surfactants, active substances, organic acids, glycosal, etc.

The use of some effective disinfectants against ASFV is limited due to their toxicity or lack of safety (e.g. formaldehyde). In practice, only some of the aforementioned chemical compounds are contained in commercial disinfectants. The disinfectants recommended in the USA by The Environmental Protection Agency (EPA) are shown in Table 11.5. Currently, there are commercially available disinfectants based on phenolic and iodine compounds, which are effective against the virus and can inactivate the ASFV at pH<4 and >11 (Gallardo *et al.*, 2015; Geering *et al.*, 2001).

11.3.6.2 Material composition

Facilities, equipment and items on pig farms are made of different materials. A primary factor to consider when selecting and performing C&D operations is the type of surface being treated to ensure the needed efficacy and minimal damage to items. The easiest surfaces for C&D are nonporous, smooth surfaces (FAD PreP and NAHEMS, 2014).

Metal surfaces (e.g. stainless steel, aluminium) are generally easier to disinfect than other materials, especially when the surfaces are smooth. However, some chemical disinfectants are incompatible with or corrosive to metal surfaces (Table 11.6). Flame guns may be a useful alternative for some metal surfaces (FAD PreP and NAHEMS, 2014).

Table 11.5. Disinfectants effective against African swine fever virus (modified from APHIS-USDA, 2011; Geering *et al.*, 2001).

Active ingredient(s)	Contact time	Application(s)
Sodium chloride Potassium peroxymonosulfate	10 min	in/on animal feeding equipment, livestock barns/pens/stalls/stables, livestock equipment, hog farrowing pen premises, hog barns/houses/ pens, animal quarters, animal feeding and watering equipment, animal transportation vehicles, agricultural premises/equipment, human footwear
Sodium dichloro-s-triazinetrione	30 min	in/on animal living quarters, farm premises, foot baths
Sodium dichloro-s-triazinetrione	30 min	in/on livestock premises, animal feeding/watering equipment, animal equipment, animal transportation vehicles, farm premises, foot baths
Sodium hypochlorite	15 min nonporous; 30 min porous	indoor or outdoor use sites, such as agricultural, transportation, quarantine, and laboratory equipment and facilities; footwear/personal protective equipment
Citric acid	15 min nonporous; 30 min porous	indoor or outdoor use sites such as agricultural and non-agricultural equipment and facilities; laboratory equipment and facilities; footwear/ personal protective equipment, personnel decontamination

Rubber and plastics should be treated as hard, nonporous surfaces; however, they may have interactions with some chemical disinfectant products. Phenols can be absorbed by plastics and 1% sodium hydroxide should be avoided. Iodophors may cause staining of these materials and can be corrosive to some plastics or rubber. Heat treatments can melt most plastics. Alcohols can swell or harden rubber or certain plastic tubing after prolonged and repeated use. Glutaraldehyde is one of the chemicals that can be used to treat rubber. In case of ASF outbreak it may be preferable to dispose of rubber and plastic tools (FAD PreP and NAHEMS, 2014).

Glass surfaces should be treated as hard, nonporous surfaces (FAD PreP and NAHEMS, 2014).

Raw concrete surfaces are porous and therefore difficult to clean. Porous surfaces should not be subjected to pre-soaking prior to washing or disinfectant application as this can cause unintended dilution. A disinfectant solution of a product registered for concrete surfaces should be applied to all surfaces once gross organic debris has been removed and the area has been washed, rinsed, and dried. High pressure washing with a disinfectant solution can be helpful for improving adequate contact but may cause damage to some surfaces. Acids and hypochlorite disinfectants can be corrosive to concrete surfaces, so flame guns may be an alternative disinfection method (FAD PreP and NAHEMS, 2014).

Wood is extremely porous and therefore difficult to disinfect. Any decaying wood surface that cannot be disinfected should be removed and disposed of appropriately (e.g. burned or buried). Wood surfaces should not be pre-soaked prior to washing or disinfectant application as this can cause unintended dilution. A disinfectant solution of a product registered for wood surfaces should be applied once gross organic debris has been removed (FAD PreP and NAHEMS, 2014).

No environmentally safe procedures exist for 'disinfecting' soil surfaces (e.g. dirt, sand, or packed clay). It is necessary to remove the layer of soil and then disinfect the area either by physical methods such as fire, boiling, steam or by a chemical method (for example caustic soda), which is the main method of disinfection.

Table 11.6. Disinfectants and effects on metal surfaces (FAD PReP and NAHEMS, 2014).

Chemical disinfectant	Effect on metal surfaces
Sodium hydroxide	corrosive to aluminium and derived alloys and galvanised metal
Sodium carbonate	corrosive to aluminium and derived alloys
Acids	highly corrosive to metals
Glutaraldehyde, potassium peroxymonosulfate, sodium chloride	mildly corrosive to metals
Iodophors, hypochlorites, formaldehyde	corrosive to some metals
Phenolics	relatively non-corrosive

11.3.6.3 Factors affecting the efficacy of disinfectants

The efficacy of disinfectants depends on the target organisms, their requirements for growth, and their resistance to environmental conditions and chemicals. The presence of interfering substances, disinfectant concentration, and the final pH of the solution as well as the contact time with surfaces and the temperature are also of importance (Holah, 1995; RUMA, 2013):

- *Interfering substances*: the efficiency of all disinfectants is reduced in the presence of organic matter for two main reasons: chemical reaction and spatial non-reaction. Organic material may form a spatial barrier (biofilm) protecting microorganisms from the effects of disinfectants. Organic material may react non-specifically with the disinfectant, which loses its biocidal potency (this is particularly true for oxidative biocides). Other interfering substances (e.g. detergents) may react chemically with the disinfectant and destroy its antimicrobial properties. Therefore, detergents should be rinsed off thoroughly before applying disinfectants. That is usually done to wash away the dirt. Therefore, it is essential to remove both soil and detergent residues prior to disinfection (Van Immerseel *et al.*; 2018; Holah, 1995).
- *Disinfectant concentration*: the relationship between microbial death and disinfectant concentration is not linear but follows a typical sigmoidal death curve. Microbial populations are difficult to kill at low biocide concentrations, but increasing concentrations leads to a point at which the majority of the population succumbs. Beyond this point, the microorganisms become more difficult to kill (through resistance or physical protection) and a proportion may survive regardless of increase in concentration. Consequently, disinfectant agents should be used at the concentration recommended by the manufacturer. Changes of this concentration will not enhance the effect (Van Immerseel *et al.*, 2018; Holah, 1995).
- *Final pH of the disinfectant solution*: efficacy of disinfectants may be affected by the pH of the water used for dilution. Only water within the pH range specified by the manufacturer's instructions should be used (Van Immerseel *et al.*; 2018; Holah, 1995).
- *Contact time*: commonly used chemical agents require several minutes in contact with pathogens to be effective. Disinfectants must find, bind to and traverse microbial cell envelopes before reaching their target site and beginning the reactions that will lead to the destruction of the microorganisms. As a consequence, sufficient contact time is critical to ensure disinfection. Most general-purpose disinfectants are formulated to require at least five minutes to reduce bacterial populations in suspension by five log reduction (Van Immerseel *et al.*, 2018; Holah, 1995; Ramirez, 2009).
- *Temperature*: the efficacy of disinfectants may also vary with temperature. In general, cold temperatures are known to reduce the effectiveness of most chemical agents (Van Immerseel *et al.*, 2018; Ramirez, 2009; RUMA, 2013).

11.3.6.4 Precautions for the use of disinfectants

Chemicals should only be used under the supervision of properly trained personnel, who must be aware of the effective spectrum of the disinfectant used, its limitations and the potential hazards to users, animals, equipment and the environment arising from the use of the product.

The staff must take appropriate steps for safe use and disposal of chemical agents. It is necessary to use personal protective equipment (PPE) for eyes and skin both during preparation and use of disinfectants. It is necessary to keep records of disinfectants used (Hurnik, 2005).

11.3.6.5 Storage of chemicals

All chemicals used on farms should be kept out of the reach of unauthorised persons. Chemicals that lose strength when exposed to light and heat should be stored in a cool, dark and dry place (OIE, 1995).

11.3.7 Drying

Cleaned and disinfected premises should have a period of downtime following the disinfection procedures. Premises should remain empty after drying of the disinfectant in order to avoid the accidental absorption of residues by the animals (FAD PReP and NAHEMS, 2014). In the event of an ASF outbreak, this period is 40 days (EC, 2002).

Final rinsing is rarely practised in pig stables, unless caustic soda or peroxide is used as a disinfectant. However, it is necessary to rinse the water system and the feeders after using disinfectants. Premises that have been cleaned and disinfected should also have a period of downtime following the procedures, to allow moisture evaporation from the building and all its surfaces. Most chemical disinfectants are not harmful to animals and therefore they do not need to be rinsed but they need to dry (FAD PReP and NAHEMS, 2014; Waddilove, 2010).

11.3.8 Testing the efficiency of the C&D procedure and evaluation of the activities carried out

Inspection of a site after C&D procedures should ensure that all tasks listed in the protocol have been performed appropriately. Factors to be addressed should include:

- All personnel are aware of and are implementing C&D measures for themselves and their equipment.
- All grossly contaminated areas have been identified and properly cleaned and disinfected.
- Gross debris (e.g. manure, unused feed or bedding) has been removed and properly disposed of.
- Any contaminated wood or items difficult to disinfect have been appraised, removed, and disposed of in a manner that minimises spread of pathogens (e.g. burned, composted, buried).
- All fixtures and fittings have been dismantled, cleaned, and disinfected.
- All infected or suspected areas have been properly washed, rinsed, and disinfected.
- Appropriate disinfectants were selected and used at the recommended concentration.
- The necessary contact time for the disinfectant was allowed.
- Effluent from the C&D procedures has been handled in a manner to minimise or avoid environmental impact.

Experienced personnel should conduct final inspection of the premises. If there is any doubt or sign of inadequate procedures, the disinfection process must be repeated (FAD PReP and NAHEMS, 2014).

11.4 Procedure for cleaning and disinfection

11.4.1 Buildings

Premises C&D will vary depending on the situation (routine, emergency) but it should always follow the previously described protocol (i.e. dry clean, pre-soak, apply detergent, rinse, dry, apply disinfectant, contact time, downtime) (Levis and Baker, 2011; Van Immerseel *et al.*, 2018).

Cleaning and disinfection of premises must be carried out using the 'all in/all out' system between each batch of pigs, followed by a ten-day resting period between batches to maintain low infection rates (Levis and Baker, 2011; Van Immerseel *et al.*, 2018;). The 'all in/all out' principle is essential in preventing cross-contamination and makes possible thorough C&D of the stables between different consecutive cycles (Van Immerseel *et al.*, 2018). It is important to apply the C&D process to the whole stable, including the ceiling, walls, floor, pipelines, feeding troughs, drinking nipples and other equipment in order to minimise potential contamination of previously cleaned areas (Levis and Baker, 2011; Van Immerseel *et al.*, 2018). After C&D measures in farrowing pens, it is necessary to clean and disinfect without leaving residual chemicals because the building will house parturient and neo-natal animals. Phenolic disinfectants should be avoided as they can be toxic to swine (FAD PReP and NAHEMS, 2014). In animal premises, disinfectants effective for ASFV include single active ingredient (sodium dichloro-s-triazinetrione) but also commercially available multi-constituent compounds (APHIS-USDA, 2011) (Table 11.5).

During the cleaning programme, it is important to pay attention to the water system. If possible, this should be drained, and drinkers removed and cleaned. Then, the system should be refilled with a suitable efficient but safe disinfectant. Finally, it needs to be thoroughly flushed with fresh water. If this is not possible due to a non-drainable system, the system should be flushed through with a suitable disinfectant, left for a certain period and flushed with water (Waddilove, 2010).

11.4.2 Equipment

All equipment that is going to come into contact with pigs must be purchased new and with no previous contact with other holdings (Carr *et al.*, 2018). In case of ASF, items in contact with infected animals that can be difficult to clean and disinfect should be inspected and then discarded. If surfaces and ambient temperature are below freezing, the same techniques used with vehicles to prevent freezing (Section 11.4.3) should be used for the equipment (FAD PReP and USDA, 2018).

A C&D station for small equipment should be established, if possible in proximity to a water supply and drainage. The most practical method of decontaminating electronic equipment (e.g. generators, motors) involves placing the equipment inside an airtight enclosure (e.g. plastic sheeting) for fumigation. When possible, equipment should be dismantled so that all parts can be fumigated (FAD PReP and USDA, 2018). Some electrical items may be inherently airtight, in which case they can be safely disinfected by wiping down with disinfectant. Exposure to ultraviolet light can be another option for disinfecting complex equipment (FAD PReP and NAHEMS, 2014; OIE, 1995). Equipment used to euthanise livestock (e.g. captive bolt guns and firearms) should be considered grossly contaminated. After use, they need to be scrubbed with disinfectant at the location where they were used, and again at the disinfection station. C&D equipment (e.g. rakes, shovels, brushes, sprayers) must be cleaned and disinfected after use and stored in a secure location. Special care should be used when cleaning and disinfecting rubber equipment because many disinfectants are corrosive to this material. For this reason, these tools should be destroyed (CERVES, 2004; FAD PReP and NAHEMS, 2014). The USA EPA-approved list of disinfectants for ASF includes several active ingredients for farm equipment such as sodium chloride and potassium peroxymonosulfate, sodium dichloro-striazinetrione and sodium hypochlorite (APHIS-USDA, 2011).

11.4.3 Vehicles

All vehicles (e.g. cars, livestock carriers, feed trucks, milk trucks, and carcass transporters) and heavy machinery (e.g. excavators, backhoes and bulldozers) that have been used on infected premises must undergo proper C&D processes because they can potentially transport pathogens from one site to another (Missouri Department of Agriculture, 2008). The cargo space of animal transport vehicles specifically needs to undergo comprehensive C&D between animal loads (FAD PReP and USDA, 2018; Neuman and Hall, 2019).

A C&D station should be established and vehicles should follow the previously described C&D protocol. During dry cleaning procedures, soiled bedding and refuse have to be removed and placed in a proper area to avoid the re-contamination of the cleaned vehicles (FAO, 2010).

Shovels, manure forks, brushes, low-pressure sprayers or mechanical scrapers can be used to remove all visible organic material from the exterior of the vehicle. During the vehicle washing procedures, detergent and warm water (32-54 °C) have to be used. Wheels and wheel wells can be a particular fomite that requires detailed attention to ensure proper C&D. Any deposits of mud and straw should be removed from the exterior of the vehicle (FAD PReP and USDA, 2018).

Items with debris that is difficult to remove should be pre-soaked. Vehicles should be rinsed with cold water, but if it is not possible, the vehicle should remain stationary for 5-10 minutes to allow the residual rinse water to drip off (FAD PReP and USDA, 2018). Cold weather is a significant constraint to vehicle disinfection; cold temperatures in fact can preserve pathogens, freeze the water and make drying difficult. Thus, in cold climates, an indoor washing facility is essential. Additionally, forced air fans and heaters can facilitate drying process (FAO, 2010). If ambient temperature is below freezing, either heating of surfaces to prevent freezing and placing heat blankets around liquid containers should be used or 40% propylene glycol in water could be added when mixing solutions (FAD PReP and USDA, 2018).

The disinfection procedure must be performed only on dried surfaces and following a precise order: the application of the disinfectant starts at the top of the vehicle and moves downward from the outside to the inside, paying particular attention to the wheels and the underlying parts of the vehicle (CERVES, 2004).

All external areas should be sprayed, and the chassis of the vehicle should be disinfected with a non-corrosive disinfectant (Table 11.4). Then, after a proper contact time, vehicles should be rinsed and dried thoroughly. Interior C&D of the vehicle is necessary if the driver leaves the cab (FAD PReP and USDA, 2018). All non-fixed items should be removed from the vehicle to be cleaned and disinfected (FAD PReP and NAHEMS, 2014). According to APHIS-USDA (2011) disinfectants effective for ASF may include single active ingredient (sodium dichloro-s-triazinetrione) but also multi-constituent compounds like combination of sodium chloride and potassium peroxymonosulfate (Table 11.5).

11.4.4 Personnel

Personnel engaged in C&D procedures should wear at the least minimum coveralls, boots and gloves. Face protection and a mask should be worn based on the product or application method used and when mixing disinfectant solutions. Personal protective equipment such as chemical-resistant suits (including both pants and jackets with hoods) or respirators may be necessary for some situations (e.g. formaldehyde or acidic disinfectants) (FAD PReP and NAHEMS, 2014). Portable electronic equipment (e.g. hand-held radios, cameras, mobile phones) should be used while protected inside plastic bags. Personnel must practise proper personal decontamination and doffing procedures for PPE before leaving an infected premise or any quarantined area to prevent the spread of pathogens (FAD PReP and USDA, 2018). For personnel body decontamination, citric acid 0.2% weight per volume (w/v) as a safe and active ingredient effective for ASFV is recommended (APHIS-USDA, 2011; Geering *et al.*, 2001; Neuman and Hall, 2019).

- At the entrance of the farm:
  - a dedicated building should be located at the entrance of the pig unit, where personnel can undress, shower and change to farm-provided clothes and boots before entering the farm areas (Capretti *et al.*, 2009; Carroll *et al.*, 2014);
  - shower installation should be designed so that people will go through the shower in both directions and will not come back into the entrance or contaminated area to dress themselves without showering first (shower in-shower out) (Carroll *et al.*, 2014; Levis and Baker, 2011);
  - personnel should not wear jewellery (Missouri Department of Agriculture, 2008);
  - individuals must thoroughly wash their hands with antibacterial soap before entering and leaving the premises (Capretti *et al.*, 2009; Carroll *et al.*, 2014).
- At the exit of the farm:
  - a container or plastic bag for collecting dirty clothing or disposable items used by personnel must be provided;
  - personnel must wash and disinfect their hands before leaving the farm;
  - farm biosecurity procedures also require personnel to shower at least 4 minutes with hot water before exiting the farm (Levis and Baker, 2011; Neuman *et al.*, 2019).

11.5 Manure

Proper manure management is an important part of the pig farm biosecurity plan (Van Immerseel *et al.*, 2018) as pig manure is known to be an important source of many infectious diseases (swine dysentery, classical swine fever, foot and mouth disease, *Salmonella* spp., *Escherichia coli,* etc.) (Venglovsky *et al.*, 2018), including ASF virus (Davies *et al.*, 2017). Pathogens can survive or even grow in manure as this substrate provides the necessary nutrients and protects them against UV radiation, desiccation and high temperature. Temperature is a critical factor in pathogen survival. Colder temperatures generally allow longer survival times for microorganisms. Besides temperature and time of storage, the survival of pathogens in slurry also depends on other factors such as moisture content, free ammonia concentration, pH, the presence of other microorganisms and other physicochemical properties (Venglovsky *et al.*, 2018). Several studies have shown the presence of ASFV in faeces and urine (Haas *et al.*, 1995; Davies *et al.*, 2017). Therefore, in case of an outbreak, slurry must be considered infected and treated properly during the C&D process (FAO, 2010).

There is considerable diversity in farm designs and flooring, which have an influence on how the waste from the pigs' housing can be removed and stored. In the context of disease control, it is advisable to use a system that avoids the build-up of faeces on the farm and limits the time that the animals are exposed to it. In principle, well designed and maintained slatted floors are the most suitable to avoid the accumulation of faeces in the pens, minimising contact with the animals (Twomey *et al.*, 2010). Contamination with pig manure poses a risk to the health of animals and farm workers if the manure is not adequately stored, treated and controlled (FAO, 2010).

To reduce the risk of disease occurrence, practical recommendations for manure management are summarised as follows:

- Pig pens should be cleaned and disinfected regularly, ensuring that manure is removed from the pens every day (FAO, 2010). Frequent removal of manure will prevent completion of the life cycles of intestinal parasites and flies (Levis and Baker, 2011).
- Quickly remove manure, urine and straw bedding from pens where sick and dead animals have been present (FAO, 2010).
- Always use separate shovels, forks, skid-steer loaders or loader bucket for handling manure and feed. This will prevent contamination of feed and water. Ensure that equipment is properly maintained and cleaned (Levis and Baker, 2011).
- If pigs are housed outdoors, maintain clean water troughs, water bowls and feed troughs (Levis and Baker, 2011).
- Control the fly population. Methods include flypaper, parasitic wasps and insecticides (baits and sprays) (Levis and Baker, 2011).
- Store manure in a way that it is inaccessible to pigs. Ideally, slurry pits should be outside the farm with channels running from different buildings (RUMA, 2013). The pipe used to carry the manure should be rodent proof (Levis and Baker, 2011).
- Another method for storing the manure is deep pits under the floor (Levis and Baker, 2011). If manure is stored in pits in buildings, care must be taken to maintain good ventilation and good air quality (RUMA, 2013).
- Slurry pits should be large enough to hold four to six months' production to reduce the need for frequent emptying (RUMA, 2013). Containers and storage basins should comply with the minimum requirements for storage capacities recommended by the Best Available Techniques (EC, 2010) and should be located far from animal facilities to avoid cross-contamination (EC, 2010).

Pig manure is a valuable plant nutrient resource, and application of manure on agricultural land is the most economical way of disposing. However, land application is a critical period in manure management, and it might pose a risk for the spread of diseases (Bailey and Buckley, 1998). In the event of an ASF outbreak, all potentially contaminated materials on a pig farm such as manure, bedding, straw, feedstuffs should be removed and disposed of in the same way as carcasses (FAO, 2010).

In case of an ASF outbreak, manure must be stacked to heat, sprayed with disinfectant and left for at least 42 days or destroyed by burning or burying. Slurry must be stored for at least 60 days after the last addition of infective material. The veterinary authority may authorise a reduced storage period for slurry if it has been effectively treated to ensure the destruction of the virus (EC, 2002). The chemicals considered effective for ASF for manure treatment are sodium hydroxide [(lye, caustic soda) 2%, 15 lt/m^3)] (CEREP and COVEPI, 2004). During the treatment with disinfectants, no fresh slurry must be introduced into the tank (OIE, 1995; FAD PReP and USDA, 2018).

Sodium hydroxide (caustic soda) along with calcium hydroxide (lime) has been shown to reduce pathogen levels in slurry and wastewater treatment processes. A 2% caustic soda solution is the strongest chemical compound (i.e. the strongest virucidal agent) inactivating ASFV (Juszkiewicz *et al.*, 2019). In slurry disinfection, it has the added benefit of being effective in the presence of organic material (FAD PReP and NAHEMS, 2014; Kalmar *et al.*, 2018).

The disinfectants should be added to the storage tank at several points simultaneously to ensure that the disinfectant is properly distributed throughout the pit, which shall be stirred vigorously until the manure is considered safe by checking the pH (<2.00 or >11.00). During exposure of slurry to disinfectants, no fresh slurry must be introduced into the tank (OIE, 1995; FAD PReP and USDA, 2018).

Acknowledgements

This publication is based on work from 'Understanding and combating African swine fever in Europe (ASF-STOP COST action 15116)' supported by COST (European Cooperation in Science and Technology).

References

Animal and Plant Health Inspection Service (APHIS) United States Department of Agriculture (USDA) (APHIS-USDA), 2011. Potential pesticides to use against the causative agents of selected foreign animal disease in farm settings. United States Department of Agriculture (USDA) Animal and Plant Health Inspection Service (APHIS), Riverdale, MD, USA. Available at: https://www.aphis.usda.gov/animal_health/emergency_management/downloads/fad_epa_disinfectants.pdf

Bailey, L. and Buckley, K., 1998. Land application of hog manure: agronomic and environmental considerations the Canadian perspective. In: Presented at the HEMS Workshop. April 27-28, 1998. Toronto, Canada, pp. 1-20.

Capretti, S., Scarcella, C., Cinotti, S., Bellini, S., Pezza, F., Bertocchi, L., Cerioli, M., Alborali, L.G., Gradassi, M., Massi, P., Lavazza, A., Grilli, G., Domenichini, A., Autorino, G., Caprioli, A., Scicluna, M.T., Abrami, S., Feltrinelli, D., Ghiglia, P., Berneri, C. and Zanardi, G., 2009. La Biosicurezza in Veterinaria. Fondazione Iniziative Zooprofilattiche e Zootecniche, Brescia, Italy.

Carr, J., Chen, S.P, Connor, J.F., Kirkwood, R. and Segales, J., 2018. Pig health. 1st ed. CRC Press, Boca Raton, FL, USA, pp. 383-468.

Carroll, C., Boyle, L., Brady, S., Clarke, S., Finn, J., Lawlor, P., Lynch, B., Martin, M., McCutcheon, G., McKeon, M., Quinn, A. and Tuite, P., 2014. Biosecurity procedures for visitors to pig units in Ireland. 2nd ed. Teagasc pig development department, Fermoy, Ireland.

Centro di Referenza Nazionale per le Pesti Suine (CEREP) in collaborazione con il Centro di referenza Nazionale per l'Epidemiologia (COVEPI), 2004. Centro Manuale operativo Peste Suina Classica e Peste Suina Africana. Available at: http://www.izsum.it/files%5CDownload%5C48%5C-1%5CManuale%20 Operativo%20PSC-PSA.pdf.

Centro di Referenza Nazionale per le Pesti Suine (CEREP) in collaborazione con il Ministero della Salute, 2019. Manuale operativo Peste Suina Classica e Peste Suina Africana. Rev. n. 1., Perugia, Italy. Available at: http://www.izsum.it/files/Download/48/-/manuale%20operativo%20pesti_2019.pdf

Centro Nazionale di Referenza per le Malattie Vescicolari (CERVES), 2004. Manuale Operativo Malattia Vescicolare del Suino, Brescia, Italy. Available at: https://www.izsler.it/izs_bs/allegati/1241/14_ManualeOperativoMVS.pdf

Costa, A., Colosio, C., Gusmara, C., Sala, V. and Guarino, M., 2014. Effects of disinfectant fogging procedure on dust, ammonia concentration, aerobic bacteria and fungal spores in a farrowing-weaning room. Annals of Agricultural and Environmental Medicine 21: 494-499. https://doi.org/10.5604/12321966.1120589

Davies, K., Goatley, L.C., Guinat, C., Netherton, C.L., Gubbins, S., Dixon, L.K. and Reis, A.L., 2017. Survival of African swine fever virus in excretions from pigs experimentally infected with the Georgia 2007/1 isolate. Transboundary Emerging Diseases 64: 425-431. https://doi.org/10.1111/tbed.12381

De Carvalho Ferreira, H.C., Weesendorp, E., Quak, S., Stegeman, J.A. and Loeffen, W.L., 2014. Suitability of faeces and tissue samples as a basis for non-invasive sampling for African swine fever in wild boar. Veterinary Microbiology 72: 449-54. https://doi.org/10.1016/j.vetmic.2014.06.016

European Commission (EC), 2002. Council Directive of 27 June 2002 laying down specific provisions for the control of African swine fever and amending Directive 92/119/EEC as regards Teschen disease and African swine fever. Official Journal of the European Communities L 192: 27-46.

European Commission (EC), 2010. Best Available Techniques (BAT). Reference Document for the Intensive Rearing of Poultry or Pigs. EUR 28674 EN. Available at: http://publications.jrc.ec.europa.eu/repository/handle/JRC107189

European Food Safety Authority (EFSA), 2014. Scientific opinion on African swine fever. Panel on Animal Health and Welfare. EFSA Journal 12: 3628. https://doi.org/10.2903/j.efsa.2014.3628.

Foreign Animal Disease Preparedness and Response Plan/National Animal Health Emergency Management System (FAD PReP/NAHEMS), 2014. NAHEMS guidelines: cleaning and disinfection. Center for Food Security and Public Health, Iowa State University of Science and Technology and U.S. Department of Agriculture Animal and Plant Health Inspection Service, Ames, IA, USA. Available at: https://www.aphis.usda.gov/animal_health/emergency_management/downloads/nahems_guidelines/cleaning_disfection.pdf

Foreign Animal Disease Preparedness and Response Plan/United States Department of Agriculture (FAD PReP/USDA), 2018. Standard Operating Procedures (SOP): 15. Cleaning and disinfection. The Foreign Animal Disease Preparedness & Response Plan/US Department of Agriculture, Riverdale, MD, USA. Available at: https://www.aphis.usda.gov/animal_health/emergency_management/downloads/sop/sop_cd.pdf

Food and Agriculture Organization of the United Nations (FAO), 2010. Good practices for biosecurity in the pig sector. FAO Animal Production and Health. Paper No. 169. FAO, Rome, Italy. Available at: http://www.fao.org/3/a-i1435e.pdf

Gallardo, C., Soler, A., Nieto, R., Carrascosa, A., De Mia, G., Bishop, R., Martins, C., Fasina, F., Couacy-Hymman E. and Heath, L., 2013. Comparative evaluation of novel African swine fever virus (ASF) antibody detection techniques derived from specific ASF viral genotypes with the OIE internationally prescribed serological tests. Veterinary Microbiology 162: 32-43. https://doi.org/10.1016/j.vetmic.2012.08.011

Gallardo, M.C., Reoyo, A.T., Fernández-Pinero, J., Iglesias, I, Muñoz M.J. and Arias M.L., 2015. African swine fever: a global view of the current challenge. Porcine Health Management 1: 21. https://doi.org/10.1186/s40813-015-0013-y

Geering, W.A., Penrith, M.L. and Nyakahuma, D., 2001. Manual on procedures for disease eradication by stamping out. Part 3. Decontamination operation procedures. Food and Agriculture Organization of the United Nation Animal Health Manual. Paper No.12. FAO, Rome, Italy. Available at: http://www.fao.org/3/y0660E03.htm#ch3

Haas, B., Ahl, R., Bohm, R. and Strauch, D., 1995. Inactivation of viruses in liquid manure. Revue Scientifique et Technique 14: 435-445.

Holah, J.T., 1992. Cleaning and disinfection. In: Dennis, C. and Stringer, M. (eds). Chilled foods: a comprehensive guide. Ellis Horwood, London, UK, pp. 319-341.

Holah, J.T., 1995. Disinfection of food production areas. Revue Scientifique et Technique 14: 343-363.

Hurnik, D., 2005. Investigations into optimal washing and disinfection techniques for pig pens. In: Murphy, J.M. (ed.) London Swine Conference Proceedings: Production at the leading edge. April 6-7, 2005. London, Ontario, Canada, pp. 135-138.

Juszkiewicz, M., Walczak, M. and Woźniakowski, G., 2019. Characteristics of selected active substances used in disinfectants and their virucidal activity against ASFV. Journal of Veterinary Research 63: 17-25. https://doi.org/10.2478/jvetres-2019-0006

Kalmar, I.D., Cay, A.B. and Tignon, M., 2018. Sensitivity of African swine fever virus (ASFV) to heat, alkalinity, and peroxide treatment in presence or absence of porcine plasma. Veterinary Microbiology 219: 144-149. https://doi.org/10.1016/j.vetmic.2018.04.025.

Kowalenko, J., Sidorow, M. and Burba, L., 1965. African swine fever and control measures. Internationale Zeitschrift der Landwirtschaft 1: 47-52.

Kolbasov, D., Tsybanov, S., Malogolovkin, A., Gazaev, I. and Mikolaychuk, S., 2011. Identification of ASF virus in pork products. Veterinaria 10: 54-56.

Levis, D. and Baker, R., 2011. Biosecurity of pigs and farm security. Nebraska Extension, Institute of Agriculture and Natural Resources at the University of Nebraska, Lincoln, NE, USA, pp. 31. Available at: http://extensionpublications.unl.edu/assets/pdf/ec289.pdf

Marriott, N., Schilling, W. and Garavani, R.B., 2018. Principles of food sanitation. Food Science Text Series. 6th ed. Springer International Publishing, Cham, Switzerland.

McKercher, P.D., Yedloutshnig, R.J., Callis, J.J., Murphy, R., Panina, G.F., Civardi, A., Bugnetti, M., Foni, E., Laddomada, A., Scarano, C. and Scatozza, F., 1987. Survival of viruses (Parma ham). Journal of the Canadian Institute of Food Science Technology 20: 267-272.

Mebus, C.A., House, C., Ruiz Gonzalvo, F., Pined, J.M., Tapiador, J., Pire, J.J., Bergada, J., Yedloutschnig, R.J., Sahu, S., Becerra, V. and Sánchez-Vizcaíno, J.M., 1993. Survival of foot-and-mouth disease, African swine fever, and hog cholera viruses in Spanish Serrano cured hams and Iberian cured hams, shoulders and loins. Food Microbiology 10: 133-143. https://doi.org/10.1006/fmic.1993.1014

Mebus, C., Arias, M., Pineda, J., Tapiador, J., House, J. and Sánchez-Vizcaíno, J.M., 1997. Survival of several porcine viruses in Spanish dry-cured meat products. Food Chemistry 59: 555-559. https://doi.org/10.1016/S0308-8146(97)00006-X

Missouri Department of Agriculture, 2008. Cleaning and disinfection: standard operative guide n° 004. Agricultural emergency response actions – Livestock disease emergency. Available at: http://agriculture.mo.gov/animals/pdf/animalag_guide4.pdf

Montgomery, R.E., 1921. On a form of swine fever occurring in British East Africa (Kenya colony). Journal of Comparative Pathology and Therapeutics 34: 159-191 (part I), 243-269 (part II).

Neuman, E.J. and Hall, W.F., 2019. Disease control, prevention and elimination. In: Zimmerman, J.J., Karriker, L.A., Ramirez, A., Schwartz, K.J., Stevenson, G.W. and Zhang, J., (eds.). Diseases of swine. 11th ed. Iowa State Press, Ames, IA, USA, pp. 144-148.

Office International des Epizooties (OIE), 1995. Disinfectants: actions and applications. Revue Scientifique et Technique 14: 257-477.

Office International des Epizooties (OIE), 2019. African swine fever: aetiology epidemiology diagnosis prevention and control references. Office International des epizooties, Paris, France. Available at: https://www.oie.int/fileadmin/Home/eng/Our_scientific_expertise/docs/pdf/AFRICAN%20SWINE%20FEVER.pdf

Ramirez, A., 2009. Four steps to effective cleaning and disinfecting. National Hog Farmer. Available at: https://www.nationalhogfarmer.com/health-diseases/1015-effective-cleaning-disinfecting-steps

Responsible Use of Medicines in Agriculture Alliance (RUMA), 2013. RUMA guidelines. Responsible use of antimicrobials in pig production. Available at: http://www.ruma.org.uk/wp-content/uploads/2014/09/RUMA_antimicrobial_long_pigs_revised_final_Nov_2013.pdf

Turner, C. and Williams, S., 1999. Laboratory-scale inactivation of African swine fever virus and swine vesicular disease virus in pig slurry. Journal of Applied Microbiology 87: 148-157.

Twomey, D.F., Miller, A.J., Snow, L.C., Armstrong, J.D., Davies, R.H., Williamson, S.M., Featherstone, C.A., Reichel, R. and Cook, A.J.C., 2010. Association between biosecurity and *Salmonella* species prevalence on English pig farms. Veterinary Record 166: 722-724. https://doi.org/10.1136/vr.b4841

Van Immerseel, F., Luyckx, K., De Reu, K. and Dewulf, J., 2018. Cleaning and disinfection. In: Dewulf, J and Van Immerseel, F. (eds.) Biosecurity in animal production and veterinary medicine. CABI, Wallingford, UK, pp. 133-152.

Venglovsky, J., Sasakova, N., Gregova, G., Papajova, I., Toth, F. and Szaboova, T., 2018. Devitalisation of pathogens in stored pig slurry and potential risk related to its application to agricultural soil. Environmental Science and Pollution Research International 25: 21412-21419. https://doi.org/10.1007/s11356-017-0557-2

Waddilove, J., 2010. Cleaning comes before disinfection. Professional pig community. Available at: https://www.pig333.com/articles/cleaning-comes-before-disinfection_3004/

12. Conclusions

D. Gavier-Widén[1,2*], S. Bellini[3], E. Chenais[1], F. Ferreira[4], F. Jori[5], M.F. Le Potier[6], M. Montoya[7], C.L. Netherton[8], P.J. Sánchez-Cordón[9,10], K. Ståhl[1] and L. Iacolina[11,12]

[1]National Veterinary Institute (SVA), 751 89, Uppsala, Sweden; [2]Department of Biomedical Sciences and Veterinary Public Health, Swedish University of Agricultural Sciences (SLU), P.O. Box 7028, 750 07 Uppsala, Sweden; [3]Istituto Zooprofilattico Sperimentale della Lombardia e dell'Emilia Romagna (IZSLER), Via A. Bianchi 7/9, 25124 Brescia, Italy; [4]Centre for Interdisciplinary Research in Animal Health, Faculty of Veterinary Medicine, University of Lisbon, 1300-477 Lisbon, Portugal; [5]ASTRE (Animal, Health, Territories, Risk and Ecosystems), CIRAD, INRAE, University of Montpellier, 34398 Montpellier, France; [6]French Agency for Food, Environmental and Occupational Health & Safety (ANSES), 22440 Ploufragan, France; [7]Centro de Investigaciones Biológicas Margarita Salas (CIB-CSIC). Ramiro de Maeztu 9, 20840 Madrid, Spain; [8]The Pirbright Institute, Ash Road, Pirbright, Woking, GU24 0NF, United Kingdom; [9]Pathology Department, Animal and Plant Health Agency, APHA-Weybridge, Addlestone, KT153NB, United Kingdom; [10]European Union Reference Laboratory for African Swine Fever (EURL), Centro de Investigación en Sanidad Animal, CISA-INIA, Valdeolmos, 28130 Madrid, Spain; [11]Department of Chemistry and Bioscience, Aalborg University, Frederik Bajers Vej 7H, 9220 Aalborg, Denmark; [12]Faculty of Mathematics, Natural Sciences and Information Technologies, University of Primorska, Glagoljaška 8, 6000 Koper, Slovenia; dolores.gavier-widen@sva.se

Abstract

The COST Action ASF-STOP brought together an extensive network of scientists of remarkable excellence on African swine fever (ASF) vaccinology, virology, immunology, diagnostics and pathology. The network also includes global leaders in wild boar ecology and management, renowned epidemiologists specialised in ASF and disease control and scientists with vast expertise in the pig sector in Europe, biosecurity, cleaning and disinfection in pig holdings. This book collects updated knowledge in these fields, with a focus on the European situation. ASF-STOP coordinated and integrated research on ASF. Despite the new knowledge generated on ASF by scientists in Europe and worldwide, many unknowns still remain. For example, many expressed ASF virus (ASFV) genes remain uncharacterised and the information is needed for efficient antiviral drug and vaccine development. Crucial interactions between ASFV and cells like macrophages or dendritic cells are not yet completely understood. Research is needed on the mechanisms of protective immunity and identifying further viral proteins for inclusion in subunit vaccines. An in-depth analysis of the impact of ASF on the structure of the pig farming

This publication is based upon work from COST Action CA15116, ASF-STOP, supported by COST (European Cooperation in Science and Technology).
www.cost.eu

COST European Cooperation in Science & Technology

Funded by the Horizon 2020 Framework Programme of the European Union

DOI 10.3920/978-90-8686-910-7_12,

system in the EU has not yet been conducted, and transmission risks from the wild boar-habitat epidemiological cycle to the domestic pig cycle are still not fully understood. The pig sector is one of the most economically significant farming sectors in the EU and pork is the most consumed meat. The EU is the world's second biggest producer of pork. In both the wild boar-habitat and the domestic pig epidemiological cycle fully implemented biosecurity can hinder or eliminate virus transmission. National legislation and EU regulations set out clear regulations for controlling ASF. However, across Europe, wild boar populations are growing in size and range and the control of ASF continues to be a major challenge. The great socio-economic impact of ASF calls for further collaborative efforts to tackle this disease.

Keywords: wild boar, pig, African swine fever, ASF-STOP, COST

12.1 Summary and conclusions

The COST Action ASF-STOP brought together an extensive network of more than 270 scientists from 42 countries covering a wide range of disciplines related to African swine fever (ASF). The excellence of the ASF-STOP network is remarkable, including scientists with outstanding international records in the fields of ASF vaccinology, virology, immunology, diagnostics and pathology. The network also includes global leaders in wild boar ecology and management, renowned epidemiologists specialised in ASF and disease control and scientists with vast expertise in the pig sector in Europe, biosecurity, cleaning and disinfection in pig holdings. This book collects updated knowledge in these fields, with a focus on the European situation. Some of the main conclusions are summarised below.

The African swine fever virus (ASFV) is a large double-stranded DNA virus and the only member of the *Asfaviridae* family of viruses. ASFV encodes ~150-170 open reading frames with a very low mutation rate, expressed in a precise and regulated fashion, with most of the genes being actively transcribed. Recent next-generation sequencing approaches enable mapping of ASFV gene transcripts from start to finish, elucidating molecular mechanisms of ASFV transcription, including promoter motifs, factors involved, and temporal gene expression patterns. However, many expressed ASFV genes remain uncharacterised, highlighting genome-wide studies as a useful tool for efficient antiviral drug and vaccine development.

Recent studies on the architecture and composition of the infectious ASFV particle have provided a comprehensive model of the ASFV architecture that integrates compositional, structural and functional information, shedding light on the huge complexity of ASFV structure and biology. ASFV infects predominantly macrophages by a very rapid dynamin and clathrin-mediated process of endocytosis, engaging unknown cell receptor(s), with movement of viral particles and endosomes along microtubules. Inside the cells, ASFV inhibits different innate immune responses.

The complex work leading to the development of an effective drug against ASFV can be greatly accelerated by using modern approaches like genome-wide CRISPR/Cas screens or computational methods. Indeed, more *in silico* approaches should be used in order to find new antiviral agents targeting ASFV proteins or host factors involved in the virus lifecycle, which alone or in combination with vaccines will stop the spread of ASFV.

The immune system is designed to detect and subsequently eliminate harmful pathogens while maintaining a homeostatic relationship with beneficial microbes in the body. Upon infection with ASFV a variety of immunological mechanisms are triggered, but these are not sufficiently broad or powerful to eliminate the virus. The first line of defence against pathogens is the innate immune system. Crucial interactions between ASFV and cells like macrophages (Mφ) or dendritic cells (DC) are not yet completely understood. Despite the central role of Mφ in ASFV infection biology, little is known about the responses of different subsets of Mφ to ASFV. The same applies to DC subsets. Remarkably, it has been shown that ASFV uses several mechanisms to counteract type I interferon (IFN) action, also subverting interferon stimulated genes, showing the key role of this family of cytokines in ASFV infection. In fact, it has been suggested that ASFV has evolved mechanisms to become 'tolerant' to the action of type I IFN.

Since early studies, the protective role of antibody responses against ASFV has been a controversial topic and their neutralisation ability has received little attention. Further studies on antibodies generated after infection with different isolates of ASFV are required to elucidate their role in ASFV infection. Conversely, new data on cellular responses show the importance of αβ T cell receptors +T cells, like CD8+ T cells, CD4+T cells or invariant natural killer T cells, or cell subsets, such as natural killer or γ-T cells within ASFV infection. The ASFV epitopes recognised by them is an area of recent discoveries that will be useful in future vaccine design studies. Recent results on regulatory T cells show that the interplay and plasticity of the immune system is subverted by ASFV infection. Unfortunately, many knowledge gaps still exist regarding immune responses to ASFV in natural infection compared to experimental infection. Moreover, identification of the correlates of immune protection is still one of the main unknowns in ASFV infection.

Pathology plays a key role in studying the pathogenesis of ASF and complements other knowledge fields to provide a broad understanding of host-virus interactions. Development of lesions over time, from early stage to terminal or late disease can be correlated with clinical signs, virus presence and immunological response. Pathology techniques, such as immunohistochemistry to label ASFV, allow characterising the co-localisation of virus with lesions, and virus loads. In this way tissue and cell targets of the virus can be identified and associated to lesions. Pathology scoring of severity and extent of lesions allows systematic semi-quantitative comparisons of outcomes of experimental infections, for example for evaluation of ASF vaccine candidates. Several host and virus-related factors, such as the virulence of the ASFV isolate, affect the clinical presentation, presence and severity of lesions and the final outcome of the infection. Highly virulent isolates are usually involved in peracute or acute forms of disease while moderately virulent isolates can induce acute, subacute and even chronic forms of ASF. Natural infection of pigs in immunologically naïve herds by highly virulent ASFV has a case fatality of up to 90-100% four to ten days post infection. The pathology presentation of ASF is variable. Fluid exudation into body cavities, such as hydrothorax and hydropericardium, and macroscopic haemorrhagic lesions are considered hallmarks of acute and subacute forms of ASF. Significant macroscopic lesions affect the lymphoid system. The spleen displays an increase in size (hyperaemic splenomegaly) and lymph nodes are often swollen, oedematous and haemorrhagic. The lungs show diffuse congestion and oedema. The kidneys characteristically show petechiae. Macroscopic lesions associated with subacute

ASF forms are similar to those in acute forms but characterised by more severe and extensive haemorrhages and oedema as a consequence of longer clinical courses induced by less virulent isolates, severe and prolonged thrombocytopenia and consumption of coagulation factors. Animals presenting chronic forms of ASF are characteristically immunocompromised and susceptible to infection with opportunistic pathogens, mainly bacteria, which ultimately cause non-specific macroscopic lesions such as fibrinous and fibrous pericarditis, pleural adhesions, lung consolidation or necrotic and caseous pneumonia with mineralisation or necrotic areas on tonsils and tongue. Joint swelling and necrotic skin lesions are also characteristic macroscopic lesions observed not only in natural chronic forms of ASF but also in experimental infections with naturally attenuated low virulent isolates used as potential vaccine candidates; the pathogenic mechanisms remain unclear.

For the diagnosis of ASF there are currently very good tools in terms of sensitivity, specificity and robustness. Several commercial kits for the detection of either antibodies (ELISA) or viral genome (PCR) have been fully validated. Inconsistent results are most often due to poor quality of the sample or to the environment (laboratory contamination) rather than to the performance of the diagnostic tools. It is highly recommended to perform PCR with an internal control to obtain reliable results particularly with poor quality samples. Either for serological or virological methods, confirmatory methods should be applied to verify positive results, especially before the declaration of the first case of ASF in a new area in the absence of an obvious epidemiological link. The overall diagnostic results depend on several factors, such as the sampling design and the population targeted. For example, in a farm with a suspicion of ASF, it is more efficacious to target sick pigs than applying random sampling of all animals. The demand for penside tests is increasing following the expansion of the disease through the Eurasian continent. Currently, a penside test for antibody detection is commercially available, while the validation of a penside test for antigen detection is still ongoing. However, the use of penside tests for ASF surveillance is still being discussed, especially considering that ASF is a notifiable disease.

Classical methods of generating vaccines for African swine fever virus have not been successful. Inactivated vaccines induce antibody responses in pigs, but do not prevent animals from developing disease after being infected. Repeated passage of ASFV through tissue culture leads to genetic changes that attenuate the virus, and these attenuated viruses can protect pigs from virulent virus. However, these tissue culture adapted viruses cause a chronic form of disease and are therefore not suitable for deployment in the field. An alternative to attenuating ASFV by passage through tissue culture is targeted gene deletion. This approach has led to the development of a number of vaccine candidates with good efficacy and promising safety profiles. A large-scale trial of one of these viruses is currently under way in China with promising initial results. A subunit vaccine would avoid the safety issues associated with live ASFV vaccines, but the complexity of the virus makes this approach challenging. Combinations of viral proteins have been tested in pigs using different vaccine platforms including viral vectors, whole proteins, and DNA plasmids with varying degrees of success. Research on the mechanisms of protective immunity and identifying further viral proteins for inclusion in subunit vaccines is required for success.

The pig sector is one of the most economically significant farming sectors in the European Union (EU) and pork is the most consumed meat. The EU is the world's second biggest producer of pork, after China, and the biggest exporter of pork and pork products. The main producer countries, Germany, Spain and France, represent about half of the EU's total production. The major production basin extends from Germany to Belgium and accounts for 30% of the sows in the EU. Other important pig producing regions are Catalonia, Murcia, Lombardy, Bretagne and some areas of central Poland and northern Croatia. Small-scale pig producers are mostly found in eastern Europe where small units rearing 3.8% of pigs account for 73.3% of the pig farms. Following the current ASF epidemic in the eastern EU, which started in 2014, the pig sector has greatly changed. The number of mainly small-sized pig holdings has decreased. An in-depth analysis of the impact of ASF on the structure of the pig farming system in the EU has, however, not yet been conducted.

Many ASF outbreak investigations have reported biosecurity shortcomings as a critical element for virus introduction and spread. The structure of the European swine industry makes it necessary to put in place differentiated biosecurity measures, in order to meet the different risk levels for introduction and spread of ASF among the diversified farming systems. The risk of exposure to ASF for the individual farm depends on the country, area and farm location in relation to the infectious status in the surroundings. This implies that biosecurity measures should take into consideration the virus persistence in the environment, routes of transmission and excretion as well as the characteristics of the farming systems and the health situation of the territory. Biosecurity programmes are normally applied to commercial holdings. However, all holdings that have access to markets, i.e. sell pigs and products, should be included in such biosecurity programmes. Backyard holdings that sell animals have an important role in the spread of ASF.

Cleaning and disinfection (C&D) procedures are fundamental for pathogen inactivation, to prevent the spread of the disease and to facilitate the repopulation after an outbreak. The completion of the C&D procedure is also one of the requirements foreseen by the OIE and by the EU legislation for the recovery of the free status after the occurrence of ASF.The choice of disinfectants and of procedures for disinfection shall take into consideration the nature of the premises, vehicles and objects to be treated. Disinfectants should further be officially authorised by the veterinary service and the conditions for their use strictly followed.

The introduction of ASFV genotype II to Georgia in 2007 and further to the EU in 2014 sparked a new era of global ASF research. This has led to the identification of specific characteristics of the epidemiology of ASF relating to wild boar, i.e the wild boar-habitat epidemiological cycle. In this epidemiological cycle wild boar carcasses and the resulting contamination of the environment play key roles for virus persistence. Transmission risks from this cycle to the domestic pig cycle are still not fully understood. Several field studies, experimental infection trials and literature reviews have confirmed that previous knowledge concerning important factors in ASF epidemiology are valid for both wild boar and domestic pigs infected with ASFV genotype II: high case fatality rate, low contagiousness, and no evidence of asymptomatic virus carriers. In both the wild boar-habitat and the domestic pig epidemiological cycle of ASF, fully implemented biosecurity can hinder or eliminate virus transmission. National legislation and EU regulations set out clear regulations for controlling ASF in this regard. However, if these are not entirely complied with, and adequate measures thus not implemented where the disease is transmitted during the daily activities of people, they are of no value for disease control and eradication. The global pattern of the current epidemic confirms the role of humans in transmitting ASF to domestic pigs and wild boar. Therefore, achieving control requires adapting biosecurity measures to local conditions to increase their implementation, as well as to improve understanding of the sociocultural, economic, and political dimensions of domestic pig production and wild boar value chains.

The positive examples from the Czech Republic and Belgium show that control and eradication of ASF in wild boar can be achieved. Early detection and swift actions combining several different efforts to first restrain wild boar from leaving the infected area, and secondly, eliminate affected populations, seem to be keys to success in this regard. Such endeavours require multi-stakeholder cooperation and communication on international, national, and local levels.

Across Europe, wild boar populations are growing in size and range, in parallel with increased damage to agricultural crops, forests and biodiversity, higher number of traffic accidents and higher levels of detected infectious pathogens maintained and disseminated by wild boar, including ASF. Currently, a high priority goal in wild boar management practice is to reduce wild boar population densities and avoid overpopulation in order to mitigate the negative impacts.

Since wild boar is one of the most popular and widespread game species in Europe, considerable expertise has been built and lessons have been learned regarding management of wild boar populations in order to curb their associated economic and environmental impact. This expertise encompasses lethal and non-lethal methods to reduce wild boar populations and methods and tools to contain and influence their movements and behaviour. During the current ASF epidemic in the EU, many of these wild boar management methods have been combined with other disease control strategies such as regionalisation, fencing and enhanced passive surveillance to attempt control or eradication of ASF from wild boar populations. Field experience shows that successful control requires collaboration of different stakeholders including animal health authorities, local authorities, hunting associations, wildlife managers, farmers, landowners, the general public using the forests, and occasionally army or police forces.

The COST Action ASF-STOP coordinated and integrated research on ASF. It also contributed to knowledge dissemination and capacity building. Despite the new knowledge generated on ASF by scientists in Europe and worldwide many unknowns still remain. The control of ASF continues to be a major challenge, which in recent years has acquired a global dimension. The great socio-economic impact of ASF calls for further collaborative efforts to tackle the disease.

Acknowledgements

This publication is based on work from 'Understanding and combating African swine fever in Europe (ASF-STOP COST action 15116)' supported by COST (European Cooperation in Science and Technology). LI received funding from the Horizon 2020 research and innovation programme under the Marie Sklodowska-Curie Action (Grant Agreement no. 656697).

www.ingramcontent.com/pod-product-compliance
Lightning Source LLC
Chambersburg PA
CBHW061201240926
51846CB00006B/1077
9789086863570